T0229766

QUANTUM COMPUTING

*From Linear Algebra
to Physical Realizations*

QUANTUM COMPUTING

From Linear Algebra to Physical Realizations

Mikio Nakahara

Department of Physics
Kinki University, Higashi-Osaka, Japan

Tetsuo Ohmi

Interdisciplinary Graduate School of Science and Engineering
Kinki University, Higashi-Osaka, Japan

CRC Press
Taylor & Francis Group
Boca Raton London New York

CRC Press is an imprint of the
Taylor & Francis Group, an **informa** business

A TAYLOR & FRANCIS BOOK

CRC Press
Taylor & Francis Group
6000 Broken Sound Parkway NW, Suite 300
Boca Raton, FL 33487-2742

© 2008 by Taylor & Francis Group, LLC
CRC Press is an imprint of Taylor & Francis Group, an Informa business

No claim to original U.S. Government works

10 9 8 7 6 5 4 3 2 1

International Standard Book Number-13: 978-0-7503-0983-7 (Hardcover)

Library of Congress Cataloging-in-Publication Data

Nakahara, Mikio.
      Quantum computing : from linear algebra to physical realizations / M.
      Nakahara and Tetsuo Ohmi.
            p. cm.
      Includes bibliographical references and index.
      ISBN 978-0-7503-0983-7 (alk. paper)
      1. Quantum computers. I. Ohmi, Tetsuo, 1942- II. Title.

QA76.889.N34 2008
621.39'1--dc22   2007044310

**Visit the Taylor & Francis Web site at
http://www.taylorandfrancis.com**

**and the CRC Press Web site at
http://www.crcpress.com**

Dedication

To our families

Contents

Preface

One of the authors (MN) had an opportunity to give a series of lectures on quantum computing at Materials Physics Laboratory, Helsinki University of Technology, Finland during the 2001-2002 Winter term. The audience included advanced undergraduate students, postgraduate students and researchers in physics, mathematics, information science, computer science and electrical engineering among others. The host scientist, Professor Martti M. Salomaa, suggested that the lectures, mostly devoted to theoretical aspects of quantum computing, be published with additional chapters on physical realization. In fact Martti himself was willing to contribute to the physical realization part, but his unexpected early death made it impossible. After Martti passed away, MN asked his longstanding collaborator, TO, to coauthor the book. This is how this book was created. Part I, the theory part, was written by MN, while Part II, the physical realization part, was written jointly by MN and TO. Both authors have reviewed the final manuscript carefully and are equally responsible for the whole content.

Quantum computing and quantum information processing are emerging disciplines in which the principles of quantum physics are employed to store and process information. We use classical digital technology at almost every moment in our lives: computers, mobile phones, mp3 players, just to name a few. Even though quantum mechanics is used in the design of devices such as LSI, the logic is purely classical. This means that an AND circuit, for example, produces *definitely* 1 when the inputs are 1 and 1. One of the most remarkable aspects of the principles of quantum physics is the *superposition principle* by which a quantum system can take several different states *simultaneously*. The input for a quantum computing device may be a superposition of many possible inputs, and accordingly the output is also a superposition of the corresponding input states. Another aspect of quantum physics, which is far beyond the classical description, is *entanglement*. Given several objects in a classical world, they can be described by specifying each object separately. Given a group of five people, for example, this group can be described by specifying the height, color of eyes, personality and so on of each constituent person. In a quantum world, however, only a very small subset of all possible states can be described by such individual specifications. In other words, most quantum states cannot be described by such individual specifications, thereby being called "entangled." Why and how these two features give rise to the enormous computational power in quantum computing will be explained in this book.

Part I is devoted to theoretical aspects of quantum computing, starting with Chapter 1 in which a brief summary of linear algebra is given. Some subjects in this chapter, such as spectral decomposition, singular value decomposition and tensor product, may not be taught in a standard physics curriculum. The principles of quantum mechanics are outlined in Chapter 2. Some examples introduced in this chapter are important for understanding some parts in Part II. Qubit, the quantum counterpart of bit in classical information processing, is introduced in Chapter 3. Here we illustrate the first application of quantum information processing, namely quantum key distribution. By making use of the theory of measurement, a cryptosystem that is 100% secure can be realized. Quantum gates, the important parameters for quantum computing and quantum information processing, are introduced in Chapter 4, where the universality theorem is proved. Quantum gates are quantum counterparts of the elementary logic gates such as AND, NOT, OR, NAND, XOR and NOR in a classical logic circuit. In fact, it will be shown that all these classical gates can be reproduced with the quantum gates as special cases. A few simple but elucidating examples of quantum algorithms are introduced in Chapter 5. They employ the principle of quantum physics to achieve outstanding efficiency compared to their classical counterparts. Chapter 6 is devoted to the explanation of quantum circuits that implement integral transforms, which play central roles in several practical quantum algorithms, such as Grover's database search algorithm (Chapter 7) and Shor's factorization algorithm (Chapter 8). Chapter 9 describes a disturbing issue of decoherence, which is one of the obstacles against the physical realization of a working quantum computer. A quantum system gradually loses its coherence through interactions with its environment, a phenomenon known as decoherence. Quantum error correcting codes (QECC) introduced in Chapter 10 are designed to overcome certain kinds of decoherence. We will illustrate QECC with several important examples.

Part II starts with Chapter 11, where the DiVincenzo criteria, the criteria that any physical system has to satisfy to be a candidate for a working quantum computer, are introduced. The subsequent chapters introduce physical systems wherein the DiVincenzo criteria are evaluated for respective realizations. Liquid state NMR, the subject of Chapter 12, is introduced first since it is one of the well-understood systems. The subject of liquid state NMR has a long history, and numerous theoretical techniques have been developed for understanding the system. The liquid state NMR system has, however, several drawbacks and cannot be the ultimate candidate for a scalable quantum computer — at least not in its current form. The molecular Hamiltonian for the liquid state NMR system is determined very precisely, and the agreement between the theory and experiments is remarkable. Chapters 13 and 14 are devoted to ionic and atomic qubits, respectively. The ion trap quantum computer is one of the most promising systems: the largest quantum register with 8 qubits has been reported. Atomic qubits trapped in an optical lattice are expected to have very small decoherence due to their charge neutrality.

Chapters 15 and 16 describe solid state realization of a quantum computer. Chapter 15 introduces several types of Josephson junction qubits. The interaction among them is analyzed in detail. Chapter 16 describes quantum dots realization of qubits. There are two types: charge qubits and spin qubits, and they are treated separately.

Suggestion to readers and instructors: The whole book may be used for a one-year course on quantum computing. Using Part I for a semester and Part II for the subsequent semester is ideal. Alternatively, Part I may be used for a single semester course for physics, mathematics or information science graduate students. It may not be a good idea to use only Part II for lectures. Instead, Chapters 1 through 4 followed by Part II may be reasonable course materials for physics graduate students. An instructor may choose chapters in Part II depending on his/her preference. Chapters in Part II are only loosely related with each other.

MN used various parts of the book for lectures at several universities including Kinki University, Helsinki University of Technology, Shizuoka University, Kyoto University, Osaka City University, Ehime University, Kobe University and Kumamoto University. He would like to thank Yoshimasa Nakano, Martti M Salomaa, Mikko Paalanen, Jukka Pekola, Akihiko Matsuyama, Takao Mizusaki, Tohru Hata, Katsuhiro Nakamura, Ayumu Sugita, Taro Kashiwa, Yukio Fukuda, Toshiro Kohmoto and Masaharu Mitsunaga for giving him opportunities to improve the manuscript and also for the warm hospitality extended to him.

Tero Heikkilä, Teemu Ojanen and Juha Voutilainen at Helsinki University of Technology worked as course assistants for MN's lectures. MN would like to thank them for their excellent course management.

We are also grateful to Takashi Aoki, Koji Chinen, Kazuyuki Fujii, Toshimasa Fujisawa, Shigeru Kanemitsu, Go Kato, Toshiyuki Kikuta, Sachiko Kitajima, Yasushi Kondo, Hiroyuki Miyoshi, Mikko Möttönen, Yumi Nakajima, Hayato Nakano, Kae Nemoto, Antti Niskanen, Manabu Ozaki, Robabeh Rahimi Darabad, Akira SaiToh, Martti Salomaa, Kouichi Semba, Fumiaki Shibata Yasuhiro Takahashi, Shogo Tanimura, Chikako Uchiyama, Juha Vartiainen, Makoto Yamashita and Paolo Zanardi for illuminating discussions and collaborations.

We would like to thank Ville Bergholm, David DiVincenzo, Kazuyuki Fujii, Toshimasa Fujisawa, Saburo Higuchi, Akio Hosoya, Hartmut Häffner, Bob Joynt, Yasuhito Kawano, Seth Lloyd, David Mermin, Masaharu Mitsunaga, Hiroyuki Miyoshi, Bill Munro, Mikko Möttönen, Yumi Nakajima, Hayato Nakano, Kae Nemoto, Harumichi Nishimura, Antti Niskanen, Izumi Ojima, Kouichi Semba, Juha Vartiainen, Frank Wilhelm, Makoto Yamashita and Paolo Zanardi for giving enlightening lectures at Kinki University.

Takashi Aoki, Shigeru Kanemitsu, Toshiyuki Kikuta, Yasushi Kondo, Yukihiro Ohta, Takayoshi Ootsuka, Juha Pirkkalainen, Robabeh Rahimi Darabad, Akira SaiToh and Hiroyuki Tomita have pointed out numerous typos and errors in the draft. Their comments helped us enormously to improve the manuscript.

Technical assistance from Akira SaiToh and Juha Vartiainen in the preparation of figures is greatly appreciated.

MN would like to thank JSPS grant (Grant Nos. 14540346 and 19540422), MEXT grant (Grant No. 13135215) and various research grants from Kinki University for supporting this book writing project.

We are grateful to Clare Brannigan, Theresa Delforn, Amber Donley, Shashi Kumar, Jay Margolis and John Navas of Taylor & Francis for their excellent editorial work. John's patience over our failure to meet deadlines is also gratefully acknowledged.

Last but not least, we would like to thank our families for patience and encouragement, to whom this book is dedicated.

Mikio Nakahara and Testuo Ohmi
Higashi-Osaka, Japan

Part I

From Linear Algebra to Quantum Computing

1

Basics of Vectors and Matrices

The set of natural numbers $\{1, 2, 3, \ldots\}$ is denoted by $\mathbb{N}$. The set of integers $\{\ldots, -2, -1, 0, 1, 2, \ldots\}$ is denoted by $\mathbb{Z}$. $\mathbb{Q}$ denotes the set of rational numbers. Finally $\mathbb{R}$ and $\mathbb{C}$ denote the sets of real numbers and complex numbers, respectively. Observe that

$$\mathbb{N} \subset \mathbb{Z} \subset \mathbb{Q} \subset \mathbb{R} \subset \mathbb{C}$$

The vector spaces encountered in physics are mostly real vector spaces and complex vector spaces. Classical mechanics and electrodynamics are formulated mainly in real vector spaces while quantum mechanics (and hence this book) is founded on complex vector spaces. In the rest of this chapter, we briefly summarize vector spaces and matrices (linear maps), taking applications to quantum mechanics into account.

The **Pauli matrices**, also known as the spin matrices, are defined by

$$\sigma_x = \begin{pmatrix} 0 & 1 \\ 1 & 0 \end{pmatrix}, \quad \sigma_y = \begin{pmatrix} 0 & -i \\ i & 0 \end{pmatrix}, \quad \sigma_z = \begin{pmatrix} 1 & 0 \\ 0 & -1 \end{pmatrix}.$$

They are also referred to as σ_1, σ_2 and σ_3, respectively.

The symbol I_n denotes the **unit matrix** of order n with ones on the diagonal and zeros off the diagonal. The subscript n will be dropped when the dimension is clear from the context. The arrow $\rightarrow$ often indicates logical implication. We use e^x and $\exp(x)$ interchangeably to denote the exponential function.

For any two matrices A and B of the same dimension, their commutator, or commutation relation, is a matrix defined as

$$[A, B] \equiv AB - BA,$$

while the anticommutator, or anticommutation relation, is

$$\{A, B\} \equiv AB + BA.$$

The symbol ∎ denotes the end of a proof.

1.1   Vector Spaces

Let K be a field, which is a set where ordinary addition, substraction, multi-plication and division are well-defined. The sets $\mathbb{R}$ and $\mathbb{C}$ are the only fields which we will be concerned with in this book. A **vector space** is a set where the addition of two vectors and a multiplication by an element of K, so-called a **scalar**, are defined.

DEFINITION 1.1 A vector space V is a set with the following properties;

(0-1) For any $u, v \in V$, their sum $u + v \in V$.

(0-2) For any $u \in V$ and $c \in K$, their scalar multiple $cu \in V$.

(1-1) $(u + v) + w = u + (v + w)$ for any $u, v, w \in V$.

(1-2) $u + v = v + u$ for any $u, v \in V$.

(1-3) There exists an element $0 \in V$ such that $u + 0 = u$ for any $u \in V$. This element 0 is called the **zero-vector**.

(1-4) For any element $u \in V$, there exists an element $v \in V$ such that $u + v = 0$. The vector v is called the **inverse** of u and denoted by $-u$.

(2-1) $c(x + y) = cx + cy$ for any $c \in K, u, v \in V$.

(2-2) $(c + d)u = cu + du$ for any $c, d \in K, u \in V$.

(2-3) $(cd)u = c(du)$ for any $c, d \in K, u \in V$.

(2-4) Let 1 be the unit element of K. Then $1u = u$ for any $u \in V$.

It is assumed that the reader is familiar with the above properties. We will be concerned mostly with the **complex vector space** $\mathbb{C}^n$ in the following. There are, however, occasional instances where the **real vector space** $\mathbb{R}^n$ is considered.

An element of $V = \mathbb{C}^n$ will be denoted by $|x\rangle$, instead of u, and expressed as a column of n complex numbers x_i $(1 \leq i \leq n)$ as

$$|x\rangle = \begin{pmatrix} x_1 \\ \vdots \\ x_n \end{pmatrix}, \quad x_i \in \mathbb{C} \tag{1.1}$$

It is often written as a **transpose** of a row vector, as $|x\rangle = (x_1, x_2, \ldots, x_n)^t$, to save space. The integer $n \in \mathbb{N}$ is called the **dimension** of the vector space. In some literature, $\mathbb{C}^n$ is denoted by $V(n, \mathbb{C})$. Similary we define the real vector space $\mathbb{R}^n = V(n, \mathbb{R})$ as the set of column vectors with real entries. An element $|x\rangle$ is also called a **ket vector** or simply a **ket**. We will later introduce another kind of vector called a bra vector, which, combined with a ket vector, yields the bracket (see Eq. (1.6)). For $|x\rangle, |y\rangle \in \mathbb{C}^n$ and $a \in \mathbb{C}$, vector addition and scalar multiplication are defined as

$$|x\rangle = \begin{pmatrix} x_1 \\ x_2 \\ \vdots \\ x_n \end{pmatrix}, \ |y\rangle = \begin{pmatrix} y_1 \\ y_2 \\ \vdots \\ y_n \end{pmatrix} \Rightarrow |x\rangle + |y\rangle = \begin{pmatrix} x_1 + y_1 \\ x_2 + y_2 \\ \vdots \\ x_n + y_n \end{pmatrix}, \ a|x\rangle = \begin{pmatrix} ax_1 \\ ax_2 \\ \vdots \\ ax_n \end{pmatrix},$$

$$(1.2)$$

respectively. All the components of the **zero-vector** $|0\rangle$ are zero. The zero-vector is also written as 0 in a less strict manner. The reader should verify that these definitions satisfy all the axioms in the definition of a vector space. Note, in particular, that any **linear combination** $c_1|x\rangle + c_2|y\rangle$ of vectors $|x\rangle, |y\rangle \in \mathbb{C}^n$ with $c_1, c_2 \in \mathbb{C}$ is also an element of $\mathbb{C}^n$.

1.2   Linear Dependence and Independence of Vectors

Let us consider a set of k vectors $\{|x_1\rangle, \dots, |x_k\rangle\}$ in $V = \mathbb{C}^n$. This set is said to be **linearly dependent** if the equation

$$\sum_{i=1}^{k} c_i |x_i\rangle = |0\rangle \qquad (1.3)$$

has a solution $c_1, \dots, c_k$, at least one of which is non-vanishing. In other words, vectors $\{|x_i\rangle\}$ are linearly dependent if one of the vectors is expressed as a linear combination of the other vectors. This definition implies that any set containing the zero-vector $|0\rangle$ is linearly dependent.

   If, in contrast, the trivial solution $c_i = 0$ $(1 \le i \le k)$ is the only solution of Eq. (1.3), the set is said to be **linearly independent**.

EXERCISE 1.1 Find the condition under which two vectors

$$|v_1\rangle = \begin{pmatrix} x \\ y \\ 3 \end{pmatrix}, \ |v_2\rangle = \begin{pmatrix} 2 \\ x - y \\ 1 \end{pmatrix} \in \mathbb{R}^3$$

are linearly independent.

THEOREM 1.1 If a set of k vectors in $\mathbb{C}^n$ is linearly independent, then the number k satisifies $k \le n$. The set is always linearly dependent if $k > n$.

The proof is left as an exercise for the readers. Suppose there are n linearly independent vectors $\{|v_i\rangle\}$ in $\mathbb{C}^n$. Then any $|x\rangle \in \mathbb{C}^n$ can be expressed uniquely as a linear combination of these n vectors;

$$|x\rangle = \sum_{i=1}^{n} c_i |v_i\rangle, \quad c_i \in \mathbb{C}.$$

The set of n linearly independent vectors is called a **basis** of $\mathbb{C}^n$ and the vectors are called **basis vectors**. The vector space spanned by a basis $\{|v_i\rangle\}$ is often denoted as $\mathrm{Span}(\{|v_i\rangle\})$.

EXERCISE 1.2 Show that a set of vectors

$$|v_1\rangle = \begin{pmatrix} 1 \\ 1 \\ 1 \end{pmatrix}, \quad |v_2\rangle = \begin{pmatrix} 1 \\ 0 \\ 1 \end{pmatrix}, \quad |v_3\rangle = \begin{pmatrix} 1 \\ -1 \\ -1 \end{pmatrix}$$

is a basis of $\mathbb{C}^3$.

1.3 Dual Vector Spaces

A function $f : \mathbb{C}^n \to \mathbb{C}$ ($f : |x\rangle \mapsto f(|x\rangle) \in \mathbb{C}$) satisfing the linearity condition

$$f(c_1|x\rangle + c_2|y\rangle) = c_1 f(|x\rangle) + c_2 f(|y\rangle),$$
$$\forall |x\rangle, |y\rangle \in \mathbb{C}^n, \forall c_1, c_2 \in \mathbb{C} \tag{1.4}$$

is called a **linear function**. To express f in a component form, let us introduce a row vector $\langle \alpha|$,

$$\langle \alpha| = (\alpha_1, \ldots, \alpha_n), \quad \alpha_i \in \mathbb{C} \tag{1.5}$$

A row vector is called a **bra vector** or simply a **bra** in the following. Let us define the **inner product** of a bra vector $\langle \alpha|$ and a ket vector $|x\rangle$ by

$$\langle \alpha|x\rangle = \sum_{i=1}^{n} \alpha_i x_i \tag{1.6}$$

Note that this product is nothing but an ordinary matrix multiplication of a $1 \times n$ matrix and an $n \times 1$ matrix.

A bra vector with the above inner product induces a linear function $\langle \alpha|(|x\rangle) = \langle \alpha|x\rangle$. In fact,

$$\langle \alpha|(c_1|x\rangle + c_2|y\rangle) = \sum_i \alpha_i(c_1 x_i + c_2 y_i) = c_1 \sum_i \alpha_i x_i + c_2 \sum_i \alpha_i y_i$$
$$= c_1 \langle \alpha|x\rangle + c_2 \langle \alpha|y\rangle.$$

Conversely, any linear function can be expressed as a linear function induced by a bra vector. The bra vector is explicitly constructed once a dual basis is introduced as we will see below.

The vector space of linear functions on a vector space V ($\mathbb{C}^n$ in the present case) is called the **dual vector space**, or simply the **dual space**, of V and denoted by V^*. The symbol $*$ here denotes the dual and should not be confused with complex conjugation. As mentioned above, we may identify the set of all bra vectors with

$$\mathbb{C}^{n*} = \{\langle\alpha| = (\alpha_1, \ldots, \alpha_n)|\alpha_i \in \mathbb{C}\}. \tag{1.7}$$

The reader is encouranged to verify directly that $\mathbb{C}^{n*}$ indeed satisfies the axioms of a vector space.

An important linear function is a bra vector obtained from a ket vector. Given a vector $|x\rangle = (x_1, \ldots, x_n)^t \in \mathbb{C}^n$, define a bra vector $\langle x|$ associated to $|x\rangle$ by

$$|x\rangle \mapsto \langle x| = (x_1^*, \ldots, x_n^*) \in \mathbb{C}^{n*}, \tag{1.8}$$

Note that each component is complex-conjugated under this correspondence. When a norm of a vector $|x\rangle$ is defined by

$$\||x\rangle\| = \sqrt{\langle x|x\rangle}, \tag{1.9}$$

it takes a non-negative real value due to this convention. In fact, observe that

$$\sqrt{\langle x|x\rangle} = \left[\sum_{i=1}^{n} x_i^* x_i\right]^{1/2} = \left[\sum_{i=1}^{n} |x_i|^2\right]^{1/2} \geq 0.$$

Given vectors $|x\rangle, |y\rangle \in \mathbb{C}^n$, their inner product is given by

$$\langle x|y\rangle = \sum_{i=1}^{n} x_i^* y_i. \tag{1.10}$$

In the mathematical literature, complex conjugation is taken rather with respect to the y_i. In the present book, however, we stick to physicists' convention (1.10), which should not be confused with Eq. (1.6).

Note the following sesquilinearity:*

$$\langle x|c_1 y_1 + c_2 y_2\rangle = c_1\langle x|y_1\rangle + c_2\langle x|y_2\rangle \tag{1.11}$$

$$\langle c_1 x_1 + c_2 x_2|y\rangle = c_1^*\langle x_1|y\rangle + c_2^*\langle x_2|y\rangle, \tag{1.12}$$

where $|c_1 y_1 + c_2 y_2\rangle \equiv c_1|y_1\rangle + c_2|y_2\rangle$.

---

*sesqui = 1.5.

EXERCISE 1.3 Let

$$|x\rangle = \begin{pmatrix} 1 \\ i \\ 2+i \end{pmatrix}, \quad |y\rangle = \begin{pmatrix} 2-i \\ 1 \\ 2+i \end{pmatrix}.$$

Find $\||x\rangle\|$, $\langle x|y\rangle$ and $\langle y|x\rangle$.

EXERCISE 1.4 Prove that

$$\langle x|y\rangle = \langle y|x\rangle^*. \tag{1.13}$$

1.4 Basis, Projection Operator and Completeness Relation

1.4.1 Orthonormal Basis and Completeness Relation

Any set of n linearly independent vectors $\{|v_1\rangle, \ldots, |v_n\rangle\}$ in $\mathbb{C}^n$ is called the **basis**, and an arbitrary vector $|x\rangle \in \mathbb{C}^n$ is expressed uniquely as a linear combination of these basis vectors as $|x\rangle = \sum_{i=1}^{n} c_i |v_i\rangle$. The n complex numbers c_i are called the **components** of $|x\rangle$ with respect to the basis $\{|v_i\rangle\}$.

A basis $\{|e_i\rangle\}$ that satisfies

$$\langle e_i|e_j\rangle = \delta_{ij} \tag{1.14}$$

is called an **orthonormal basis**. Clearly the choice of $\{|e_i\rangle\}$ which satisfies the above condition is far from unique. It turns out that orthonormal bases are convenient for many purposes.

Let $|x\rangle = \sum_{i=1}^{n} c_i |e_i\rangle$. The inner product of $|x\rangle$ and $\langle e_j|$ yields

$$\langle e_j|x\rangle = \sum_{i=1}^{n} c_i \langle e_j|e_i\rangle = \sum_{i=1}^{n} c_i \delta_{ji} = c_j \to c_j = \langle e_j|x\rangle.$$

Substituting this result into the expansion of $|x\rangle$, we obtain

$$|x\rangle = \sum_{i=1}^{n} \langle e_i|x\rangle |e_i\rangle = \sum_{i=1}^{n} |e_i\rangle\langle e_i|x\rangle = \left(\sum_{i=1}^{n} |e_i\rangle\langle e_i|\right) |x\rangle.$$

Since $|x\rangle$ is arbitrary, we finally obtain the **completeness relation**

$$\sum_{i=1}^{n} |e_i\rangle\langle e_i| = I. \tag{1.15}$$

The completeness relation is quite useful and will be frequently made use of in the following.

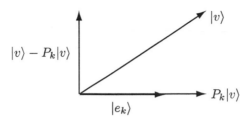

FIGURE 1.1

A vector $|v\rangle$ is projected to the direction defined by a unit vector $|e_k\rangle$ by the action of $P_k = |e_k\rangle\langle e_k|$. The difference $|v\rangle - P_k|v\rangle$ is orthogonal to $|e_k\rangle$.

1.4.2   Projection Operators

The *matrix*

$$P_k \equiv |e_k\rangle\langle e_k| \tag{1.16}$$

introduced above is called a **projection operator** in the direction defined by $|e_k\rangle$. This projects a vector $|v\rangle$ to a vector parallel to $|e_k\rangle$ in such a way that $|v\rangle - P_k|v\rangle$ is orthogonal to $|e_k\rangle$ (see Fig. 1.1).

The set $\{P_k = |e_k\rangle\langle e_k|\}$ satisfies the conditions

(i) $\qquad P_k^2 = P_k,$ $\hspace{5cm}$ (1.17)

(ii) $\qquad P_k P_j = 0 \quad (k \neq j),$ $\hspace{3.5cm}$ (1.18)

(iii) $\quad \displaystyle\sum_k P_k = I \quad$ (completeness relation). $\hspace{1cm}$ (1.19)

The conditions (i) and (ii) are obvious from the orthonormality $\langle e_j|e_k\rangle = \delta_{jk}$.

EXAMPLE 1.1 Let

$$|e_1\rangle = \frac{1}{\sqrt{2}}\begin{pmatrix} 1 \\ 1 \end{pmatrix}, |e_2\rangle = \frac{1}{\sqrt{2}}\begin{pmatrix} 1 \\ -1 \end{pmatrix}.$$

They define an orthonormal basis as is easily verified. Projection operators are

$$P_1 = |e_1\rangle\langle e_1| = \frac{1}{2}\begin{pmatrix} 1 & 1 \\ 1 & 1 \end{pmatrix}, \quad P_2 = |e_2\rangle\langle e_2| = \frac{1}{2}\begin{pmatrix} 1 & -1 \\ -1 & 1 \end{pmatrix}.$$

They satisfy the completeness relation

$$\sum_k P_k = \begin{pmatrix} 1 & 0 \\ 0 & 1 \end{pmatrix} = I$$

and the orthogonality condition

$$P_1 P_2 = \begin{pmatrix} 0 & 0 \\ 0 & 0 \end{pmatrix}.$$

The reader should verify that $P_k^2 = P_k$.

EXERCISE 1.5 Let $\{|e_k\rangle\}$ be as in Example 1.1 and let

$$|v\rangle = \begin{pmatrix} 3 \\ 2 \end{pmatrix} = \sum c_k |e_k\rangle.$$

Find the coefficients c_1 and c_2.

1.4.3   Gram-Schmidt Orthonormalization

Let us construct a set of k orthonormal vectors, given a linearly independent set of k vectors $\{|v_i\rangle\}$ in $\mathbb{C}^n$ ($k \leq n$). The first step is to define a vector

$$|e_1\rangle = \frac{|v_1\rangle}{\||v_1\rangle\|},$$

which is clearly normalized; $\||e_1\rangle\| = 1$. Before we proceed further, let us recall that the component of a vector $|u\rangle$ along $|e_k\rangle$ is given by $\langle e_k|u\rangle$. Then we define, in the next step, a vector

$$|f_2\rangle = |v_2\rangle - |e_1\rangle\langle e_1|v_2\rangle,$$

which is clearly orthogonal to $|e_1\rangle$; $\langle e_1|f_2\rangle = \langle e_1|v_2\rangle - \langle e_1|e_1\rangle\langle e_1|v_2\rangle = 0$. This vector must be normalized as

$$|e_2\rangle = \frac{|f_2\rangle}{\||f_2\rangle\|}.$$

Similarly we find, in the jth step, the vector

$$|e_j\rangle = \frac{|v_j\rangle - \sum_{i=1}^{j-1}\langle e_i|v_j\rangle|e_i\rangle}{\||v_j\rangle - \sum_{i=1}^{j-1}\langle e_i|v_j\rangle|e_i\rangle\|} \quad (1 \leq j \leq k).$$

By construction, $\{|e_1\rangle, |e_2\rangle, \ldots, |e_k\rangle\}$ is an orthonormal set, which spans a k-dimensional subspace in $\mathbb{C}^n$. This is called the **Gram-Schmidt orthonormalization**. When $k = n$, it spans the whole vector space $\mathbb{C}^n$.

EXAMPLE 1.2 Let

$$|v_1\rangle = \begin{pmatrix} 1 \\ i \end{pmatrix}, \quad |v_2\rangle = \begin{pmatrix} 2i \\ 4 \end{pmatrix}.$$

Then we obtain

$$|e_1\rangle = \frac{1}{\sqrt{2}} \begin{pmatrix} 1 \\ i \end{pmatrix}.$$

Moreover

$$|f_2\rangle = \begin{pmatrix} 2i \\ 4 \end{pmatrix} - \frac{1}{2} \begin{pmatrix} 1 \\ i \end{pmatrix} (1, -i) \begin{pmatrix} 2i \\ 4 \end{pmatrix} = \begin{pmatrix} 3i \\ 3 \end{pmatrix},$$

from which we find

$$|e_2\rangle = \frac{1}{\sqrt{2}} \begin{pmatrix} i \\ 1 \end{pmatrix}.$$

EXERCISE 1.6 (1) Use the Gram-Schmidt orthonormalization to find an orthonormal basis $\{|e_k\rangle\}$ from a linearly independent set of vectors

$$|v_1\rangle = (-1, 2, 2)^t, \quad |v_2\rangle = (2, -1, 2)^t, \quad |v_3\rangle = (3, 0, -3)^t.$$

(2) Let

$$|u\rangle = (1, -2, 7)^t = \sum_k c_k |e_k\rangle.$$

Find the coefficients c_k.

EXERCISE 1.7 Let

$$|v_1\rangle = (1, i, 1)^t, \quad |v_2\rangle = (3, 1, i)^t.$$

Find an orthonormal basis for a two-dimensional subspace spanned by $\{|v_1\rangle, |v_2\rangle\}$.

1.5 Linear Operators and Matrices

A map $A : \mathbb{C}^n \to \mathbb{C}^n$ is a **linear operator** if

$$A(c_1|x\rangle + c_2|y\rangle) = c_1 A|x\rangle + c_2 A|y\rangle \tag{1.20}$$

is satified for arbitrary $|x\rangle, |y\rangle \in \mathbb{C}^n$ and $c_k \in \mathbb{C}$. Let us choose an arbitrary orthonormal basis $\{|e_k\rangle\}$. It is shown below that A is expressed as an $n \times n$ matrix.

Let $|v\rangle = \sum_{k=1}^n v_k|e_k\rangle$ be an arbitrary vector in $\mathbb{C}^n$. Linearity implies that $A|v\rangle = \sum_k v_k A|e_k\rangle$. Therefore, the action of A on an arbitrary vector is fixed provided that its action on the basis vectors is given. Since $A|e_k\rangle \in \mathbb{C}^n$, it can be expanded as

$$A|e_k\rangle = \sum_{i=1}^n |e_i\rangle A_{ik}.$$

By taking the inner product between $\langle e_j|$ and the above equation, we obtain

$$A_{jk} = \langle e_j|A|e_k\rangle. \tag{1.21}$$

This is the matrix element of A given an orthonormal basis $\{|e_k\rangle\}$.

It is easy then to show that

$$A = \sum_{j,k} A_{jk}|e_j\rangle\langle e_k| \tag{1.22}$$

since by multiplying the completeness relation $I = \sum_{i=1}^n |e_i\rangle\langle e_i|$ from the left and the right on A simultaneously, we obtain

$$A = IAI = \sum_{j,k} |e_j\rangle\langle e_j|A|e_k\rangle\langle e_k| = \sum_{j,k} A_{jk}|e_j\rangle\langle e_k|.$$

1.5.1 Hermitian Conjugate, Hermitian and Unitary Matrices

Hermitian matrices play important role in many areas in mathematics and physics. To define a Hermitian matrix, we need to introduce the **Hermitian conjugate** operation, denoted †.[†]

DEFINITION 1.2 (Hermitian conjugate) Given a linear operator $A :$ $\mathbb{C}^n \to \mathbb{C}^n$, its Hermitian conjugate $A^\dagger$ is defined by

$$\langle u|A|v \rangle \equiv \langle A^\dagger u|v \rangle = \langle v|A^\dagger|u \rangle^*, \tag{1.23}$$

where $|u\rangle, |v\rangle$ are arbitrary vectors in $\mathbb{C}^n$.

The above definition shows that $\langle e_j|A|e_k \rangle = \langle e_k|A^\dagger|e_j \rangle^*$. Therefore, we find the relation $A_{jk} = (A^\dagger)^*_{kj}$, namely

$$(A^\dagger)_{jk} = A^*_{kj}. \tag{1.24}$$

In other words, the matrix elements of $A^\dagger$ are obtained by the transpose and the complex conjugation of A.

This definition also applies to a ket vector $|x\rangle$. We have

$$|x\rangle^\dagger = (x^*_1, \ldots, x^*_n) = \langle x|.$$

Namely, the procedure to produce a bra vector from a ket vector is regarded as a Hermitian conjugation of the ket vector.

EXERCISE 1.8 Let A and B be $n \times n$ matrices and $c \in \mathbb{C}$. Show that

$$(cA)^\dagger = c^*A^\dagger, \quad (A+B)^\dagger = A^\dagger + B^\dagger, \quad (AB)^\dagger = B^\dagger A^\dagger. \tag{1.25}$$

DEFINITION 1.3 (Hermitian matrix) A matrix $A : \mathbb{C}^n \to \mathbb{C}^n$ is said to be a **Hermitian matrix** if it satisifies $A^\dagger = A$.

Let $\{|e_1\rangle, \ldots, |e_n\rangle\}$ be an orthonormal basis in $\mathbb{C}^n$. Suppose a matrix $U :$ $\mathbb{C}^n \to \mathbb{C}^n$ satisifes $U^\dagger U = I$. By operating U on $\{|e_k\rangle\}$, we obtain a vector $|f_k\rangle = U|e_k\rangle$. These vectors are again orthonormal since

$$\langle f_j|f_k \rangle = \langle e_j|U^\dagger U|e_k \rangle = \langle e_j|e_k \rangle = \delta_{jk}. \tag{1.26}$$

Note that $|\det U| = 1$ since $\det U^\dagger U = \det U^\dagger \det U = |\det U|^2 = 1$.

[†]Mathematicians tend to use $*$ to denote Hermitian conjugate. We will follow the physicists' convention here.

DEFINITION 1.4 (Unitary matrix) Let $U : \mathbb{C}^n \to \mathbb{C}^n$ be a matrix which satisfies $U^\dagger = U^{-1}$. Then U is called a **unitary matrix**. Moreover, if U is unimodular, namely $\det U = 1$, U is said to be a **special unitary matrix**.

The set of unitary matrices is a group called the **unitary group**, while that of the special unitary matrices is a group called the **special unitary group**. They are denoted by $\mathrm{U}(n)$ and $\mathrm{SU}(n)$, respectively.

Remarks: If a real matrix $A : \mathbb{R}^n \to \mathbb{R}^n$ satisfies $A^t = A^{-1}$, A is called an orthogonal matrix. From $\det(AA^t) = \det A \det A^t = (\det A)^2 = \det I = 1$, we find that $\det A = \pm 1$. If A is unimodular, $\det A = 1$, it is called a special orthogonal matrix. The set of orthogonal (special orthogonal) matrices is a group called the **orthogonal group** (**special orthogonal group**) and denoted by $\mathrm{O}(n)$ ($\mathrm{SO}(n)$).

1.6 Eigenvalue Problems

Suppose we operate a matrix A on a vector $|v\rangle \in \mathbb{C}^n$, where $|v\rangle \neq |0\rangle$. The result $A|v\rangle$ is not proportional to $|v\rangle$ in general. If, however, $|v\rangle$ is properly chosen, we may end up with $A|v\rangle$, which is a scalar multiple of $|v\rangle$;

$$A|v\rangle = \lambda|v\rangle, \quad \lambda \in \mathbb{C}. \tag{1.27}$$

Then λ is called an **eigenvalue** of A, while $|v\rangle$ is called the corresponding **eigenvector**. The above equation being a linear equation, the norm of the eigenvector cannot be fixed. Of course, it is always possible to normalize $|v\rangle$ such that $\||v\rangle\| = 1$. We often use the symbol $|\lambda\rangle$ for an eigenvector corresponding to an eigenvalue λ to save symbols.

Let $\{|e_k\rangle\}$ be an orthonormal basis in $\mathbb{C}^n$ and let $\langle e_i|A|e_j\rangle = A_{ij}$ and $v_i = \langle e_i|v\rangle$ be the components of A and $|v\rangle$ with respect to the basis. Then the component expression for the above equation is obtained from

$$A|v\rangle = \sum_{i,j} |e_i\rangle\langle e_i|A|e_j\rangle\langle e_j|v\rangle = \sum_{i,j} A_{ij} v_j |e_i\rangle$$

as

$$\sum_j A_{ij} v_j = \lambda v_i. \tag{1.28}$$

Let us find the eigenvalue λ next. Note first that the eigenvalue equation is rewritten as

$$\sum_j (A - \lambda I)_{ij} v_j = 0.$$

This equation in v_j has nontrivial solutions if and only if the matrix $A - \lambda I$ has no inverse, namely

$$D(\lambda) \equiv \det(A - \lambda I) = 0. \tag{1.29}$$

If it had the inverse, then $|v\rangle = (A - \lambda I)^{-1}|0\rangle = 0$ would be the unique solution. This equation (1.29) is called the **characteristic equation** or the **eigen equation** of A.

Let A be an $n \times n$ matrix. Then the characteristic equation has n solutions, including the multiplicity, which we write as $\{\lambda_1, \lambda_2, \ldots, \lambda_n\}$. The function $D(\lambda)$ is also written as

$$
\begin{aligned}
D(\lambda) &= \prod_{i=1}^{n}(\lambda_i - \lambda) \\
&= (-\lambda)^n + \sum_i \lambda_i(-\lambda)^{(n-1)} + \ldots + \prod_{i=1}^{n}\lambda_i \\
&= (-\lambda)^n + \operatorname{tr} A(-\lambda)^{(n-1)} + \ldots + \det A, \tag{1.30}
\end{aligned}
$$

where use has been made of the facts $\operatorname{tr} A = \sum_i \lambda_i$ and $\det A = \prod_i \lambda_i$.

1.6.1 Eigenvalue Problems of Hermitian and Normal Matrices

The eigenvalue problems of Hermitian matrices and unitary matrices are particularly important in practical applications.

THEOREM 1.2 All the eigenvalues of a Hermitian matrix are real numbers. Moreover, two eigenvectors corresponding to different eigenvalues are orthogonal.

Proof. Let A be a Hermitian matrix and let $A|\lambda\rangle = \lambda|\lambda\rangle$. The Hermitian conjugate of this equation is $\langle\lambda|A = \lambda^*\langle\lambda|$. From these equations we obtain $\langle\lambda|A|\lambda\rangle = \lambda\langle\lambda|\lambda\rangle = \lambda^*\langle\lambda|\lambda\rangle$, which proves $\lambda = \lambda^*$ since $\langle\lambda|\lambda\rangle \neq 0$.

Let $A|\mu\rangle = \mu|\mu\rangle$ $(\mu \neq \lambda)$, next. Then $\langle\mu|A = \mu\langle\mu|$ since $\mu \in \mathbb{R}$. From $\langle\mu|A|\lambda\rangle = \lambda\langle\mu|\lambda\rangle$ and $\langle\mu|A|\lambda\rangle = \mu\langle\mu|\lambda\rangle$, we obtain $0 = (\lambda - \mu)\langle\mu|\lambda\rangle$. Since $\mu \neq \lambda$, we must have $\langle\mu|\lambda\rangle = 0$. ∎

Suppose λ is k-fold degenerate. Then there are k independent eigenvectors corresponding to λ. We may invoke to the Gram-Schmidt orthonormalization, for example, to obtain an orthonormal basis in this k-dimensional space. Accordingly, the set of eigenvectors of a Hermitian matrix is always chosen to be orthonormal. Therefore, the set of eigenvectors $\{|\lambda_k\rangle\}$ of a Hermitian matrix A may be made into a complete set

$$\sum_{k=1}^{n} |\lambda_k\rangle\langle\lambda_k| = I$$

EXAMPLE 1.3 The Pauli matrix

$$\sigma_y = \begin{pmatrix} 0 & -i \\ i & 0 \end{pmatrix}$$

is Hermitian. Let us find its eigenvalues and corresponding eigenvectors. From

$$\det(\sigma_y - \lambda I) = \lambda^2 - 1 = 0,$$

we find the eigenvalues $\lambda_1 = 1$ and $\lambda_2 = -1$. Let $|\lambda_1\rangle = (x, y)^t$ be an eigenvector corresponding to λ_1;

$$\sigma_y|\lambda_1\rangle = |\lambda_1\rangle \; \rightarrow \; \begin{pmatrix} 0 & -i \\ i & 0 \end{pmatrix} \begin{pmatrix} x \\ y \end{pmatrix} = \begin{pmatrix} x \\ y \end{pmatrix} \; \rightarrow \; \begin{cases} -iy = x \\ ix = y \end{cases}.$$

These relations are satified with $x = 1, y = i$. Thus the normalized eigenvector is

$$|\lambda_1\rangle = \frac{1}{\sqrt{2}} \begin{pmatrix} 1 \\ i \end{pmatrix}.$$

Similarly we obtain

$$|\lambda_2\rangle = \frac{1}{\sqrt{2}} \begin{pmatrix} i \\ 1 \end{pmatrix}.$$

It is easy to verify that they are orthogonal

$$\langle \lambda_1 | \lambda_2 \rangle = \frac{1}{2}(1, -i) \begin{pmatrix} i \\ 1 \end{pmatrix} = 0$$

and satisfy the completeness relation

$$\sum_k |\lambda_k\rangle\langle\lambda_k| = \frac{1}{2} \begin{pmatrix} 1 & -i \\ i & 1 \end{pmatrix} + \frac{1}{2} \begin{pmatrix} 1 & i \\ -i & 1 \end{pmatrix} = I.$$

Finally let us find a unitary matrix U which diagonalizes σ_y as

$$U^\dagger \sigma_y U = \begin{pmatrix} \lambda_1 & 0 \\ 0 & \lambda_2 \end{pmatrix}.$$

Let us consider a matrix

$$U = (|\lambda_1\rangle, |\lambda_2\rangle) = \frac{1}{\sqrt{2}} \begin{pmatrix} i & 1 \\ 1 & i \end{pmatrix}.$$

Then

$$\sigma_y U = (\sigma_y|\lambda_1\rangle, \sigma_y|\lambda_2\rangle) = (\lambda_1|\lambda_1\rangle, \lambda_2|\lambda_2\rangle)$$

from which we find

$$U^\dagger \sigma_y U = \begin{pmatrix} \langle\lambda_1| \\ \langle\lambda_2| \end{pmatrix} (\lambda_1|\lambda_1\rangle, \lambda_2|\lambda_2\rangle) = \begin{pmatrix} 1 & 0 \\ 0 & -1 \end{pmatrix}$$

as promised. Note that the unitarity of U is attributed to the orthonormality of $\{|\lambda_k\rangle\}$.

EXAMPLE 1.4 (1) The eigenvalues and the corresponding eigenvectors of σ_x are found in a similar way as the above example as $\lambda_1 = 1, \lambda_2 = -1$ and

$$|\lambda_1\rangle = \frac{1}{\sqrt{2}} \begin{pmatrix} 1 \\ 1 \end{pmatrix}, \quad |\lambda_2\rangle = \frac{1}{\sqrt{2}} \begin{pmatrix} 1 \\ -1 \end{pmatrix}.$$

(2) Let us consider the eigenvalue problem of a matrix

$$A = \begin{pmatrix} 1 & 0 & 0 & 0 \\ 0 & 1 & 0 & 0 \\ 0 & 0 & 0 & 1 \\ 0 & 0 & 1 & 0 \end{pmatrix}.$$

Note that this matrix is block diagonal with diagonal blocks I and σ_x. It is found from this observation that the eigenvalues are $1, 1, 1$ and -1. The corresponding eigenvectors are obtained by making use of the result of (1) as

$$\begin{pmatrix} 1 \\ 0 \\ 0 \\ 0 \end{pmatrix}, \begin{pmatrix} 0 \\ 1 \\ 0 \\ 0 \end{pmatrix}, \frac{1}{\sqrt{2}} \begin{pmatrix} 0 \\ 0 \\ 1 \\ 1 \end{pmatrix}, \frac{1}{\sqrt{2}} \begin{pmatrix} 0 \\ 0 \\ 1 \\ -1 \end{pmatrix}.$$

(3) Let us consider the eigenvalue problem of a matrix

$$B = \begin{pmatrix} 0 & 0 & 0 & 1 \\ 0 & 1 & 0 & 0 \\ 0 & 0 & 1 & 0 \\ 1 & 0 & 0 & 0 \end{pmatrix}.$$

Although this matrix is not block diagonal, change of the order of basis vectors from $|e_1\rangle, |e_2\rangle, |e_3\rangle, |e_4\rangle$ to $|e_3\rangle, |e_2\rangle, |e_1\rangle, |e_4\rangle$ maps the matrix B to A in (2). Therefore the eivenvalues of B are the same as those of A. (Note that the characteristic equation is left unchanged under a permutation of basis vectors.) By putting back the order of the basis vectors, the eigenvectors of A are mapped to those of B as

$$\begin{pmatrix} 0 \\ 0 \\ 1 \\ 0 \end{pmatrix}, \begin{pmatrix} 0 \\ 1 \\ 0 \\ 0 \end{pmatrix}, \frac{1}{\sqrt{2}} \begin{pmatrix} 1 \\ 0 \\ 0 \\ 1 \end{pmatrix}, \frac{1}{\sqrt{2}} \begin{pmatrix} 1 \\ 0 \\ 0 \\ -1 \end{pmatrix}.$$

EXERCISE 1.9 Let

$$A = \frac{1}{\sqrt{2}} \begin{pmatrix} 0 & 1+i \\ 1-i & 0 \end{pmatrix}.$$

Find the eigenvalues and the corresponding normalized eigenvectors. Show that the eigenvectors are mutually orthogonal and that they satisfy the completeness relation. Find a unitary matrix which diagonalizes A.

It has been shown above that eigenvalues of a Hermitian matrix are real. Note that converse is not true. For example,

$$A = \begin{pmatrix} a & b \\ -b & -a \end{pmatrix} \qquad a, b \in \mathbb{R}$$

has real eigenvalues $\pm\sqrt{a^2 - b^2}$ for $|a| \geq |b|$. How about the orthonormality of the eigenvectors?

A matrix A is **normal** if it satisfies

$$AA^\dagger = A^\dagger A. \tag{1.31}$$

THEOREM 1.3 Let A be a normal matrix. Then its eigenvectors corresponding to different eigenvalues are orthogonal.

Proof. Let us write the eigenvalue equation as $(A - \lambda_j)|\lambda_j\rangle = 0$. Then we find, from the assumed condition $[A, A^\dagger] = 0$, that

$$\langle \lambda_j |(A^\dagger - \lambda_j^*)(A - \lambda_j)|\lambda_j\rangle = \langle \lambda_j |(A - \lambda_j)(A^\dagger - \lambda_j^*)|\lambda_j\rangle = 0,$$

which implies $\langle \lambda_j |A = \lambda_j \langle \lambda_j|$. Then it follows that

$$\langle \lambda_k |A|\lambda_j\rangle = \lambda_k \langle \lambda_k |\lambda_j\rangle = \lambda_j \langle \lambda_k |\lambda_j\rangle,$$

which proves that $\langle \lambda_k |\lambda_j\rangle = 0$ for $\lambda_j \neq \lambda_k$. ∎

If some of the eigenvalues are degenerate, we may use the Gram-Schmidt procedure to make the corresponding eigenvectors orthonormal. Therefore it is always possible to assume the set of eigenvectors of a normal matrix satisfies the completeness relation.

Important examples of normal matrices are Hermitian matrices, unitary matrices and skew-Hermitian matrices; see the next exercise.

EXERCISE 1.10 (1) Suppose A is skew-Hermitian, namely $A^\dagger = -A$. Show that all the eigenvalues are pure imaginary.

(2) Let U be a unitary matrix. Show that all the eigenvalues are unimodular, namely $|\lambda_j| = 1$.

(3) Let A be a normal matrix. Show that A is Hermitian if and only if all the eigenvalues of A are real.

EXERCISE 1.11 Let

$$U = \begin{pmatrix} 0 & 0 & i \\ 0 & i & 0 \\ i & 0 & 0 \end{pmatrix}.$$

Find the eigenvalues (without calculation if possible) and the corresponding eigenvectors.

EXERCISE 1.12 Let H be a Hermitian matrix. Show that

$$U = (I + iH)(I - iH)^{-1}$$

is unitary. This transformation is called the **Cayley transformation**.

1.7   Pauli Matrices

Let us consider spin $1/2$ particles, such as an electron or a proton. These particles have an internal degree of freedom: the spin-up and spin-down states. (To be more precise, these are expressions that are relevant when the z-component of an angular momentum S_z is diagonalized. If S_x is diagonalized, for example, these two quantum states can be either "spin-right" or "spin-left.") Since the spin-up and spin-down states are orthogonal, we can take their components to be

$$|\uparrow\rangle = \begin{pmatrix} 1 \\ 0 \end{pmatrix}, \quad |\downarrow\rangle = \begin{pmatrix} 0 \\ 1 \end{pmatrix}. \tag{1.32}$$

Verify that they are eigenvectors of σ_z satisfying $\sigma_z|\uparrow\rangle = |\uparrow\rangle$ and $\sigma_z|\downarrow\rangle = -|\downarrow\rangle$. In quantum information, we often use the notations $|0\rangle = |\uparrow\rangle$ and $|1\rangle = |\downarrow\rangle$. Moreover, the states $|0\rangle$ and $|1\rangle$ are not necessarily associated with spins. They may represent any two mutually orthogonal states, such as horizontally and vertically polarized photons. Thus we are free from any physical system, even though the terminology of spin algebra may be employed.

For electrons and protons, the spin angular momentum operator is conveniently expressed in terms of the Pauli matrices σ_k as $S_k = (\hbar/2)\sigma_k$. We often employ natural units in which $\hbar = 1$. Note the tracelessness property $\text{tr}\,\sigma_k = 0$ and the Hermiticity $\sigma_k^\dagger = \sigma_k$.[‡] In addition to the Pauli matrices, we introduce the unit matrix I in the algebra, which amounts to expanding the Lie algebra $\mathfrak{su}(2)$ to $\mathfrak{u}(2)$. The Pauli matrices satisfy the anticommutation relations

$$\{\sigma_i, \sigma_j\} = \sigma_i\sigma_j + \sigma_j\sigma_i = 2\delta_{ij}I. \tag{1.33}$$

Therefore, the eigenvalues of σ_k are found to be ± 1.

The commutation relations between the Pauli matrices are

$$[\sigma_i, \sigma_j] = \sigma_i\sigma_j - \sigma_j\sigma_i = 2i\sum_k \varepsilon_{ijk}\sigma_k, \tag{1.34}$$

[‡]Mathematically speaking, these two properties imply that $i\sigma_k$ are generators of the $\mathfrak{su}(2)$ Lie algebra associated with the Lie group SU(2).

where ε_{ijk} is the totally antisymmetric tensor of rank 3, also known as the **Levi-Civita symbol**,

$$\varepsilon_{ijk} = \begin{cases} 1, & (i,j,k) = (1,2,3), (2,3,1), (3,1,2) \\ -1 & (i,j,k) = (2,1,3), (1,3,2), (3,2,1) \\ 0 & \text{otherwise.} \end{cases}$$

The commutation relations, together with the anticommutation relations, yield

$$\sigma_i \sigma_j = i \sum_{k=1}^{3} \varepsilon_{ijk} \sigma_k + \delta_{ij} I. \tag{1.35}$$

The spin-flip ("ladder") operators are defined by

$$\sigma_+ = \frac{1}{2}(\sigma_x + i\sigma_y) = \begin{pmatrix} 0 & 1 \\ 0 & 0 \end{pmatrix}, \quad \sigma_- = \frac{1}{2}(\sigma_x - i\sigma_y) = \begin{pmatrix} 0 & 0 \\ 1 & 0 \end{pmatrix}. \tag{1.36}$$

Verify that $\sigma_+| \uparrow \rangle = \sigma_-| \downarrow \rangle = 0$, $\sigma_+| \downarrow \rangle = | \uparrow \rangle$, $\sigma_-| \uparrow \rangle = | \downarrow \rangle$. The projection operators to the eigenspaces of σ_z with the eigenvalues ± 1 are

$$P_+ = | \uparrow \rangle \langle \uparrow | = \tfrac{1}{2}(I + \sigma_z) = \begin{pmatrix} 1 & 0 \\ 0 & 0 \end{pmatrix},$$

$$P_- = | \downarrow \rangle \langle \downarrow | = \tfrac{1}{2}(I - \sigma_z) = \begin{pmatrix} 0 & 0 \\ 0 & 1 \end{pmatrix}. \tag{1.37}$$

In fact, it is straightforward to show

$$P_+| \uparrow \rangle = | \uparrow \rangle, \quad P_+| \downarrow \rangle = 0, \quad P_-| \uparrow \rangle = 0, \quad P_-| \downarrow \rangle = | \downarrow \rangle.$$

Finally, we note the following identities:

$$\sigma_\pm^2 = 0, \quad P_\pm^2 = P_\pm, \quad P_+ P_- = 0. \tag{1.38}$$

1.8 Spectral Decomposition

Spectral decomposition of a normal matrix is quite a powerful technique in several applications.

THEOREM 1.4 Let A be a normal matrix with eigenvalues $\{\lambda_i\}$ and eigenvectors $\{|\lambda_i\rangle\}$, which are assumed to be orthonormal. Then A is decomposed as

$$A = \sum_i \lambda_i |\lambda_i\rangle \langle \lambda_i|,$$

which is called the **spectral decomposition** of A.

Proof. This is a straightforward consequence of the completeness relation

$$I = \sum_{i=1}^{n} |\lambda_i\rangle\langle\lambda_i|.$$

If we operate A on the above equation from the left, we obtain

$$A = AI = \sum_{i=1}^{n} A|\lambda_i\rangle\langle\lambda_i| = \sum_{i=1}^{n} \lambda_i|\lambda_i\rangle\langle\lambda_i|,$$

which proves the theorem.   ∎

Let us recall that $P_i = |\lambda_i\rangle\langle\lambda_i|$ is a projection operator onto the direction of $|\lambda_i\rangle$. Then the spectral decomposition claims that the operation of A in the one-dimensional subspace spanned by $|\lambda_i\rangle$ is equivalent with a multiplication by a scalar λ_i. This observation reveals a neat way to obtain the spectral decomposition of a normal matrix. Let A be a normal matrix and let $\{\lambda_\alpha\}$ and $\{|\lambda_{\alpha,p}\rangle \ (1 \leq p \leq g_\alpha)\}$ be the sets of eigenvalues and eigenvectors, respectively. Here we use subscripts $\alpha, \beta, \ldots$ to denote distinct eigenvalues, while g_α denotes the degeneracy of the eigenvalue λ_α, namely λ_α has g_α linearly independent eigenvectors, which are indexed by p. Therefore we have

$$\sum_\alpha 1 \leq n, \quad \sum_\alpha g_\alpha = \sum_i 1 = n.$$

Now consider the following expression:

$$P_\alpha = \frac{\prod_{\beta \neq \alpha}(A - \lambda_\beta I)}{\prod_{\gamma \neq \alpha}(\lambda_\alpha - \lambda_\gamma)}. \tag{1.39}$$

This is a projection operator onto the g_α-dimensional space corresponding to the eigenvalue λ_α. In fact, it is straightforward to verify that

$$P_\alpha|\lambda_{\alpha,p}\rangle = \frac{\prod_{\beta \neq \alpha}(\lambda_\alpha - \lambda_\beta)}{\prod_{\gamma \neq \alpha}(\lambda_\alpha - \lambda_\gamma)}|\lambda_{\alpha,p}\rangle = |\lambda_{\alpha,p}\rangle \quad (1 \leq p \leq g_\alpha)$$

and

$$P_\alpha|\lambda_{\delta,q}\rangle = \frac{\prod_{\beta \neq \alpha}(\lambda_\delta - \lambda_\beta)}{\prod_{\gamma \neq \alpha}(\lambda_\alpha - \lambda_\gamma)}|\lambda_{\delta,q}\rangle = 0 \quad (\delta \neq \alpha, \ 1 \leq q \leq g_\delta)$$

since one of $\beta(\neq \alpha)$ is equal to $\delta(\neq \alpha)$ in the numerator. Therefore, we conclude that P_α is a projection operator

$$P_\alpha = \sum_{p=1}^{g_\alpha} |\lambda_{\alpha,p}\rangle\langle\lambda_{\alpha,p}| \tag{1.40}$$

onto the g_α-dimensional subspace correcponding to the eigenvalue λ_α. It follows from Eq. (1.40) that rank $P_\alpha = g_\alpha$. Note also that

$$AP_\alpha = \lambda_\alpha P_\alpha. \tag{1.41}$$

The above method is particularly suitable when the eigenvalues are degenerate. It is also useful when eigenvectors are difficult to obtain or unnecessary.

EXAMPLE 1.5 Let us take σ_y as an example. We found in Example 1.3 that the eigenvalues are $\lambda_1 = +1$ and $\lambda_2 = -1$, from which we obtain the projection operators directly by using Eq. (1.39) as

$$P_1 = \frac{(\sigma_y - (-I))}{(1 - (-1))} = \frac{1}{2}\begin{pmatrix} 1 & -i \\ i & 1 \end{pmatrix}, \quad P_2 = \frac{(\sigma_y - I)}{(-1 - 1)} = \frac{1}{2}\begin{pmatrix} 1 & i \\ -i & 1 \end{pmatrix}.$$

We find the spectral decomposition of σ_y as

$$\sigma_y = \sum_i \lambda_i P_i = \frac{1}{2}\begin{pmatrix} 1 & -i \\ i & 1 \end{pmatrix} + (-1)\frac{1}{2}\begin{pmatrix} 1 & i \\ -i & 1 \end{pmatrix}.$$

One of the advantages of the spectral decomposition is that a function of a matrix is evaluated quite easily. Let us prove the following formula.

PROPOSITION 1.1 Let A be a normal matrix in the above theorem. Then for an arbitrary $n \in \mathbb{N}$, we obtain

$$A^n = \sum_\alpha \lambda_\alpha^n P_\alpha. \tag{1.42}$$

If, furthermore, A^{-1} exists, the above formula may be extended to $n \in \mathbb{Z}$ by noting that λ_α^{-1} is an eigenvalue of A^{-1}.

Proof. Let $n \in \mathbb{N}$. Then

$$A^n P_\alpha = \lambda_\alpha A^{n-1} P_\alpha = \ldots = \lambda_\alpha^{n-1} AP_\alpha = \lambda_\alpha^n P_\alpha,$$

from which we obtain

$$A^n = A^n \sum_\alpha P_\alpha = \sum_\alpha A^n P_\alpha = \sum_\alpha \lambda_\alpha^n P_\alpha.$$

To prove the second half of the proposition, we only need to show that A^{-1} has an eigenvalue λ_α^{-1}, provided that A^{-1} exists (and hence $\lambda_\alpha \neq 0$), and the corresponding projection operator is P_α. We find

$$|\lambda_{\alpha,p}\rangle = A^{-1}A|\lambda_{\alpha,p}\rangle = \lambda_\alpha A^{-1}|\lambda_{\alpha,p}\rangle \to A^{-1}|\lambda_{\alpha,p}\rangle = \lambda_\alpha^{-1}|\lambda_{\alpha,p}\rangle.$$

Therefore the projection operator corresponding to the eivengalue λ_α^{-1} is P_α. The case $n = 0$, $I = \sum_\alpha P_\alpha$, is nothing but the completeness relation. Now we have proved that Eq. (1.42) applies to an arbitrary $n \in \mathbb{Z}$. ∎

From the above proposition, we obtain for a normal matrix A and an arbitrary analytic function $f(x)$,

$$f(A) = \sum_\alpha f(\lambda_\alpha) P_\alpha. \tag{1.43}$$

Even when $f(x)$ does not admit a series expansion, we may still formally define $f(A)$ by Eq. (1.43). Let $f(x) = \sqrt{x}$ and $A = \sigma_y$, for example. Then we obtain from Example 1.5 that

$$\sqrt{\sigma_y} = (\pm 1) P_1 + (\pm i) P_2.$$

It is easy to show that the RHS squares to σ_y. However, there are four possible $\sqrt{\sigma}$ depending on the choice of $\pm$ for each eigenvalue. Therefore the spectral decomposition is not unique in this case. Of course this ambiguity originates in the choice of the branch in the definition of $\sqrt{x}$.

EXAMPLE 1.6 Let us consider σ_y again. It follows directly from Example 1.5 that

$$\exp(i\alpha\sigma_y) \equiv \sum_{k=0}^{\infty} \frac{(i\alpha\sigma_y)^k}{k!} = e^{i\alpha} P_1 + e^{-i\alpha} P_2 = \begin{pmatrix} \cos\alpha & \sin\alpha \\ -\sin\alpha & \cos\alpha \end{pmatrix}.$$

EXERCISE 1.13 Suppose a 2×2 matrix A has eigenvalues $-1, 3$ and the corresponding eigenvectors

$$|e_1\rangle = \frac{1}{\sqrt{2}} \begin{pmatrix} -1 \\ i \end{pmatrix}, \quad |e_2\rangle = \frac{1}{\sqrt{2}} \begin{pmatrix} 1 \\ i \end{pmatrix},$$

respectively. Find A.

EXERCISE 1.14 Let

$$A = \begin{pmatrix} 2 & 1 \\ 1 & 2 \end{pmatrix}.$$

(1) Find the eigenvalues and the corresponding normalized eigenvectors of A.
(2) Write down the spectral decomposition of A.
(3) Find $\exp(i\alpha A)$.

EXERCISE 1.15 Let

$$A = \begin{pmatrix} 5 & -2 & -4 \\ -2 & 2 & 2 \\ -4 & 2 & 5 \end{pmatrix}.$$

(1) Find the eigenvalues and the corresponding eigenvectors of A.
(2) Find the spectral decomposition of A.
(3) Find the inverse of A by making use of the spectral decomposition.

Now we prove a formula which will turn out to be very useful in the following. This is a generalization of Example 1.6

PROPOSITION 1.2 Let $\hat{\boldsymbol{n}} \in \mathbb{R}^3$ be a unit vector and $\alpha \in \mathbb{R}$. Then

$$\exp\left(i\alpha\hat{\boldsymbol{n}} \cdot \boldsymbol{\sigma}\right) = \cos\alpha I + i(\hat{\boldsymbol{n}} \cdot \boldsymbol{\sigma})\sin\alpha, \tag{1.44}$$

where $\boldsymbol{\sigma} = (\sigma_x, \sigma_y, \sigma_z)$.

Proof. Let

$$A = \boldsymbol{n} \cdot \boldsymbol{\sigma} = \begin{pmatrix} n_z & n_x - in_y \\ n_x + in_y & -n_z \end{pmatrix}.$$

The eigenvalues of A are $\lambda_1 = +1$ and $\lambda_2 = -1$. It then follows that

$$P_1 = \frac{(A+I)}{2} = \frac{1}{2}\begin{pmatrix} 1+n_z & n_x - in_y \\ n_x + in_y & 1 - n_z \end{pmatrix},$$

$$P_2 = \frac{(A-I)}{-2} = \frac{1}{2}\begin{pmatrix} 1-n_z & -n_x + in_y \\ -n_x - in_y & 1 + n_z \end{pmatrix},$$

from which we readily find

$$e^{i\alpha A} = \frac{e^{i\alpha}}{2}\begin{pmatrix} 1+n_z & n_x - in_y \\ n_x + in_y & 1 - n_z \end{pmatrix} + \frac{e^{-i\alpha}}{2}\begin{pmatrix} 1-n_z & -n_x + in_y \\ -n_x - in_y & 1 + n_z \end{pmatrix}$$

$$= \cos\alpha I + i(\boldsymbol{n} \cdot \boldsymbol{\sigma})\sin\alpha.$$

∎

EXERCISE 1.16 Let $f : \mathbb{C} \to \mathbb{C}$ be an analytic function. Let $\hat{\boldsymbol{n}}$ be a real three-dimensional unit vector and α be a real number. Show that

$$f(\alpha\hat{\boldsymbol{n}} \cdot \boldsymbol{\sigma}) = \frac{f(\alpha) + f(-\alpha)}{2}I + \frac{f(\alpha) - f(-\alpha)}{2}\hat{\boldsymbol{n}} \cdot \boldsymbol{\sigma}. \tag{1.45}$$

(c.f., Proposition 1.2.)

1.9 Singular Value Decomposition (SVD)

A subject somewhat related to the eigenvalue problem is the singular value decomposition. In a sense, it is a generalization of the eigenvalue problem to arbitrary matrices.

THEOREM 1.5 Let A be an $m \times n$ matrix with complex entries. Then it is possible to decompose A as

$$A = U\Sigma V^\dagger, \tag{1.46}$$

where $U \in U(m), V \in U(n)$ and Σ is an $m \times n$ matrix whose diagonals are nonnegative real numbers, called the **singular values**, while all the off diagonal components are zero. The matrix Σ is called the **singular value matrix**.

The decompostion (1.46) is called the **singular value decomosition** and is often abbreviated as SVD.

We now sketch the proof of the decomposition. Let us assume $m > n$ for definiteness. Consider the eigenvalue problem of an $n \times n$ Hermitian matrix $A^\dagger A$;

$$A^\dagger A |\lambda_i\rangle = \lambda_i |\lambda_i\rangle \quad (1 \le i \le n),$$

where λ_i is a nonnegative real number, where nonnegativity follows from the observation $\lambda_i = \langle \lambda_i | \lambda_i | \lambda_i \rangle = \langle \lambda_i | A^\dagger A | \lambda_i \rangle = \|A|\lambda_i\rangle\|^2 \ge 0$. Note that the set $\{|\lambda_i\rangle\}$ satisfies the completeness relation

$$\sum_{i=1}^{n} |\lambda_i\rangle\langle\lambda_i| = I_n$$

if they are made orthonormal by the Gram-Schmidt orthonormalization. We assume r of the eigenvalues are strictly positive and $n - r$ are zero. The set $\{\lambda_i\}$ is arranged in nonincreasing order $\lambda_1 \ge \lambda_2 \ge \ldots \ge \lambda_r > 0$ while $\lambda_{r+1} = \ldots = \lambda_n = 0$. Now define

$$V = (|\lambda_1\rangle, |\lambda_2\rangle, \ldots, |\lambda_r\rangle, |\lambda_{r+1}\rangle, \ldots, |\lambda_n\rangle),$$

$$\Sigma = \begin{pmatrix} \sqrt{\lambda_1} & & & & & & \\ & \sqrt{\lambda_2} & & & & & \\ & & \ddots & & & & \\ & & & \sqrt{\lambda_r} & & & \\ & & & & 0 & & \\ & & & & & 0 & \\ & & & & & & \ddots \end{pmatrix},$$

and

$$U = (|\mu_1\rangle, |\mu_2\rangle, \ldots, |\mu_r\rangle, |\mu_{r+1}\rangle, \ldots, |\mu_m\rangle),$$

where

$$|\mu_k\rangle = \frac{1}{\sqrt{\lambda_k}} A|\lambda_k\rangle \quad (1 \le k \le r),$$

while other orthonormal vectors $|\mu_{r+1}\rangle, \ldots, |\mu_m\rangle$ are taken to be orthogonal to $|\mu_k\rangle$ $(1 \le k \le r)$. Note that $V \in U(n)$ and $U \in U(m)$ by construction. Then we find

$$U\Sigma V^\dagger$$

$$= \left(\frac{1}{\sqrt{\lambda_1}} A|\lambda_1\rangle, \frac{1}{\sqrt{\lambda_2}} A|\lambda_2\rangle, \ldots \frac{1}{\sqrt{\lambda_r}} A|\lambda_r\rangle, |\mu_{r+1}\rangle, \ldots, |\mu_m\rangle \right) \begin{pmatrix} \sqrt{\lambda_1}\langle\lambda_1| \\ \sqrt{\lambda_2}\langle\lambda_2| \\ \vdots \\ \sqrt{\lambda_r}\langle\lambda_r| \\ 0 \\ \vdots \\ 0 \end{pmatrix}$$

$$= A \sum_{i=1}^{r} |\lambda_i\rangle\langle\lambda_i| = A \sum_{i=1}^{n} |\lambda_i\rangle\langle\lambda_i| = A,$$

where we noted that $A|\lambda_i\rangle = 0$ for $r + 1 \le i \le n$ and use has been made of the completeness relation of $\{|\lambda_i\rangle\}$. The reader should examine the case in which $m \le n$. ∎

EXAMPLE 1.7 Let

$$A = \begin{pmatrix} 1 & 1 \\ 0 & 0 \\ i & i \end{pmatrix}$$

for which

$$A^\dagger A = \begin{pmatrix} 2 & 2 \\ 2 & 2 \end{pmatrix}.$$

The eigenvalues of $A^\dagger A$ are $\lambda_1 = 4$ and $\lambda_2 = 0$ with the corresponding eigenvectors

$$|\lambda_1\rangle = \frac{1}{\sqrt{2}} \begin{pmatrix} 1 \\ 1 \end{pmatrix}, \quad |\lambda_2\rangle = \frac{1}{\sqrt{2}} \begin{pmatrix} -1 \\ 1 \end{pmatrix}.$$

Unitary matrix V and the singular value matrix Σ are found from these data as

$$V = \frac{1}{\sqrt{2}} \begin{pmatrix} 1 & -1 \\ 1 & 1 \end{pmatrix} \text{ and } \Sigma = \begin{pmatrix} 2 & 0 \\ 0 & 0 \\ 0 & 0 \end{pmatrix}.$$

To construct U, we need

$$|\mu_1\rangle = \frac{1}{2} A|\lambda_1\rangle = \frac{1}{\sqrt{2}} (1, 0, i)^t$$

and two other vectors orthogonal to $|\mu_1\rangle$. By inspection, we find

$$|\mu_2\rangle = (0, 1, 0)^t \text{ and } |\mu_3\rangle = \frac{1}{\sqrt{2}}(1, 0, -i)^t,$$

for example. From these vectors we construct U as

$$U = \frac{1}{\sqrt{2}} \begin{pmatrix} 1 & 0 & 1 \\ 0 & \sqrt{2} & 0 \\ i & 0 & -i \end{pmatrix}.$$

The reader should verify that $U\Sigma V^\dagger$ really reproduces A.

EXERCISE 1.17 Find the SVD of

$$A = \begin{pmatrix} 1 & 0 & i \\ i & 0 & 1 \end{pmatrix}.$$

1.10 Tensor Product (Kronecker Product)

DEFINITION 1.5 Let A be an $m \times n$ matrix and let B be a $p \times q$ matrix. Then

$$A \otimes B = \begin{pmatrix} a_{11}B, a_{12}B, \ldots, a_{1n}B \\ a_{21}B, a_{22}B, \ldots, a_{2n}B \\ \cdots \\ a_{m1}B, a_{m2}B, \ldots, a_{mn}B \end{pmatrix} \tag{1.47}$$

is an $(mp) \times (nq)$ matrix called the **tensor product (Kronecker product)** of A and B.

It should be noted that not all $(mp) \times (nq)$ matrices are tensor products of an $m \times n$ matrix and a $p \times q$ matrix. In fact, an $(mp) \times (np)$ matrix has $mnpq$ degrees of freedom, while $m \times n$ and $p \times q$ matrices have $mn + pq$ in total. Observe that $mnpq \gg mn + pq$ for large enough m, n, p and q. This fact is ultimately related to the power of quantum computing compared to its classical counterpart.

EXAMPLE 1.8

$$\sigma_x \otimes \sigma_z = \begin{pmatrix} 0 & \sigma_z \\ \sigma_z & 0 \end{pmatrix} = \begin{pmatrix} 0 & 0 & 1 & 0 \\ 0 & 0 & 0 & -1 \\ 1 & 0 & 0 & 0 \\ 0 & -1 & 0 & 0 \end{pmatrix}.$$

EXAMPLE 1.9 We can also apply the tensor product to vectors as a special case. Let

$$|u\rangle = \begin{pmatrix} a \\ b \end{pmatrix}, \quad |v\rangle = \begin{pmatrix} c \\ d \end{pmatrix}.$$

Then we obtain

$$|u\rangle \otimes |v\rangle = \begin{pmatrix} a|v\rangle \\ b|v\rangle \end{pmatrix} = \begin{pmatrix} ac \\ ad \\ bc \\ bd \end{pmatrix}.$$

The tensor product $|u\rangle \otimes |v\rangle$ is often abbreviated as $|u\rangle|v\rangle$ or $|uv\rangle$ when it does not cause confusion.

EXERCISE 1.18 Let A and B be as above and let C be an $n \times r$ matrix and D be a $q \times s$ matrix. Show that

$$(A \otimes B)(C \otimes D) = (AC) \otimes (BD). \tag{1.48}$$

It similarly holds that

$$(A_1 \otimes B_1)(A_2 \otimes B_2)(A_3 \otimes B_3) = (A_1 A_2 A_3) \otimes (B_1 B_2 B_3),$$

and its generalizations whenever the dimensions of the matrices match so that the products make sense.

EXERCISE 1.19 Show that

$$A \otimes (B + C) = A \otimes B + A \otimes C \tag{1.49}$$
$$(A \otimes B)^\dagger = A^\dagger \otimes B^\dagger \tag{1.50}$$
$$(A \otimes B)^{-1} = A^{-1} \otimes B^{-1} \tag{1.51}$$

whenever the matrix operations are well-defined.

Show, from the above observations, that the tensor product of two unitary matrices is also unitary and that the tensor product of two Hermitian matrices is also Hermitian.

EXERCISE 1.20 Let A and B be an $m \times m$ matrix and a $p \times p$ matrix, respectively. Show that

$$\mathrm{tr}(A \otimes B) = (\mathrm{tr}A)(\mathrm{tr}B),$$
$$\det(A \otimes B) = (\det A)^p (\det B)^m.$$

EXERCISE 1.21 Let $|a\rangle, |b\rangle, |c\rangle, |d\rangle \in \mathbb{C}^n$. Show that

$$(|a\rangle\langle b|) \otimes (|c\rangle\langle d|) = (|a\rangle \otimes |c\rangle)(\langle b| \otimes \langle d|) = |ac\rangle\langle bd|.$$

THEOREM 1.6 Let A be an $m \times m$ matrix and B be a $p \times p$ matrix. Let A have the eigenvalues $\lambda_1, \ldots, \lambda_m$ with the corresponding eigenvectors $|u_1\rangle, \ldots, |u_m\rangle$ and let B have the eigenvalues $\mu_1, \ldots, \mu_p$ with the corresponding eigenvectors $|v_1\rangle, \ldots, |v_p\rangle$. Then $A \otimes B$ has mp eigenvalues $\{\lambda_j \mu_k\}$ with the corresponding eigenvectors $\{|u_j v_k\rangle\}$.

Proof. We show that $|u_j v_k\rangle$ is an eigenvector. In fact,

$$(A \otimes B)(|u_j v_k\rangle) = (A|u_j\rangle) \otimes (B|v_k\rangle) = (\lambda_j |u_j\rangle) \otimes (\mu_k |v_k\rangle)$$
$$= \lambda_j \mu_k (|u_j v_k\rangle) \ .$$

Therefore, the eigenvalue is $\lambda_j \mu_k$ with the corresponding eigenvector $|u_j v_k\rangle$. Since there are mp eigenvectors, the vectors $|u_j v_k\rangle$ exhaust all of them. ∎

EXERCISE 1.22 Let A and B be as above. Show that $A \otimes I_p + I_m \otimes B$ has the eigenvalues $\{\lambda_j + \mu_k\}$ with the corresponding eigenvectors $\{|u_j v_k\rangle\}$, where I_p is the $p \times p$ unit matrix.

2

Framework of Quantum Mechanics

Quantum mechanics is founded on several postulates, which cannot be proven theoretically. They are justified only through an empirical fact that they are consistent with all the known experimental results. The choice of the postulates depends heavily on authors' taste. Here we give one that turns out to be the most convenient in the study of quantum information and computation. For a general introduction to quantum mechanics, we recommend [1, 2, 3, 4], for example. [5] and [6] contain more advanced subjects than those treated in this chapter.

2.1 Fundamental Postulates

Quantum mechanics was discovered roughly a century ago. In spite of its long history, the interpretation of the wave function remains an open question. Here we adopt the most popular one, called the **Copenhagen interpretation**.

A 1 A pure state in quantum mechanics is represented in terms of a normalized vector $|\psi\rangle$ in a Hilbert space $\mathcal{H}$ (a complex vector space with an inner product): $\langle\psi|\psi\rangle = 1$. Suppose two states $|\psi_1\rangle$ and $|\psi_2\rangle$ are physical states of the system. Then their linear superposition $c_1|\psi_1\rangle + c_2|\psi_2\rangle$ ($c_k \in \mathbb{C}$) is also a possible state of the same system. This is called the **superposition principle**.

A 2 For any physical quantity (i.e., **observable**) a, there exists a corresponding Hermitian operator A acting on the Hilbert space $\mathcal{H}$. When we make a measurement of a, we obtain one of the eigenvalues λ_j of the operator A. Let λ_1 and λ_2 be two eigenvalues of A: $A|\lambda_i\rangle = \lambda_i|\lambda_i\rangle$. Suppose the system is in a superposition state $c_1|\lambda_1\rangle + c_2|\lambda_2\rangle$. If we measure a in this state, then the state undergoes an abrupt change to one of the eigenstates corresponding to the observed eigenvalue: If the observed eigenvalue is λ_1 (λ_2), the system undergoes a **wave function collapse** as follows: $c_1|\lambda_1\rangle + c_2|\lambda_2\rangle \rightarrow |\lambda_1\rangle$ ($|\lambda_2\rangle$), and the state immediately after the measurement is $|\lambda_1\rangle$ ($|\lambda_2\rangle$). Suppose we prepare many copies of the state $c_1|\lambda_1\rangle + c_2|\lambda_2\rangle$. The probability of collapsing to the state $|\lambda_k\rangle$ is given by $|c_k|^2$ ($k = 1, 2$). In this sense, the complex coefficient c_i is called the **probability amplitude**. It should be noted that a measurement produces one outcome λ_i and the probability of obtaining it is experimentally evaluated only after repeating measurements with many copies of the same state. These statements are easily generalized to states in a superposition of more than two states.

A 3 The time dependence of a state is governed by the **Schrödinger equation**

$$i\hbar\frac{\partial|\psi\rangle}{\partial t} = H|\psi\rangle, \tag{2.1}$$

where $\hbar$ is a physical constant known as the **Planck constant** and H is a Hermitian operator (matrix) corresponding to the energy of the system and is called the **Hamiltonian**.

Several comments are in order.

- In Axiom A 1, the phase of the vector may be chosen arbitrarily; $|\psi\rangle$ in fact represents the "ray" $\{e^{i\alpha}|\psi\rangle \,|\alpha \in \mathbb{R}\}$. This is called the **ray representation**. In other words, we can totally ignore the phase of a vector since it has no observable consequence. Note, however, that the *relative* phase of two different states is meaningful. Although $|\langle\phi|e^{i\alpha}\psi\rangle|^2$ is independent of α, $|\langle\phi|\psi_1 + e^{i\alpha}\psi_2\rangle|^2$ does depend on α.

- Axiom A 2 may be formulated in a different but equivalent way as follows. Suppose we would like to measure an observable a. Let $A = \sum_i \lambda_i|\lambda_i\rangle\langle\lambda_i|$ be the corresponding operator, where $A|\lambda_i\rangle = \lambda_i|\lambda_i\rangle$. Then the expectation value $\langle A\rangle$ of a after measurements with respect to many copies of a state $|\psi\rangle$ is

$$\langle A\rangle = \langle\psi|A|\psi\rangle. \tag{2.2}$$

Let us expand $|\psi\rangle$ in terms of $|\lambda_i\rangle$ as $|\psi\rangle = \sum_i c_i|\lambda_i\rangle$ to show the equivalence between two formalisms. According to A 2, the probability of observing λ_i upon measurement of a is $|c_i|^2$, and therefore the expectation value after many measurements is $\sum_i \lambda_i|c_i|^2$. If, conversely, Eq. (2.2) is employed, we will obtain the same result since

$$\langle\psi|A|\psi\rangle = \sum_{i,j} c_j^* c_i\langle\lambda_j|A|\lambda_i\rangle = \sum_{i,j} c_j^* c_i\lambda_i\delta_{ij} = \sum_i \lambda_i|c_i|^2.$$

This measurement is called the **projective measurement**. Any particular outcome λ_i will be found with the probability

$$|c_i|^2 = \langle\psi|P_i|\psi\rangle, \tag{2.3}$$

where $P_i = |\lambda_i\rangle\langle\lambda_i|$ is the projection operator, and the state immediately after the measurement is $|\lambda_i\rangle$ or equivalently

$$\frac{P_i|\psi\rangle}{\sqrt{\langle\psi|P_i|\psi\rangle}}, \tag{2.4}$$

where the overall phase has been ignored.

• The Schrödinger equation (2.1) in Axiom A 3 is formally solved to yield

$$|\psi(t)\rangle = e^{-iHt/\hbar}|\psi(0)\rangle, \tag{2.5}$$

if the Hamiltonian H is time-independent, while

$$|\psi(t)\rangle = \mathcal{T}\exp\left[-\frac{i}{\hbar}\int_0^t H(t)dt\right]|\psi(0)\rangle \tag{2.6}$$

if H depends on t, where $\mathcal{T}$ is the time-ordering operator defined by

$$\mathcal{T}[A(t_1)B(t_2)] = \begin{cases} A(t_1)B(t_2), & t_1 > t_2 \\ B(t_2)A(t_1), & t_2 \geq t_1 \end{cases},$$

for a product of two operators. Generalization to products of more than two operators should be obvious. We write Eqs. (2.5) and (2.6) as $|\psi(t)\rangle = U(t)|\psi(0)\rangle$. The operator $U(t) : |\psi(0)\rangle \mapsto |\psi(t)\rangle$, which we call the **time-evolution operator**, is unitary. Unitarity of $U(t)$ guarantees that the norm of $|\psi(t)\rangle$ is conserved:

$$\langle\psi(0)|U^\dagger(t)U(t)|\psi(0)\rangle = \langle\psi(0)|\psi(0)\rangle = 1.$$

EXERCISE 2.1 (Uncertainty Principle)
(1) Let A and B be Hermitian operators and $|\psi\rangle$ be some quantum state on which A and B operate. Show that

$$|\langle\psi|[A,B]|\psi\rangle|^2 + |\langle\psi|\{A,B\}|\psi\rangle|^2 = 4|\langle\psi|AB|\psi\rangle|^2.$$

(2) Prove the Cauchy-Schwarz inequality

$$|\langle\psi|AB|\psi\rangle|^2 \leq \langle\psi|A^2|\psi\rangle\langle\psi|B^2|\psi\rangle.$$

(3) Show that

$$|\langle\psi|[A,B]|\psi\rangle|^2 \leq 4\langle\psi|A^2|\psi\rangle\langle\psi|B^2|\psi\rangle.$$

(4) Show that

$$\Delta(A)\Delta(B) \geq \frac{1}{2}|\langle\psi|[A,B]|\psi\rangle|, \tag{2.7}$$

where $\Delta(A) \equiv \sqrt{\langle\psi|A^2|\psi\rangle - \langle\psi|A|\psi\rangle^2}$.
(5) Suppose $A = Q$ and $B = P \equiv \dfrac{\hbar}{i}\dfrac{d}{dQ}$. Deduce from the above arguments that

$$\Delta(Q)\Delta(P) \geq \frac{\hbar}{2}.$$

The uncertaintly principle in terms of standard deviation has been formulated first in [7] and [8].

2.2   Some Examples

We now give some examples to clarify the axioms introduced in the previous section. They turn out to have relevance to certain physical realizations of a quantum computer.

EXAMPLE 2.1 Let us consider a time-independent Hamiltonian

$$H = -\frac{\hbar}{2}\omega\sigma_x. \tag{2.8}$$

Suppose the system is in the eigenstate of σ_z with the eigenvalue $+1$ at time $t = 0$;

$$|\psi(0)\rangle = \begin{pmatrix} 1 \\ 0 \end{pmatrix}.$$

The wave function $|\psi(t)\rangle$ $(t > 0)$ is then found from Eq. (2.5) to be

$$|\psi(t)\rangle = \exp\left(i\frac{\omega}{2}\sigma_x t\right)|\psi(0)\rangle. \tag{2.9}$$

The matrix exponential function in this equation is evaluated with the help of Eq. (1.44) and we find

$$|\psi(t)\rangle = \begin{pmatrix} \cos\omega t/2 & i\sin\omega t/2 \\ i\sin\omega t/2 & \cos\omega t/2 \end{pmatrix}\begin{pmatrix} 1 \\ 0 \end{pmatrix} = \begin{pmatrix} \cos\omega t/2 \\ i\sin\omega t/2 \end{pmatrix}. \tag{2.10}$$

Suppose we measure the observable σ_z. Note that $|\psi(t)\rangle$ is expanded in terms of the eigenvectors of σ_z as

$$|\psi(t)\rangle = \cos\frac{\omega}{2}t|\sigma_z = +1\rangle + i\sin\frac{\omega}{2}t|\sigma_z = -1\rangle.$$

Therefore we find the spin is in the spin-up state with the probability $P_\uparrow(t) = \cos^2(\omega t/2)$ and in the spin-down state with the probability $P_\downarrow(t) = \sin^2(\omega t/2)$ as depicted in Fig. 2.1. Of course, the total probability is independent of time since $\cos^2(\omega t/2) + \sin^2(\omega t/2) = 1$. This result is consistent with classical spin dynamics. The Hamiltonian (2.8) depicts a spin under a magnetic field along the x-axis. Our initial condition signifies that the spin points the z-direction at $t = 0$. Then the spin starts precession around the x-axis, and the z-component of the spin oscillates sinusoidally as is shown above.

Next let us take the initial state

$$|\psi(0)\rangle = \frac{1}{\sqrt{2}}\begin{pmatrix} 1 \\ 1 \end{pmatrix},$$

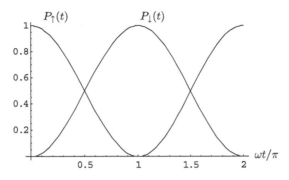

FIGURE 2.1
Probability $P_\uparrow(t)$ with which a spin is observed in the $\uparrow$-state and $P_\downarrow(t)$ observed in the $\downarrow$-state.

which is an eigenvector of σ_x (and hence the Hamiltonian) with the eigenvalue $+1$. We find $|\psi(t)\rangle$ in this case as

$$|\psi(t)\rangle = \begin{pmatrix} \cos\omega t/2 & i\sin\omega t/2 \\ i\sin\omega t/2 & \cos\omega t/2 \end{pmatrix} \frac{1}{\sqrt{2}}\begin{pmatrix} 1 \\ 1 \end{pmatrix} = \frac{e^{i\omega t/2}}{\sqrt{2}}\begin{pmatrix} 1 \\ 1 \end{pmatrix}. \tag{2.11}$$

Therefore the state remains in its initial state at an arbitrary $t > 0$. This is an expected result since the system at $t = 0$ is an eigenstate of the Hamiltonian.

EXERCISE 2.2 Let us consider a Hamiltonian

$$H = -\frac{\hbar}{2}\omega\sigma_y. \tag{2.12}$$

Suppose the initial state of the system is

$$|\psi(0)\rangle = \begin{pmatrix} 0 \\ 1 \end{pmatrix}. \tag{2.13}$$

(1) Find the wave function $|\psi(t)\rangle$ at later time $t > 0$.
(2) Find the probability for the system to have the outcome $+1$ upon measurement of σ_z at $t > 0$.
(3) Find the probability for the system to have the outcome $+1$ upon measurement of σ_x at $t > 0$.

Now let us formulate Example 2.1 and Exercise 2.2 in the most general form. Consider a Hamiltonian

$$H = -\frac{\hbar}{2}\omega\hat{n}\cdot\boldsymbol{\sigma}, \tag{2.14}$$

where $\hat{n}$ is a unit vector in $\mathbb{R}^3$. The time-evolution operator is readily obtained, by making use of the result of Proposition 1.2, as

$$U(t) = \exp(-iHt/\hbar) = \cos\frac{\omega}{2}t\, I + i(\hat{n}\cdot\boldsymbol{\sigma})\sin\frac{\omega}{2}t. \tag{2.15}$$

Suppose the initial state is

$$|\psi(0)\rangle = \begin{pmatrix} 1 \\ 0 \end{pmatrix},$$

for example. Then we find

$$|\psi(t)\rangle = U(t)|\psi(0)\rangle = \begin{pmatrix} \cos(\omega t/2) + in_z \sin(\omega t/2) \\ i(n_x + in_y)\sin(\omega t/2) \end{pmatrix}. \tag{2.16}$$

The reader should verify that $|\psi(t)\rangle$ is normalized at any instant of time $t > 0$.

EXAMPLE 2.2 (Rabi oscillation) This example is often employed for a quantum gate implementation as will be shown later. We will take the natural unit $\hbar = 1$ to simplify our notation throughout this example. Let us consider a spin-1/2 particle in a magnetic field along the z-axis, whose Hamiltonian is given by

$$H_0 = -\frac{\omega_0}{2}\sigma_z. \tag{2.17}$$

Suppose the particle is irradiated by an oscillating magnetic field of angular frequency ω, which introduces transitions between two energy eigenstates of H_0. Then the perturbed Hamiltonian is modelled as

$$H = -\frac{\omega_0}{2}\sigma_z + \frac{\omega_1}{2}\begin{pmatrix} 0 & e^{i\omega t} \\ e^{-i\omega t} & 0 \end{pmatrix} = \frac{1}{2}\begin{pmatrix} -\omega_0 & \omega_1 e^{i\omega t} \\ \omega_1 e^{-i\omega t} & \omega_0 \end{pmatrix}, \tag{2.18}$$

where $\omega_1 > 0$ is a parameter proportional to the amplitude of the oscillating field. Let us evaluate the wave function $|\psi(t)\rangle$ at time $t > 0$ assuming that the system is in the ground state of the unperturbed Hamiltonian

$$|\psi(0)\rangle = \begin{pmatrix} 1 \\ 0 \end{pmatrix} \tag{2.19}$$

at $t = 0$. Note that we cannot simply exponentiate the Hamiltonian since it is time-dependent. Surprisingly, however, the following trick makes it time-independent. Let us consider the following "gauge transformation":

$$|\phi(t)\rangle = e^{-i\omega\sigma_z t/2}|\psi(t)\rangle. \tag{2.20}$$

A straightforward calculation shows that $|\phi(t)\rangle$ satisfies

$$i\frac{d}{dt}|\phi(t)\rangle = \tilde{H}|\phi(t)\rangle, \tag{2.21}$$

where

$$\tilde{H} = e^{-i\omega\sigma_z t/2}He^{i\omega\sigma_z t/2} - ie^{-i\omega\sigma_z t/2}\frac{d}{dt}e^{i\omega\sigma_z t/2} = \frac{1}{2}\begin{pmatrix} -\omega_0 + \omega & \omega_1 \\ \omega_1 & \omega_0 - \omega \end{pmatrix}$$

$$= -\frac{\delta}{2}\sigma_z + \frac{\omega_1}{2}\sigma_x \tag{2.22}$$

is, in fact, time-independent. Here $\delta = \omega_0 - \omega$ stands for the "detuning" between ω and ω_0. Note that the Hamiltonian $\tilde{H}$ can be put into the form (2.14) as

$$\tilde{H} = \frac{\Delta}{2}\left(\frac{\omega_1}{\Delta}\sigma_x - \frac{\delta}{\Delta}\sigma_z\right), \quad \Delta \equiv \sqrt{\delta^2 + \omega_1^2}. \tag{2.23}$$

Now it is easy to solve Eq. (2.21). The time evolution operator is obtained using Eq. (2.15) as

$$\begin{aligned}
\tilde{U}(t) &= \cos\frac{\Delta t}{2}I - i\left(\frac{\omega_1}{\Delta}\sigma_x - \frac{\delta}{\Delta}\sigma_z\right)\sin\frac{\Delta t}{2} \\
&= \begin{pmatrix} \cos\dfrac{\Delta t}{2} + i\dfrac{\delta}{\Delta}\sin\dfrac{\Delta t}{2} & -i\dfrac{\omega_1}{\Delta}\sin\dfrac{\Delta t}{2} \\[2mm] -i\dfrac{\omega_1}{\Delta}\sin\dfrac{\Delta t}{2} & \cos\dfrac{\Delta t}{2} - i\dfrac{\delta}{\Delta}\sin\dfrac{\Delta t}{2} \end{pmatrix}.
\end{aligned} \tag{2.24}$$

The wave function $|\phi(t)\rangle$ with the initial condtion $|\phi(0)\rangle = (1,0)^t$ is

$$|\phi(t)\rangle = \tilde{U}(t)|\phi(0)\rangle = \begin{pmatrix} \cos\dfrac{\Delta t}{2} + i\dfrac{\delta}{\Delta}\sin\dfrac{\Delta t}{2} \\[2mm] -i\dfrac{\omega_1}{\Delta}\sin\dfrac{\Delta t}{2} \end{pmatrix}. \tag{2.25}$$

We find $|\psi(t)\rangle$ from Eq. (2.20) as

$$|\psi(t)\rangle = e^{i\omega\sigma_z t/2}|\phi(t)\rangle = \begin{pmatrix} e^{i\omega t/2}\left(\cos\dfrac{\Delta t}{2} + i\dfrac{\delta}{\Delta}\sin\dfrac{\Delta t}{2}\right) \\[2mm] -ie^{-i\omega t/2}\dfrac{\omega_1}{\Delta}\sin\dfrac{\Delta t}{2} \end{pmatrix}. \tag{2.26}$$

Suppose the applied field is in resonance with the energy difference of two levels, namely $\omega = \omega_0$. We obtain $\delta = 0$ and $\Delta = \omega_1$ in this case. The wave function $|\psi(t)\rangle$ at later time $t > 0$ is

$$|\psi(t)\rangle = e^{i\omega\sigma_z t/2}|\phi(t)\rangle = \begin{pmatrix} e^{i\omega_0 t/2}\cos\dfrac{\omega_1 t}{2} \\[2mm] -ie^{-i\omega_0 t/2}\sin\dfrac{\omega_1 t}{2} \end{pmatrix}. \tag{2.27}$$

The probability with which the system is found in the ground (excited) state of H_0 is given by

$$P_0 = \cos^2 \omega_1 t/2 \quad (P_1 = \sin^2 \omega_1 t/2). \tag{2.28}$$

This oscillatory behavior is called the **Rabi oscillation**. The frequency ω_1 is called the **Rabi frequency**, while Δ in Eq. (2.23) is called the **generalized Rabi frequency**

2.3   Multipartite System, Tensor Product and Entangled State

So far, we have assumed implictly that the system is made of a single component. Suppose a system is made of two components; one lives in a Hilbert space $\mathcal{H}_1$ and the other in another Hilbert space $\mathcal{H}_2$. A system composed of two separate components is called **bipartite**. Then the system as a whole lives in a Hilbert space $\mathcal{H} = \mathcal{H}_1 \otimes \mathcal{H}_2$, whose general vector is written as

$$|\psi\rangle = \sum_{i,j} c_{ij} |e_{1,i}\rangle \otimes |e_{2,j}\rangle, \tag{2.29}$$

where $\{|e_{a,i}\rangle\}$ $(a = 1, 2)$ is an orthonormal basis in $\mathcal{H}_a$ and $\sum_{i,j} |c_{ij}|^2 = 1$.

A state $|\psi\rangle \in \mathcal{H}$ written as a tensor product of two vectors as $|\psi\rangle = |\psi_1\rangle \otimes |\psi_2\rangle$, $(|\psi_a\rangle \in \mathcal{H}_a)$ is called a **separable state** or a **tensor product state**. A separable state admits a classical interpretation such as "The first system is in the state $|\psi_1\rangle$, while the second system is in $|\psi_2\rangle$." It is clear that the set of separable states has dimension $\dim\mathcal{H}_1 + \dim\mathcal{H}_2$. Note however that the total space $\mathcal{H}$ has different dimensions since we find, by counting the number of coefficients in (2.29), that $\dim\mathcal{H} = \dim\mathcal{H}_1\dim\mathcal{H}_2$. This number is considerably larger than the dimension of the sparable states when $\dim\mathcal{H}_a$ $(a = 1, 2)$ are large. What are the missing states then? Let us consider a spin state

$$|\psi\rangle = \frac{1}{\sqrt{2}} (|\uparrow\rangle \otimes |\uparrow\rangle + |\downarrow\rangle \otimes |\downarrow\rangle) \tag{2.30}$$

of two separated electrons. Suppose $|\psi\rangle$ may be decomposed as

$$\begin{aligned}|\psi\rangle &= (c_1|\uparrow\rangle + c_2|\downarrow\rangle) \otimes (d_1|\uparrow\rangle + d_2|\downarrow\rangle) \\ &= c_1 d_1|\uparrow\rangle \otimes |\uparrow\rangle + c_1 d_2|\uparrow\rangle \otimes |\downarrow\rangle + c_2 d_1|\downarrow\rangle \otimes |\uparrow\rangle + c_2 d_2|\downarrow\rangle \otimes |\downarrow\rangle.\end{aligned}$$

However this decomposition is not possible since we must have

$$c_1 d_2 = c_2 d_1 = 0, \; c_1 d_1 = c_2 d_2 = \frac{1}{\sqrt{2}}$$

simultaneously, and it is clear that the above equations have no common solution. Therefore the state $|\psi\rangle$ is not separable.

Such non-separable states are called **entangled** in quantum theory [9]. The fact

$$\dim\mathcal{H}_1\dim\mathcal{H}_2 \gg \dim\mathcal{H}_1 + \dim\mathcal{H}_2$$

tells us that most states in a Hilbert space of a bipartite system are entangled when the constituent Hilbert spaces are higher dimensional. These entangled states refuse classical descriptions. Entanglement will be used extensively as a powerful computational resource in quantum information processing and quantum computation.

Suppose a bipartite state (2.29) is given. We are interested in when the state is separable and when entangled. The criterion is given by the Schmidt decomposition of $|\psi\rangle$.

PROPOSITION 2.1 Let $\mathcal{H} = \mathcal{H}_1 \otimes \mathcal{H}_2$ be the Hilbert space of a bipartite system. Then a vector $|\psi\rangle \in \mathcal{H}$ admits the **Schmidt decomposition**

$$|\psi\rangle = \sum_{i=1}^{r} \sqrt{s_i}|f_{1,i}\rangle \otimes |f_{2,i}\rangle \text{ with } \sum_i s_i = 1, \tag{2.31}$$

where $s_i > 0$ are called the **Schmidt coefficients** and $\{|f_{a,i}\rangle\}$ is an orthonormal set of $\mathcal{H}_a$. The number $r \in \mathbb{N}$ is called the **Schmidt number** of $|\psi\rangle$.

Proof. This is a direct consequence of SVD introduced in §1.9. Let $|\psi\rangle$ be expanded as in Eq. (2.29). Note that the coefficients c_{ij} form a $\dim\mathcal{H}_1 \times \dim\mathcal{H}_2$ matrix C. We apply the SVD to obtain $C = U\Sigma V^\dagger$, where U and V are unitary matrices and Σ is a matrix whose diagonal elements are nonnegative real numbers while all the off-diagonal elements vanish. Now $|\psi\rangle$ of Eq. (2.29) is put in the form

$$|\psi\rangle = \sum_{i,j,k} U_{ik}\Sigma_{kl}V_{jl}^*|e_{1,i}\rangle \otimes |e_{2,j}\rangle.$$

Now define $|f_{1,k}\rangle = \sum_i U_{ik}|e_{1,i}\rangle$ and $|f_{2,k}\rangle = \sum_j V_{jk}^*|e_{2,j}\rangle$. Unitarity of U and V guarantees that they are orthonormal bases of $\mathcal{H}_1$ and $\mathcal{H}_2$, respectively. By noting that $\Sigma_{kl} = d_k\delta_{kl}$, we obtain

$$|\psi\rangle = \sum_{i=1}^{r} d_i|f_{1,i}\rangle \otimes |f_{2,i}\rangle,$$

where r is the number of nonvanishing diagonal elements in Σ. The wave function (2.31) is obtained by replacing the positive number d_i by $d_i = \sqrt{s_i}$. Moreover, the normalization condition implies $\langle\psi|\psi\rangle = \sum_i s_i = 1$. ∎

It follows from the above proposition that a bipartite state $|\psi\rangle$ is separable if and only if its Schmidt number r is 1.

EXAMPLE 2.3 Consider a bipartite state

$$|\psi\rangle = \frac{1}{2}(|e_{1,1}\rangle|e_{2,1}\rangle + |e_{1,1}\rangle|e_{2,2}\rangle + i|e_{1,3}\rangle|e_{2,1}\rangle + i|e_{1,3}\rangle|e_{2,2}\rangle),$$

whose coefficients form a matrix

$$C = \frac{1}{2}\begin{pmatrix} 1 & 1 \\ 0 & 0 \\ i & i \end{pmatrix}.$$

Note that this is essentially the same matrix whose SVD was analyzed in Example 1.7. By making use of the result obtained there, we find

$$|\psi\rangle = |f_{1,1}\rangle|f_{2,1}\rangle,$$

where

$$|f_{1,1}\rangle = \sum_{i=1}^{3} U_{i1}|e_{1,i}\rangle = \frac{1}{\sqrt{2}}(|e_{1,1}\rangle + i|e_{1,3}\rangle)$$

and

$$|f_{2,1}\rangle = \sum_{j=1}^{2} V_{j1}^{*}|e_{2,j}\rangle = \frac{1}{\sqrt{2}}(|e_{2,1}\rangle + |e_{2,2}\rangle).$$

Therefore the Schmidt number is 1 and the state is separable.

Generalization to a system with more components, i.e., a **multipartite system**, should be obvious. A system composed of N components has a Hilbert space

$$\mathcal{H} = \mathcal{H}_1 \otimes \mathcal{H}_2 \otimes \ldots \otimes \mathcal{H}_N, \tag{2.32}$$

where $\mathcal{H}_a$ is the Hilbert space to which the ath component belongs. Classification of entanglement in a multipartite system is far from obvious, and an analogue of the Schmidt decompostion is not known to date for $N \geq 3$.*

2.4   Mixed States and Density Matrices

It might happen in some cases that a quantum system under considertation is in the state $|\psi_i\rangle$ with a probability p_i. In other words, we cannot say definitely which state the system is in. Therefore some random nature comes into the description of the system. This random nature should not be confused with a probabilistic behavior of a quantum system. Such a system is said to be in a **mixed state**, while a system whose vector is uniquely specified is in a **pure state**. A pure state is a special case of a mixed state in which $p_i = 1$ for some i and $p_j = 0$ $(j \neq i)$.

Mixed states may happen in the following cases, for example.

- Suppose we observe a beam of totally unpolarized light and measure whether photons are polarized vertically or horizontally. The measurement outcome of a particular photon is *either* horizontal *or* vertical. Therefore when the beam passes through a linear polarizer, the intensity is halved. The beam is a mixture of horizontally polarized photons and vertically polarized photons.

---

*See, however, [10, 11].

- A particle source emits a particle in a state $|\psi_i\rangle$ with a probability p_i $(1 \leq i \leq N)$.

- Let us consider a canonical ensemble. If we pick up one of the members in the ensemble, it is in a state $|\psi_i\rangle$ with energy E_i with a probability $p_i = e^{-E_i/k_B T}/Z(T)$, where $Z(T) = \text{Tr}\, e^{-H/k_B T}$ is the partition function.

In each of these examples, a particular state $|\psi_i\rangle \in \mathcal{H}$ appears with probability p_i, in which case the expectation value of the observable a is $\langle \psi_i | A | \psi_i \rangle$, where we assume $|\psi_i\rangle$ is normalized; $\langle \psi_i | \psi_i \rangle = 1$. The mean value of a is then given by

$$\langle A \rangle = \sum_{i=1}^{N} p_i \langle \psi_i | A | \psi_i \rangle, \tag{2.33}$$

where N is the number of available states. Let us introduce the **density matrix** by

$$\rho = \sum_{i=1}^{N} p_i |\psi_i\rangle\langle\psi_i|. \tag{2.34}$$

Then Eq. (2.33) is rewritten in a compact form as

$$\langle A \rangle = \text{Tr}(\rho A). \tag{2.35}$$

EXERCISE 2.3 Let A be a Hermitian matrix. A is called positive-semidefinite if $\langle \psi | A | \psi \rangle \geq 0$ for any $|\psi\rangle$ in the relevant Hilbert space $\mathcal{H}$. Show that all the eigenvalues of a positive-semidefinite Hermitian matrix are non-negative.

Conversely, show that a Hermitian matrix A, whose eigenvalues are all non-negative, satisfies $\langle \psi | A | \psi \rangle \geq 0$ for any $|\psi\rangle \in \mathcal{H}$.

Properties which a density matrix ρ satisfies are very much like axioms for pure states.

A 1′ A physical state of a system, whose Hilbert space is $\mathcal{H}$, is completely specified by its associated density matrix $\rho : \mathcal{H} \to \mathcal{H}$. A density matrix is a positive semi-definite Hermitian operator with $\text{tr}\,\rho = 1$ (see remarks below).

A 2′ The mean value of an observable a is given by

$$\langle A \rangle = \text{tr}\,(\rho A). \tag{2.36}$$

A 3′ The temporal evolution of the density matrix is given by the **Liouville-von Neumann equation**,

$$i\hbar \frac{d}{dt}\rho = [H, \rho], \tag{2.37}$$

where H is the system Hamiltonian (see remarks below).

Several remarks are in order.

- We assume $\{|\psi_i\rangle\}$ is not necessarily an orthogonal basis of $\mathcal{H}$, although it is assumed $\langle\psi_i|\psi_i\rangle = 1$. The density matrix (2.34) is Hermitian since $p_i \in \mathbb{R}$. It is positive semi-definite since

$$\langle\phi|\rho|\phi\rangle = \sum_i p_i\langle\phi|\psi_i\rangle\langle\psi_i|\phi\rangle = \sum_i p_i|\langle\psi_i|\phi\rangle|^2 \geq 0,$$

where $|\phi\rangle$ is an arbitrary vector. We also find

$$\operatorname{tr}\rho = \sum_k \langle e_k|\rho|e_k\rangle = \sum_{i,k}\langle e_k|p_i|\psi_i\rangle\langle\psi_i|e_k\rangle$$

$$= \sum_i p_i\langle\psi_i|\left(\sum_k |e_k\rangle\langle e_k|\right)|\psi_i\rangle = \sum_i p_i\langle\psi_i|\psi_i\rangle = 1,$$

where $\{|e_k\rangle\}$ is an orthonormal basis of $\mathcal{H}$.

- Each $|\psi_i\rangle$ follows the Schrödinger equation

$$i\hbar\frac{d}{dt}|\psi_i\rangle = H|\psi_i\rangle$$

in a closed quantum system. Its Hermitian conjugate is

$$-i\hbar\frac{d}{dt}\langle\psi_i| = \langle\psi_i|H.$$

We prove the Liouville-von Neumann equation from these equations as

$$i\hbar\frac{d}{dt}\rho = i\hbar\frac{d}{dt}\sum_i p_i|\psi_i\rangle\langle\psi_i| = \sum_i p_i H|\psi_i\rangle\langle\psi_i| - \sum_i p_i|\psi_i\rangle\langle\psi_i|H$$

$$= [H, \rho].$$

We denote the set of all possible density matrices as $\mathcal{S}(\mathcal{H})$, where $\mathcal{H}$ is the Hilbert space associated with a system under consideration. It is easy to verify that $t\rho_1 + (1-t)\rho_2$ for $\rho_{1,2} \in \mathcal{S}(\mathcal{H})$ is also a density matrix, which shows that $\mathcal{S}(\mathcal{H})$ is a convex set.

EXAMPLE 2.4 A pure state $|\psi\rangle$ is a special case in which the corresponding density matrix is

$$\rho = |\psi\rangle\langle\psi|. \tag{2.38}$$

Therefore ρ in this case is nothing but the projection operator onto the state $|\psi\rangle$. Observe that

$$\langle A\rangle = \operatorname{tr}\rho A = \sum_i \langle\psi_i|\psi\rangle\langle\psi|A|\psi_i\rangle = \sum_i \langle\psi|A|\psi_i\rangle\langle\psi_i|\psi\rangle = \langle\psi|A|\psi\rangle.$$

A general density matrix is a convex combination of pure states.

Let us consider a beam of photons. We take a horizontally polarized state $|e_1\rangle = |\leftrightarrow\rangle$ and a vertically polarized state $|e_2\rangle = |\updownarrow\rangle$ as orthonormal basis vectors. If the photons are a totally uniform mixture of two polarized states, the density matrix is given by

$$\rho = \frac{1}{2}|e_1\rangle\langle e_1| + \frac{1}{2}|e_2\rangle\langle e_2| = \frac{1}{2}\begin{pmatrix} 1 & 0 \\ 0 & 1 \end{pmatrix} = \frac{1}{2}I.$$

This state is a uniform mixture of $|\updownarrow\rangle$ and $|\leftrightarrow\rangle$ and called a **maximally mixed state**.

If photons are in a pure state $|\psi\rangle = (|e_1\rangle + |e_2\rangle)/\sqrt{2}$, the density matrix, with $\{|e_i\rangle\}$ as basis, is

$$\rho = |\psi\rangle\langle\psi| = \frac{1}{2}\begin{pmatrix} 1 & 1 \\ 1 & 1 \end{pmatrix}.$$

If $|\psi\rangle$ itself is used as a basis vector, the other vector being $|\phi\rangle = (|e_1\rangle - |e_2\rangle)/\sqrt{2}$, the density matrix with respect to the basis $\{|\psi\rangle, |\phi\rangle\}$ has a component expression

$$\rho = \begin{pmatrix} 1 & 0 \\ 0 & 0 \end{pmatrix}.$$

Verify that they all satisfy Hermiticity, positive semi-definitness and $\operatorname{tr}\rho = 1$.

Let $A = \sum_a \lambda_a |\lambda_a\rangle\langle\lambda_a|$ be the spectral decomposition of an observable A and let $\rho = \sum_i p_i |\psi_i\rangle\langle\psi_i|$ be an arbitrary state. Then the measurement outcome of A is λ_a with the probability

$$p(a) = \sum_i p_i |\langle\lambda_a|\psi_i\rangle|^2 = \langle\lambda_a|\rho|\lambda_a\rangle = \operatorname{tr}(P_a\rho), \tag{2.39}$$

where $P_a = |\lambda_a\rangle\langle\lambda_a|$ is the projection operator. The state changes to a pure state $|\lambda_a\rangle\langle\lambda_a|$ immediately after the measurement with the outcome λ_a. This change is written as $\rho \mapsto P_a\rho P_a/p(a)$.

Now, we are interested in when ρ represents a pure state or a mixed state.

THEOREM 2.1 A state ρ is pure if and only if $\rho^2 = \rho$.

Proof. Since ρ is Hermitian, all its eigenvalues λ_i $(1 \leq i \leq \dim\mathcal{H})$ are real and the corresponding eigenvectors $\{|\lambda_i\rangle\}$ are made orthonormal. Let $\rho = \sum_i \lambda_i |\lambda_i\rangle\langle\lambda_i|$ be the spectral decomposition of ρ. Suppose $\rho^2 = \sum_i \lambda_i^2 |\lambda_i\rangle\langle\lambda_i| = \rho$. Then the eigenvalue λ_i satisfies $\lambda_i^2 = \lambda_i$ for any i. Therefore λ_i is either 0 or 1. It follows from $\operatorname{tr}\rho = \sum_i \lambda_i = 1$ that $\lambda_p = 1$ for some p and $\lambda_i = 0$ for $i \neq p$, namely, ρ is a pure state $|\lambda_p\rangle\langle\lambda_p|$.

The converse is trivial. ∎

EXERCISE 2.4 Show that a state ρ is pure if and only if $\operatorname{tr}\rho^2 = 1$.

We classify mixed states into three classes, similar to the classification of pure states into separable states and entangled states. We use a bipartite system in the definition, but generalization to multipartitle systems should be obvious.

DEFINITION 2.1 A state ρ is called **uncorrelated** if it is written as

$$\rho = \rho_1 \otimes \rho_2. \tag{2.40}$$

It is called **separable** if it is written in the form

$$\rho = \sum_i p_i \rho_{1,i} \otimes \rho_{2,i}, \tag{2.41}$$

where $0 \leq p_i \leq 1$ and $\sum_i p_i = 1$. It is called **inseparable** if ρ does not admit the decompostion (2.41),

It is important to realize that only inseparable states have quantum correlations analogous to that of an entangled pure state. However, it does not necessarily imply separable states have no non-classical correlation. It was pointed out that useful non-classical correlation exists in the subset of separable states [12].

In the next subsection, we discuss how to find whether a given bipartite density matrix is separable or inseparable.

2.4.1   Negativity

Let ρ be a bipartite state and define the **partial transpose** ρ^{pt} of ρ with respect to the second Hilbert space as

$$\rho_{ij,kl} \rightarrow \rho_{il,kj}, \tag{2.42}$$

where

$$\rho_{ij,kl} = (\langle e_{1,i}| \otimes \langle e_{2,j}|) \, \rho \, (|e_{1,k}\rangle \otimes |e_{2,l}\rangle).$$

Here $\{|e_{1,k}\rangle\}$ is the basis of the first system, while $\{|e_{2,k}\rangle\}$ is the basis of the second system. Suppose ρ takes a separable form (2.41). Then the partial transpose yields

$$\rho^{\mathrm{pt}} = \sum_i p_i \rho_{1,i} \otimes \rho_{2,i}^t. \tag{2.43}$$

Note here that ρ^t for any density matrix ρ is again a density matrix since it is still a positive semi-definite Hermitian with unit trace. Therefore the partial transposed density matrix (2.43) is another density matrix. It was conjectured by Peres [6] and subsequently proved by the Hordecki family [14] that positivity of the partially transposed density matrix is a necessary and sufficient condition for ρ to be separable in the cases of $\mathbb{C}^2 \otimes \mathbb{C}^2$ systems and $\mathbb{C}^2 \otimes \mathbb{C}^3$ systems. Conversely, if the partial transpose of ρ of these systems is not a density matrix, then ρ is inseparable. Instead of giving the proof, we look at the following example.

EXAMPLE 2.5 Let us consider the Werner state

$$\rho = \begin{pmatrix} \frac{1-p}{4} & 0 & 0 & 0 \\ 0 & \frac{1+p}{4} & -\frac{p}{2} & 0 \\ 0 & -\frac{p}{2} & \frac{1+p}{4} & 0 \\ 0 & 0 & 0 & \frac{1-p}{4} \end{pmatrix}, \tag{2.44}$$

where $0 \le p \le 1$. Here the basis vectors are arranged in the order

$$|e_{1,1}\rangle|e_{2,1}\rangle, |e_{1,1}\rangle|e_{2,2}\rangle, |e_{1,2}\rangle|e_{2,1}\rangle, |e_{1,2}\rangle|e_{2,2}\rangle.$$

Partial transpose of ρ yields

$$\rho^{\mathrm{pt}} = \begin{pmatrix} \frac{1-p}{4} & 0 & 0 & -\frac{p}{2} \\ 0 & \frac{1+p}{4} & 0 & 0 \\ 0 & 0 & \frac{1+p}{4} & 0 \\ -\frac{p}{2} & 0 & 0 & \frac{1-p}{4} \end{pmatrix}.$$

Note that we need to consider off-diagonal matrix elements only when we partically transpose the elements. We have, for example,

$$\rho_{12,21} = (\langle e_{1,1}| \otimes \langle e_{2,2}|) \, \rho \, (|c_{1,2}\rangle \otimes |c_{2,1}\rangle)$$
$$\rightarrow (\langle e_{1,1}| \otimes \langle e_{2,1}|) \, \rho \, (|e_{1,2}\rangle \otimes |e_{2,2}\rangle) = \rho^{\mathrm{pt}}_{11,22}.$$

For ρ^{pt} to be a physically acceptable state, it must have non-negative eigenvalues. The characteristic equation of ρ^{pt} is

$$D(\lambda) = \det(\rho^{\mathrm{pt}} - \lambda I) = \left(\lambda - \frac{p+1}{4}\right)^3 \left(\lambda - \frac{1-3p}{4}\right) = 0.$$

There are threefold degenerate eigenvalues $\lambda = (1+p)/4$ and a nondegenerate eigenvalue $\lambda = (1 - 3p)/4$. This shows that ρ^{pt} is an unphysical state for $1/3 < p \le 1$. If this is the case, ρ is inseparable.

From the above observation, inseparable states are characterized by non-vanishing **negativity** defined as

$$N(\rho) \equiv \frac{\sum_i |\lambda_i| - 1}{2}. \tag{2.45}$$

Note that negativity vanishes if and only if all the eigenvalues of ρ^{pt} are nonnegative.

Negativity is one of the so-called entanglement monotones [15], which also include concurrence, entanglement of formation and entropy of entanglement.

EXAMPLE 2.6 It was mentioned above that vanishing negativity is equivalent with separability only for $\mathbb{C}^2 \otimes \mathbb{C}^2$ systems and $\mathbb{C}^2 \otimes \mathbb{C}^3$ systems. A counter example in a $\mathbb{C}^2 \otimes \mathbb{C}^4$ system has been given in Horodecki [16]. Let us consider

$$
\rho = \frac{1}{7b+1}
\begin{pmatrix}
b & 0 & 0 & 0 & 0 & b & 0 & 0 \\
0 & b & 0 & 0 & 0 & 0 & b & 0 \\
0 & 0 & b & 0 & 0 & 0 & 0 & b \\
0 & 0 & 0 & b & 0 & 0 & 0 & 0 \\
0 & 0 & 0 & 0 & \frac{1+b}{2} & 0 & 0 & \frac{\sqrt{1-b^2}}{2} \\
b & 0 & 0 & 0 & 0 & b & 0 & 0 \\
0 & b & 0 & 0 & 0 & 0 & b & 0 \\
0 & 0 & b & 0 & \frac{\sqrt{1-b^2}}{2} & 0 & 0 & \frac{1+b}{2}
\end{pmatrix}
\qquad (0 \le b \le 1), \qquad (2.46)
$$

which is known to be inseparable. The partial transposed matrix with respect to the second system is

$$
\rho^{\mathrm{pt}} = \frac{1}{7b+1}
\begin{pmatrix}
b & 0 & 0 & 0 & 0 & 0 & 0 & 0 \\
0 & b & 0 & 0 & b & 0 & 0 & 0 \\
0 & 0 & b & 0 & 0 & b & 0 & 0 \\
0 & 0 & 0 & b & 0 & 0 & b & 0 \\
0 & b & 0 & 0 & \frac{1+b}{2} & 0 & 0 & \frac{\sqrt{1-b^2}}{2} \\
0 & 0 & b & 0 & 0 & b & 0 & 0 \\
0 & 0 & 0 & b & 0 & 0 & b & 0 \\
0 & 0 & 0 & 0 & \frac{\sqrt{1-b^2}}{2} & 0 & 0 & \frac{1+b}{2}
\end{pmatrix}.
\qquad (2.47)
$$

The eigenvalues of ρ^{pt} are

$$
0, 0, 0, \frac{b}{7b+1}, \frac{2b}{7b+1}, \frac{2b}{7b+1}
$$

$$
\frac{1 + 14b^2 + 9b - \sqrt{98b^4 - 70b^3 + 23b^2 + 12b + 1}}{2\left(49b^2 + 14b + 1\right)},
$$

$$
\frac{1 + 14b^2 + 9b + \sqrt{98b^4 - 70b^3 + 23b^2 + 12b + 1}}{2\left(49b^2 + 14b + 1\right)}.
$$

It can be shown that the seventh eigenvalue takes the maximum value $(25 - 2\sqrt{10})/130$ at $b = (47 - 10\sqrt{10})/31$ and the minimum value 0 at $b = 0$, and hence all the eigenvalues are non-negative for $0 \le b \le 1$ in spite of inseparability of ρ.

EXERCISE 2.5 Verify that

$$
\rho_1 =
\begin{pmatrix}
\frac{1+p}{4} & 0 & 0 & \frac{p}{2} \\
0 & \frac{1-p}{4} & 0 & 0 \\
0 & 0 & \frac{1-p}{4} & 0 \\
\frac{p}{2} & 0 & 0 & \frac{1+p}{4}
\end{pmatrix}
\qquad (0 \le p \le 1) \qquad (2.48)
$$

is a density matrix. Show that the negativity does not vanish for $p > 1/3$.

EXERCISE 2.6 Verify that

$$\rho_2 = \begin{pmatrix} \frac{p}{2} & 0 & 0 & \frac{p}{2} \\ 0 & \frac{1-p}{2} & \frac{1-p}{2} & 0 \\ 0 & \frac{1-p}{2} & \frac{1-p}{2} & 0 \\ \frac{p}{2} & 0 & 0 & \frac{p}{2} \end{pmatrix} \qquad (0 \le p \le 1) \qquad (2.49)$$

is a density matrix. Show that the negativity vanishes only for $p = 1/2$.

2.4.2 Partial Trace and Purification

Let $\mathcal{H} = \mathcal{H}_1 \otimes \mathcal{H}_2$ be a Hilbert space of a bipartite system made of components 1 and 2 and let A be an arbitrary operator acting on $\mathcal{H}$. The **partial trace** of A over $\mathcal{H}_2$ is an operator acting on $\mathcal{H}_1$ defined as

$$A_1 = \text{tr}_2 A \equiv \sum_k (I \otimes \langle k|)A(I \otimes |k\rangle). \qquad (2.50)$$

Let $\rho = |\psi\rangle\langle\psi| \in \mathcal{S}(\mathcal{H})$ be a density matrix of a pure state $|\psi\rangle$. Suppose we are interested only in the first system and have no access to the second system. Then the partial trace allows us to "forget" about the second system. In other words, the partial trace quantifies our ignorance of the second sytem.

To be more concrete, let us consider a pure state of a two-dimensional system

$$|\psi\rangle = \frac{1}{\sqrt{2}}(|e_1\rangle|e_1\rangle + |e_2\rangle|e_2\rangle),$$

where $\{|e_i\rangle\}$ is an orthonormal basis. The corresponding density matrix is

$$\rho = \frac{1}{2} \begin{pmatrix} 1 & 0 & 0 & 1 \\ 0 & 0 & 0 & 0 \\ 0 & 0 & 0 & 0 \\ 1 & 0 & 0 & 1 \end{pmatrix},$$

where the basis vectors are ordered as $\{|e_1\rangle|e_1\rangle, |e_1\rangle|e_2\rangle, |e_2\rangle|e_1\rangle, |e_2\rangle|e_2\rangle\}$. The partial trace of ρ is

$$\rho_1 = \text{tr}_2\rho = \sum_{i=1,2} (I \otimes \langle e_i|)\rho(I \otimes |e_i\rangle) = \frac{1}{2}\begin{pmatrix} 1 & 0 \\ 0 & 1 \end{pmatrix}. \qquad (2.51)$$

Note that a pure state $|\psi\rangle$ is mapped to a maximally mixed state ρ_1.

Observe that

$$\text{tr}\,(\rho_1 A) = \text{tr}\,(\rho(A \otimes I)) \qquad (2.52)$$

for an observable A acting on the first Hilbert space. The expectation value of A under that state ρ is equally obtained by using ρ_1.

EXERCISE 2.7 Let

$$|\psi'\rangle = \frac{1}{\sqrt{2}}(|e_1\rangle|e_2\rangle - |e_2\rangle|e_1\rangle).$$

Find the corresponding density matrix. Then partial-trace it over the first Hilbert space to find a density matrix of the second system.

We have seen above that the partial trace of a pure state density matrix of a bipartite system over one of the constituent Hilbert spaces yields a mixed state. How about the converse? Given a mixed state density matrix, is it always possible to find a pure state density matrix whose partial trace over the extra Hilbert space yields the given density matrix? The answer is yes and the process to find the pure state is called the **purification**. Let $\rho_1 = \sum_k p_k |\psi_k\rangle\langle\psi_k|$ be a general density matrix of a system 1 with the Hilbert space $\mathcal{H}_1$. Now let us introduce the second Hilbert space $\mathcal{H}_2$ whose dimension is the same as that of $\mathcal{H}_1$. Then formally introduce a normalized vector

$$|\Psi\rangle = \sum_k \sqrt{p_k}|\psi_k\rangle \otimes |\phi_k\rangle, \tag{2.53}$$

where $\{|\phi_k\rangle\}$ is an orthonormal basis of $\mathcal{H}_2$. We find

$$\mathrm{tr}_2|\Psi\rangle\langle\Psi| = \sum_{i,j,k}(I \otimes \langle\phi_i|)\left[\sqrt{p_j p_k}|\psi_j\rangle|\phi_j\rangle\langle\psi_k|\langle\phi_k|\right](I \otimes |\phi_i\rangle)$$

$$= \sum_k p_k|\psi_k\rangle\langle\psi_k| = \rho_1. \tag{2.54}$$

Thus it is always possible to purify a mixed state by tensoring an extra Hilbert space of the same dimension as that of the original Hilbert space. It is easy to see, by construction, that purification is far from unique. In fact, there are an infinite number of purifications of a given mixed state density matrix; see Exercise 2.9.

EXERCISE 2.8 Let

$$\rho_1 = \frac{1}{4}\begin{pmatrix} 1 & 0 \\ 0 & 3 \end{pmatrix}$$

be a density matrix with a basis $\{|\psi_i\rangle\}$. Find a purification of ρ_1.

EXERCISE 2.9 Let

$$|\Psi\rangle = \sum_k \sqrt{p_k}|\psi_k\rangle \otimes |\phi_k\rangle$$

be a purification of $\rho_1 = \sum_k p_k|\psi_k\rangle\langle\psi_k| \in \mathcal{S}(\mathcal{H})$. Show that

$$|\Psi'\rangle = \sum_k \sqrt{p_k}|\psi_k\rangle \otimes U|\phi_k\rangle$$

is another purification of ρ_1, where U is an arbitrary unitary matrix in $U(\dim\mathcal{H})$.

2.4.3 Fidelity

It often happens that one has to compare two density matrices and tell how much they are close to each other. An experimentalist, for example, conducts an experiment and then wants to compare the resulting quantum state with the theoretical prediction. A good measure for this purpose is **fidelity**, which we now define [17].

DEFINITION 2.2 Let ρ_1 and ρ_2 be two density matrices belonging to the same state space $\mathcal{S}(\mathcal{H})$. Then the fidelity is defined by

$$F(\rho_1, \rho_2) = \mathrm{tr}\left(\sqrt{\sqrt{\rho_1}\rho_2\sqrt{\rho_1}}\right), \qquad (2.55)$$

where $\sqrt{\rho_1}$ is chosen such that all the squre-roots of the eievnvalues are positive-semidefinite.

A few comments are in order.

- Let $\rho_1 = \sum_i p_i |p_i\rangle\langle p_i|$ be the spectral decomposition of ρ_1. Then the requirement in the definition claims that $\sqrt{\rho_1} = \sum_i \sqrt{p_i}|p_i\rangle\langle p_i|$.

- $F(\rho, \rho) = 1$ since

$$F(\rho, \rho) = \mathrm{tr}\left(\sqrt{\sqrt{\rho}\rho\sqrt{\rho}}\right) = \mathrm{tr}\,\rho = 1.$$

- F is non-negative by definition and $F(\rho_1, \rho_2) < 1$ for $\rho_1 \neq \rho_2$. See [17] for the proof.

EXAMPLE 2.7 Let $\rho_1 = |\psi_1\rangle\langle\psi_1|$ and $\rho_2 = |\psi_2\rangle\langle\psi_2|$. We note $\sqrt{\rho_i} = |\psi_i\rangle\langle\psi_i| = \rho_i$ by definition. Then the fidelity for them is

$$F(\rho_1, \rho_2) = \mathrm{tr}\sqrt{|\psi_1\rangle\langle\psi_1||\psi_2\rangle\langle\psi_2||\psi_1\rangle\langle\psi_1|}$$
$$= |\langle\psi_1|\psi_2\rangle|\mathrm{tr}\,|\psi_1\rangle\langle\psi_1| = |\langle\psi_1|\psi_2\rangle|.$$

Let $[\rho_1, \rho_2] = 0$. Then they are simultaneously diagonalizable. Let $\rho_i = \sum_i p_i |i\rangle\langle i|$ and $\rho_2 = \sum_i q_i |i\rangle\langle i|$ be their spectral decompositions, where $\{|i\rangle\}$ is the set of simultaneous eigenvectors, which is assumed to be an orthrnormal set. Then the fidelity is

$$F(\rho_1, \rho_2) = \mathrm{tr}\sqrt{\sum_{ijk}\sqrt{p_i p_k}q_j|i\rangle\langle i|j\rangle\langle j|k\rangle\langle k|} = \sum_j \sqrt{p_j q_j}.$$

EXERCISE 2.10 Let U be a unitary operator acting on ρ_1 and ρ_2. Show that

$$F(U\rho_1 U^\dagger, U\rho_2 U^\dagger) = F(\rho_1, \rho_2). \qquad (2.56)$$

EXERCISE 2.11 Let

$$\rho_1 = \frac{1}{2} \begin{pmatrix} 1 & 0 & 0 & 0 \\ 0 & 0 & 0 & 0 \\ 0 & 0 & 0 & 0 \\ 0 & 0 & 0 & 1 \end{pmatrix}, \quad \rho_2 = \frac{1}{2} \begin{pmatrix} 1 & 0 & 0 & 1 \\ 0 & 0 & 0 & 0 \\ 0 & 0 & 0 & 0 \\ 1 & 0 & 0 & 1 \end{pmatrix}.$$

Find the fidelity $F(\rho_1, \rho_2)$.

Now we are ready to proceed to the world of quantum information and quantum computation. Variations on the themes introduced here and in the previous chapter will appear repeatedly in the following chapters.

References

[1] P. A. M. Dirac, *Principles of Quantum Mechanics* (4th ed.), Clarendon Press (1981).

[2] L. I. Shiff, *Quantum Mechanics* (3rd ed.), McGraw-Hill (1968).

[3] A. Messiah, *Quantum Mechanics*, Dover (2000).

[4] J. J. Sakurai, *Modern Quantum Mechanics* (2nd Edition), Addison Wesley, Boston (1994).

[5] L. E. Ballentine, *Quantum Mechanics*, World Scientific, Singapore (1998).

[6] A. Peres, *Quantum Theory: Concepts and Methods*, Springer (2006).

[7] E. H. Kennard, Z. Phys. **44**, 326 (1927).

[8] H. P. Robertson, Phys. Rev. **34** 163, (1929).

[9] R. Horodecki *et al.*, eprint, quant-ph/0702225 (2007).

[10] A. Acín *et al.*, Phys. Rev. Lett. **85**, 1560 (2000).

[11] A. Acín *et al.*, Phys. Rev. Lett. **87**, 040401 (2001).

[12] C. H. Bennett *et al.*, Phys. Rev. A **59**, 1070 (1999) and D. P. DiVincenzo, D. W. Leung and B. M. Terhal, IEEE Trans. Info. Theory **48**, 580 (2002). See also A. SaiToh, R. Rahimi and M. Nakahara, e-print quant-ph/0703133.

[13] A. Peres, Phys. Rev. Lett. **77**, 1413 (1996).

[14] M. Horodecki *et al.*, Phys. Lett. A **223**, 1 (1996).

[15] G. Vidal, J. Mod. Opt. **47**, 355 (2000).

[16] P. Horodecki, Phys. Lett. A **232**, 333 (1997).

[17] R. Jozsa, J. Mod. Opt. **41**, 2315 (1994).

3

Qubits and Quantum Key Distribution

3.1 Qubits

A (Boolean) **bit** assumes two distinct values, 0 and 1. Bits constitute the building blocks of the classical information theory founded by C. Shannon. Quantum information theory, on the other hand, is based on **qubits**.

General references for this chapter are [1] and [2].

3.1.1 One Qubit

A qubit is a (unit) vector in the vector space $\mathbb{C}^2$, whose basis vectors are denoted as

$$|0\rangle = \begin{pmatrix} 1 \\ 0 \end{pmatrix} \text{ and } |1\rangle = \begin{pmatrix} 0 \\ 1 \end{pmatrix}. \tag{3.1}$$

What these vectors physically mean depends on the physical realization employed for quantum-information processing.

- In some cases, $|0\rangle$ stands for a vertically polarized photon $|\updownarrow\rangle$, while $|1\rangle$ represents a horizontally polarized photon $|\leftrightarrow\rangle$. Alternatively they might correspond to photons polarized in different directions. For example, $|0\rangle$ may represent a polarization state

$$|\nearrow\rangle = \frac{1}{\sqrt{2}}(|\updownarrow\rangle + |\leftrightarrow\rangle),$$

while $|1\rangle$ represents a state

$$|\searrow\rangle = \frac{1}{\sqrt{2}}(|\updownarrow\rangle - |\leftrightarrow\rangle).$$

Note that if $|\updownarrow\rangle$ $(|\leftrightarrow\rangle)$ corresponds to an eigenstate of σ_z with the eigenvalue $+1$ (-1), respectively, then $|\nearrow\rangle$ $(|\searrow\rangle)$ corresponds to an eigenstate of σ_x with the eigenvalue $+1$ (-1), respectively.

Similarly, the states

$$|\sigma^+\rangle = \frac{1}{\sqrt{2}}(|\updownarrow\rangle + i|\leftrightarrow\rangle), \quad |\sigma^-\rangle = \frac{1}{\sqrt{2}}(|\updownarrow\rangle - i|\leftrightarrow\rangle)$$

correspond to the eigenstates of σ_y with the eigenvalues ± 1 and represent circularly polarized photons.

- They may represent spin states of an electron, $|0\rangle = |\uparrow\rangle$ and $|1\rangle = |\downarrow\rangle$. Electrons are replaced by nuclei with spin $1/2$ in NMR quantum computing.

- Truncated two states from many levels may also be employed as a qubit. Take the ground state and the first excited state of ionic energy levels or atomic energy levels, for example. We may assign $|0\rangle$ to the ground state and $|1\rangle$ to the first excited state.

In any case, we have to fix a set of basis vectors when we carry out quantum information processing. All the physics should be described with respect to this basis. In the following, the basis is written in an abstract form as $\{|0\rangle, |1\rangle\}$, unless otherwise stated.

A remark is in order. The third example of a qubit above suggests that a quantum system with more than two states may be employed for information storage and information processing. If a quantum system admits three different states, it is called a **qutrit**, while if it takes d different states, it is called a **qudit**. A spin S particle, for example, takes $d = 2S + 1$ spin states and works as a qudit. The significance of qutrits and qudits in information processing is still to be explored.

It is convenient to assume the vector $|0\rangle$ corresponds to the classical value 0, while $|1\rangle$ to 1 in quantum computation. Moreover it is possible for a qubit to be in a superposition state:

$$|\psi\rangle = a|0\rangle + b|1\rangle \text{ with } a, b \in \mathbb{C}, |a|^2 + |b|^2 = 1. \tag{3.2}$$

The fundamental requirement of quantum mechanics is that if we make measurement on $|\psi\rangle$ to see whether it is in $|0\rangle$ or $|1\rangle$, the outcome will be 0 (1) with the probability $|a|^2$ ($|b|^2$), and the state immediately after the measurement is $|0\rangle$ ($|1\rangle$).

Although a qubit may take infinitely many different states, it should be kept in mind that we can extract from it as the same amount of information as that of a classical bit. Information can be extracted only through measurements. When we make measurement on a qubit, the state vector "collapses" to the eigenvector that corresponds to the eigenvalue observed. Suppose that a spin is in the state $a|0\rangle + b|1\rangle$. If we observe that the z-component of the spin is $+1/2$, the system immediately after the measurement is *definitely* in the state $|0\rangle$. This happens with probability $\langle\psi|0\rangle\langle0|\psi\rangle = |a|^2$. The outcome of a measurement on a qubit is always one of the eigenvalues, which we call abstractly 0 and 1, just like for a classical bit. We are tempted to think that by making measurements of a large number of copies of this system, we may be able to determine the coefficients a and b (or, at least, $|a|$ and $|b|$) of the wavefunction. But this is not the case due to the "no-cloning theorem" proved later. It is impossible to duplicate an unknown quantum system with a unitary transformation.

3.1.2 Bloch Sphere

It is useful, for many purposes, to express a state of a single qubit graphically. Let us parameterize a one-qubit state $|\psi\rangle$ with θ and ϕ as

$$|\psi(\theta, \phi)\rangle = \cos\frac{\theta}{2}|0\rangle + e^{i\phi}\sin\frac{\theta}{2}|1\rangle. \tag{3.3}$$

We are not interested in the overall phase, and the phase of $|\psi\rangle$ is fixed in such a way that the coefficient of $|0\rangle$ is real. Now we show that $|\psi(\theta, \phi)\rangle$ is an eigenstate of $\hat{n}(\theta, \phi) \cdot \boldsymbol{\sigma}$ with the eigenvalue $+1$. Here $\boldsymbol{\sigma} = (\sigma_x, \sigma_y, \sigma_z)$ and $\hat{n}(\theta, \phi)$ is a real unit vector called the **Bloch vector** with components

$$\hat{n}(\theta, \phi) = (\sin\theta\cos\phi, \sin\theta\sin\phi, \cos\theta)^t.$$

In fact, a straightforward calculation shows that

$$\begin{aligned}
\hat{n}(\theta, \phi) \cdot \boldsymbol{\sigma}|\psi(\theta, \phi)\rangle &= \begin{pmatrix} \cos\theta & \sin\theta e^{-i\phi} \\ \sin\theta e^{i\phi} & -\cos\theta \end{pmatrix} \begin{pmatrix} \cos\frac{\theta}{2} \\ e^{i\phi}\sin\frac{\theta}{2} \end{pmatrix} \\
&= \begin{pmatrix} \cos\frac{\theta}{2}\cos\theta + \sin\frac{\theta}{2}\sin\theta \\ e^{i\phi}\left(\cos\frac{\theta}{2}\sin\theta - \cos\theta\sin\frac{\theta}{2}\right) \end{pmatrix} \\
&= \begin{pmatrix} \cos\frac{\theta}{2} \\ e^{i\phi}\sin\frac{\theta}{2} \end{pmatrix} = |\psi(\theta, \phi)\rangle.
\end{aligned}$$

It is therefore natural to assign a unit vector $\hat{n}(\theta, \phi)$ to a state vector $|\psi(\theta, \phi)\rangle$. Namely, a state $|\psi(\theta, \phi)\rangle$ is expressed as a unit vector $\hat{n}(\theta, \phi)$ on the surface of the unit sphere, called the **Bloch sphere**. This correspondence is one-to-one if the ranges of θ and ϕ are restricted to $0 \leq \theta \leq \pi$ and $0 \leq \phi < 2\pi$.

EXERCISE 3.1 Let $|\psi(\theta, \phi)\rangle$ be the state given by Eq. (3.3). Show that

$$\langle\psi(\theta, \phi)|\boldsymbol{\sigma}|\psi(\theta, \phi)\rangle = \hat{n}(\theta, \phi), \tag{3.4}$$

where $\hat{n}$ is the unit vector defined above.

It is possible to express a density matrix ρ of a qubit using a unit ball this time. Since ρ is a positive semi-definite Hermitian matrix with unit trace, its most general form is

$$\rho = \frac{1}{2}\left(I + \sum_{i=x,y,z} u_i\sigma_i\right), \tag{3.5}$$

where u_i are components of a real vector $\boldsymbol{u}$ with $|\boldsymbol{u}| \leq 1$. The reality follows from the Hermiticity requirement, and $\operatorname{Tr}\rho = 1$ is easy to check. The eigenvalues of ρ are

$$\lambda_+ = \frac{1}{2}\left(1 + \sqrt{|\boldsymbol{u}|}\right), \quad \lambda_- = \frac{1}{2}\left(1 - \sqrt{|\boldsymbol{u}|}\right) \tag{3.6}$$

and therefore non-negative. In case $|\boldsymbol{u}| = 1$, the eigenvalue λ_- vanishes and rank $\rho = 1$. Therefore the surface of the unit ball corresponds to pure states. The converse is also shown easily. In contrast, all the points $\boldsymbol{u}$ inside a unit ball correspond to mixed states. The ball is called the **Bloch ball**, also called the Bloch sphere in a mathematically less strict sense, and the vector $\boldsymbol{u}$ is also called the Bloch vector. The normalized vector $\boldsymbol{n}$ of the Bloch sphere is a special case of $\boldsymbol{u}$ restricted in pure states.

EXERCISE 3.2 Find the density matrix of a pure state (3.3) and write it in the form of Eq. (3.5).

EXERCISE 3.3 Let ρ be given by Eq. (3.5). Show that

$$\langle \boldsymbol{\sigma} \rangle = \mathrm{tr}\,(\rho\boldsymbol{\sigma}) = \boldsymbol{u}. \tag{3.7}$$

3.1.3   Multi-Qubit Systems and Entangled States

Let us consider a group of many (n) qubits next. Such a system behaves quite differently from a classical one, and this difference gives a distinguishing aspect to quantum information theory. An n-qubit system is often called a (quantum) **register** in the context of quantum computing.

Consider a classical system made of several components. The state of this system is completely determined by specifying the state of each component. This is *not* the case for a quantum system. A quantum system made of many components is not necessarily described by specifying the state of each component as we have learned in §2.3.

As an example, let us consider an n-qubit register. Suppose we specify the state of each qubit separately in analogy with a classical case. Each of the qubits is then described by a two-dimensional complex vector of the form $a_i|0\rangle + b_i|1\rangle$, and we need $2n$ complex numbers $\{a_i, b_i\}_{1 \leq i \leq n}$ to specify the state. This corresponds the the tensor product state

$$(a_1|0\rangle + b_1|1\rangle) \otimes (a_2|0\rangle + b_2|1\rangle) \otimes \ldots \otimes (a_n|0\rangle + b_n|1\rangle)$$

introduced in §2.3. If the system is treated in a fully quantum-mechanical way, however, we have to include superposition of such tensor product states, which is not necessarily decomposable into a tensor product form. Such a state is **entangled** (see §2.3). A general state vector of the register is represented as

$$|\psi\rangle = \sum_{i_k=0,1} a_{i_1 i_2 \ldots i_n} |i_1\rangle \otimes |i_2\rangle \otimes \ldots \otimes |i_n\rangle$$

and lives in a 2^n-dimensional complex vector space. Note that $2^n \gg 2n$ for a large number n. The ratio $2^n/2n$ is $\sim 6.3 \times 10^{27}$ for $n = 100$ and $\sim 5.4 \times 10^{297}$ for $n = 1000$. These astronomical numbers tell us that most quantum states in a Hilbert space with large n are entangled, i.e., they do not have classical analogy which tensor product states have. Entangled states that have no classical counterparts are extremely powerful resources for quantum computation and quantum communication as we will show later.

Let us consider a system of two qubits for definiteness. The combined system has a basis $\{|00\rangle, |01\rangle, |10\rangle, |11\rangle\}$. More generally, a basis for a system of n qubits may be taken to be $\{|b_{n-1}b_{n-2}\ldots b_0\rangle\}$, where $b_{n-1}, b_{n-2}, \ldots, b_0 \in \{0, 1\}$. It is also possible to express the basis in terms of the decimal system. We write $|x\rangle$, instead of $|b_{n-1}b_{n-2}\ldots b_0\rangle$, where $x = b_{n-1}2^{n-1} + b_{n-2}2^{n-2} + \ldots + b_0$ is the decimal expression of the binary number $b_{n-1}b_{n-2}\ldots b_0$. Thus the basis for a two-qubit system may be written also as $\{|0\rangle, |1\rangle, |2\rangle, |3\rangle\}$ with this decimal notation. Whether the binary system or the decimal system is employed should be clear from the context. An n-qubit system has $2^n = \exp(n \ln 2)$ basis vectors.

The set

$$\{|\Phi^+\rangle = \frac{1}{\sqrt{2}}(|00\rangle + |11\rangle), \quad |\Phi^-\rangle = \frac{1}{\sqrt{2}}(|00\rangle - |11\rangle),$$
$$|\Psi^+\rangle = \frac{1}{\sqrt{2}}(|01\rangle + |10\rangle), \quad |\Psi^-\rangle = \frac{1}{\sqrt{2}}(|01\rangle - |10\rangle)\} \tag{3.8}$$

is an orthonormal basis of a two-qubit system and is called the **Bell basis**. Each vector is called the **Bell state** or the **Bell vector**. Note that all the Bell states are entangled.

EXERCISE 3.4 The Bell basis is obtained from the binary basis $\{|00\rangle, |01\rangle, |10\rangle, |11\rangle\}$ by a unitary transformation. Write down the unitary transformation explicitly.

Among three-qubit entangled states, the following two states are important for various reasons and hence deserve special names. The state

$$|\text{GHZ}\rangle = \frac{1}{\sqrt{2}}(|000\rangle + |111\rangle) \tag{3.9}$$

is called the **Greenberger-Horne-Zeilinger state** and is often abbreviated as the **GHZ state**[3]. Another important three-qubit state is the **W state** [4],

$$|\text{W}\rangle = \frac{1}{\sqrt{3}}(|100\rangle + |010\rangle + |001\rangle). \tag{3.10}$$

EXERCISE 3.5 Find the expectation value of $\sigma_x \otimes \sigma_z$ measured in each of the Bell states.

3.1.4   Measurements

Classical information theory is formulated independently of measurements of the system under consideration. This is because the readout of the result is always the same for anyone and at any time, provided that the system processes the same information. This is completely different in quantum information processing. Measurement is an essential part of the theory as we see below.

By making a measurement on a system, we *project* the state vector to one of the basis vectors that the measurement equipment defines.* Suppse we have a state vector $|\psi\rangle = a|0\rangle + b|1\rangle$ and measure it to see if it is in the state $|0\rangle$ or $|1\rangle$. Depending on the system, this means if a spin points up or down or a photon is polarized horizontally or vertically, for example. The result is either 0 or 1. In the first case, the state "collapses" to $|0\rangle$ while in the second case, to $|1\rangle$. We find, after many measurements, the probability of obtaining outcome 0 (1) is $|a|^2$ ($|b|^2$).

To be more formal, we construct a **measurement operator** M_m such that the probability of obtaining the outcome m in the state $|\psi\rangle$ is

$$p(m) = \langle\psi|M_m^\dagger M_m|\psi\rangle, \tag{3.11}$$

and the state immediately after the measurement is

$$|m\rangle = \frac{M_m|\psi\rangle}{\sqrt{p(m)}}. \tag{3.12}$$

In the above example, the measurement operators are nothing but projection operators; $M_0 = |0\rangle\langle0|$ and $M_1 = |1\rangle\langle1|$. In fact, we have

$$p(0) = \langle\psi|M_0^\dagger M_0|\psi\rangle = \langle\psi|0\rangle\langle0|\psi\rangle = |a|^2,$$

and

$$\frac{M_0|\psi\rangle}{\sqrt{p(0)}} = \frac{a}{|a|}|0\rangle \simeq |0\rangle,$$

and similarly for the other case M_1. It should be noted that a quantum state is defined up to a phase and hence $a/|a|$ does not play any role. [*Remark:* See [2] for the difference between a general measurement operator and a projective measurement operator.]

Suppose we are given many copies of a particular state $|\psi\rangle$. If we measure an observable M in each of the copies, the expectation value of M is given, in terms of the projection operators, by

$$\begin{aligned} E(M) &= \sum_m m\,p(m) = \sum_m m\langle\psi|P_m|\psi\rangle \\ &= \langle\psi|\sum_m mP_m|\psi\rangle = \langle\psi|M|\psi\rangle, \end{aligned} \tag{3.13}$$

---

*This is called a projective measurement as was noted in Chapter 2. We will be concerned only with projective measurements, unless stated otherwise.

where use has been made of the spectral decomposition $M = \sum m P_m$. The standard deviation is given by

$$\Delta(M) = \sqrt{\langle(M - \langle M\rangle)^2\rangle} = \sqrt{\langle M^2\rangle - \langle M\rangle^2}. \tag{3.14}$$

Let us analyze measurements in a two-qubit system in some detail. An arbitrary state is written as

$$|\psi\rangle = a|00\rangle + b|01\rangle + c|10\rangle + d|11\rangle, \quad |a|^2 + |b|^2 + |c|^2 + |d|^2 = 1,$$

where $a, b, c, d \in \mathbb{C}$. We make a measurement of the first qubit with respect to the basis $\{|0\rangle, |1\rangle\}$. To this end, we rewrite the state as

$$a|00\rangle + b|01\rangle + c|10\rangle + d|11\rangle$$
$$= |0\rangle \otimes (a|0\rangle + b|1\rangle) + |1\rangle \otimes (c|0\rangle + d|1\rangle)$$
$$= u|0\rangle \otimes \left(\frac{a}{u}|0\rangle + \frac{b}{u}|1\rangle\right) + v|1\rangle \otimes \left(\frac{c}{v}|0\rangle + \frac{d}{v}|1\rangle\right),$$

where $u = \sqrt{|a|^2 + |b|^2}$ and $v = \sqrt{|c|^2 + |d|^2}$. The measurement operators acting on the first qubit are

$$M_0 = |0\rangle\langle 0| \otimes I, \quad M_1 = |1\rangle\langle 1| \otimes I. \tag{3.15}$$

Note that we need to specify $\otimes I$ explicitly since we are working in a two-qubit Hilbert space $\mathbb{C}^4$. Upon a measurement of the first qubit, we obtain 0 with the probability

$$\langle\psi|M_0|\psi\rangle = u^2 = |a|^2 + |b|^2,$$

projecting the state to

$$\frac{M_0|\psi\rangle}{\sqrt{p(0)}} = |0\rangle \otimes \left(\frac{a}{u}|0\rangle + \frac{b}{u}|1\rangle\right),$$

while we obtain $|1\rangle$ with the probability $v^2 = |c|^2 + |d|^2$, projecting the state to $|1\rangle \otimes \left(\frac{c}{v}|0\rangle + \frac{d}{v}|1\rangle\right)$. Note that the state after the measurement has unit norm in both cases. The measurement of the second qubit can be carried out similarly. Measurements on an n-qubit system can be carried out by repeating one-qubit measurement n times.

In the two-qubit example above, the Hilbert space for the system is separated into a direct sum of $\mathcal{H}_0$, where the first qubit is in the state $|0\rangle$, and $\mathcal{H}_1$, where it is in $|1\rangle$: $\mathcal{H} = \mathcal{H}_0 \oplus \mathcal{H}_1$. An arbitrary two-qubit state $|\psi\rangle$ is uniquely decomposed into two vectors, each of which belongs to $\mathcal{H}_0$ or $\mathcal{H}_1$ as

$$(|0\rangle\langle 0| \otimes I)|\psi\rangle \in \mathcal{H}_0, \quad (|1\rangle\langle 1| \otimes I)|\psi\rangle \in \mathcal{H}_1,$$

where normalization has been ignored. More generally, an observation of k qubits in an n-qubit system yields 2^k possible outcomes m_i $(1 \leq i \leq 2^k)$.

Accordingly, the 2^n-dimensional Hilbert space of the system is separated into the direct sum of mutually orthogonal subspaces $\mathcal{H}_{m_1}, \mathcal{H}_{m_2}, \ldots, \mathcal{H}_{m_{2^k}}$ as $\mathcal{H} = \mathcal{H}_{m_1} \oplus \mathcal{H}_{m_2} \oplus \ldots \oplus \mathcal{H}_{m_{2^k}}$. When the result of the observation of the k qubits is m_i, the state after the observation is projected to the subspace $\mathcal{H}_{m_i}$. It should be clear from the construction that each subspace $\mathcal{H}_{m_i}$ has dimension $2^n/2^k = 2^{n-k}$. The measurement device projects the state before the observation

$$|\psi\rangle = c_{m_1}|\psi_{m_1}\rangle + c_{m_2}|\psi_{m_2}\rangle + \ldots + c_{m_{2^k}}|\psi_{m_{2^k}}\rangle, \quad (|\psi_{m_i}\rangle \in \mathcal{H}_{m_i})$$

into one of the subspaces $\mathcal{H}_{m_i}$ randomly with the probability $|c_{m_i}|^2$.

Measurement gives an alternative viewpoint to entangled states. A state is not entangled if a measurement of a qubit does not affect the state of the other qubits. Suppose the first qubit of the state

$$\frac{1}{\sqrt{2}}(|00\rangle + |11\rangle)$$

was measured to be 0 (1). Then the outcome of the measurement of the second qubit is *definitely* 0 (1). Therefore the measurement of the first qubit affects the outcome of the measurement on the second qubit, which shows that the initial state is an entangled state. In other words, there exists a strong correlation between the two qubits. This correlation may be used for information processing as will be shown later. In contrast with this, the state $\frac{1}{\sqrt{2}}(|00\rangle + |01\rangle)$ is not entangled since it can be written as

$$\frac{1}{\sqrt{2}}(|00\rangle + |01\rangle) = |0\rangle \otimes \frac{1}{\sqrt{2}}(|0\rangle + |1\rangle).$$

Irrespectively of the measurement of the second qubit, the measurement of the first qubit definitely yields 0. Moreover, the second qubit is measured to be 0 (1) with the probability $1/2$, independently of whether the first qubit is measured or not.

EXERCISE 3.6 In many quantum algorithms, the result of an action of a function f on x is encoded into the form

$$U_f : \mathcal{N} \sum_x |x\rangle|0\rangle \to \mathcal{N} \sum_x |x\rangle|f(x)\rangle,$$

where $|x\rangle$ stands for the tensor product state $|b_{n-1}b_{n-2}\ldots b_0\rangle$ with $x = b_{n-1}2^{n-1} + b_{n-2}2^{n-2} + \ldots b_0$ and $\mathcal{N}$ is the normalization constant. The first register is for the input x, while the second one is for the corresponding output $f(x)$. Note that U_f acts on *all* possible states simultaneously.

Let $f(x) = a^x \mod N$, where a and N are coprime, and consider the state

$$U_f \left[\frac{1}{\sqrt{512}} \sum_{x=0}^{511} |x\rangle|0\rangle \right] = \frac{1}{\sqrt{512}} \sum_{x=0}^{511} |x\rangle|a^x \mod N\rangle$$

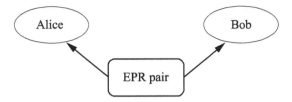

FIGURE 3.1
EPR pair produced by a source in the middle. One particle is sent to Alice and the other to Bob.

with $a = 6$ and $N = 91$. Suppose the measurement of the first register results in (1) $x = 11$, (2) $x = 23$ and (3) $x = 35$. What is the state immediately after each measurement?

3.1.5 Einstein-Podolsky-Rosen (EPR) Paradox

Einstein, Podolsky and Rosen (EPR) proposed a Gedanken experiment which, at first glance, shows that an entangled state violates an axiom of the special theory of relativity [5]. Suppose a particle source produces the so-called EPR pair in the state

$$|\Psi^-\rangle = \frac{1}{\sqrt{2}}(|01\rangle - |10\rangle)$$

and it sends one particle to Alice and the other to Bob, who may be separated far away (see Fig. 3.1).[†] Alice measures her particle and obtains her reading $|0\rangle$ or $|1\rangle$. Depending on her reading, the EPR state is projected to $|01\rangle$ ($|10\rangle$), and Bob will *definitely* observe $|1\rangle$ ($|0\rangle$) in his measurement. The change of the state

$$\frac{1}{\sqrt{2}}(|01\rangle - |10\rangle) \rightarrow |01\rangle \quad \text{or} \quad |10\rangle \tag{3.16}$$

takes place instantaneously even when they are separated by a large distance. It seems that Alice's measurement propageted to Bob's qubit instantaneously, and it violates the special theory of relativity. This is the very point EPR proposed to defeat quantum mechanics.

Note, however, that nothing has propagated from Alice to Bob and *vice versa*, upon Alice's measurement. Clearly no energy has propagated. What about information? It is impossible for Alice to control her and hence Bob's readings. Therefore it is impossible to use EPR pairs to send a sensible message from Alice to Bob. If they could, the message would be sent instantaneously, which certainly violates the special theory of relativity. If a large number of EPR pairs are sent to Alice and Bob and they independently

---

[†]Alice and Bob are names frequently used in information theory.

measure their qubits, they will observe random sequences of 0 and 1. They notice that their readings are strongly correlated only after they exchange their sequences by means of classical communication, which can be done at most with the speed of light.

3.2   Quantum Key Distribution (BB84 Protocol)

A large number of qubits and gate operations on them are involved in most practical applications of quantum computing. We will postpone these applications in the subsequent chapters. Is there any practical use of single qubits then? There is a suprisingly secure way of distributing a cryptographic key using a sequence of individual qubits [6] called BB84 protocol, which is already available commercially [7]. **Quantum key distribution** (QKD) is a secure way of distributing an encryption and decryption key by making use of qubits. The sender and the receiver can detect a possible third party eavesdropping their communication by comparing the sequence sent with that of the received one.

One-time pad is an absolutely secure cryptosystem if and only if the key for encoding and decoding is shared only between the sender and the receiver and used once for all. Suppose we want to send a message 1100101001 in a binary form using a key 1001010011, for example. The message is encrypted by adding the message to the key bitwise modulo 2, which we denote by $i \oplus j$. We have explicitly $0 \oplus 0 = 0$, $0 \oplus 1 = 1$, $1 \oplus 0 = 1$, $1 \oplus 1 = 0$. For the above case, we have the encrypted message 0101111010. For decryption, the receiver is required only to add the same key bitwise again since $(i \oplus j) \oplus j = i$. Decryption of an encrypted message is impossible without the key since there are 2^n possible keys for an n-bit string and many of them yield sensible messages. This cryptosystem is not secure any more if the same key is used many times.

A key must be sent from the sender to the receiver, or in the opposite direction, each time this cryptosystem is used. If the key is sent through a classical communication channel, there always exists a possibility of eavesdropping. However, this problem is completely solved if a quantum channel is employed as we show now. Suppose Alice wants to send Bob a one-time pad key to encode and decode her secret message. They can communicate with each other using a bidirectional classical channel. There also exists a quantum channel that is unidirectional from Alice to Bob. See Fig. 3.2. There is a possibility that their communication is being eavesdropped by a third party, which we call Eve. Alice sends Bob many qubits, one by one, and Bob measures the states of each of the qubits he receives. To make our discussion concrete, we assume qubits are made of polarized photons.

Alice employs two coding systems when she sends photons to Bob. The

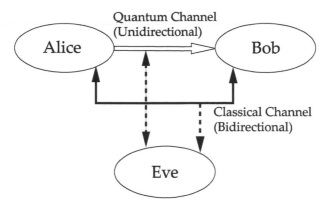

FIGURE 3.2
Quantum key distribution protocol BB84.

coding systems are

$$\text{coding system (1)} \quad 0 \mapsto |\updownarrow\rangle, \quad 1 \mapsto |\leftrightarrow\rangle,$$
$$\text{coding system (2)} \quad 0 \mapsto |\nwarrow\rangle, \quad 1 \mapsto |\nearrow\rangle.$$

They are chosen *at random* for each photon. Bob also chooses coding systems (1) and (2) randomly, independently of Alice, to measure the polarization of each photon Alice sends. Suppose $4N$ photons are sent from Alice to Bob. After all the photons have been sent, Alice and Bob exchange the sequence of the coding systems they employed using the classical communication channel (so this is not 100 % secure), without disclosing the bits (0 or 1) Alice sent and Bob received. They will know, as a result, for which photons they employed the same coding systems. They discard all the cases for which they employed different coding systems since the sent bits and the received bits agree only with probability $1/2$ in these cases. Now $\sim 2N$ photons, on average, should be correctly transmitted and they share $\sim 2N$ bits of binary numbers in their hands. To make sure that no one eavesdrops their quantum channel, they choose N cases randomly out of $2N$ cases with the same coding systems employed and exchange N bits (0 or 1) associated with these N cases over the classical channel. If there are no eavesdroppers operating, they should have the same bits for all the N cases. After verifying that they are free from eavesdroppers, they discard these N cases (since the classical channel may be eavesdropped) and use the remaining N bits to generate a one-time pad key.

Suppose Eve is in action. After eavesdropping the photons, she immediately sends Bob her results in order to hide her presence. Note that Bob will immediately notice the presence of an eavesdropper from missing photons unless Eve sends some photons to Bob. Eve's coding system is different from Alice's also with probability $1/2$, and she sends Bob the results of her measurement with the same coding system as she used to eavesdrop. Then there exist cases in which the photon Alice sends disagrees with the one Bob receives, even when Alice and Bob employ the same coding system. This happens with probability $1/4$ as is shown now. Suppose both Alice and Bob happen to employ the coding system (1) and Alice sends Bob 0. Eve will use the coding system (1) with probability $1/2$, in which case Eve measures 0 and sends Bob $|\uparrow\rangle$. Bob, also employing the coding system (1), will obtain 0 with probability 1. If, in contrast, Eve employs coding system (2), which happens with probability $1/2$, then Eve measures 0 or 1 with probability $1/2$ for each photon and sends Bob her result with coding system (2). Then Bob, with coding system (1), will obtain 0 or 1 both with probability $1/2$. In the end of the day, Bob obtains 0 with probability $3/4$ and 1 with probability $1/4$, even though Alice and Bob employ the same coding system. Suppose $4N$ photons are sent from Alice to Bob, as before. They find their codings agree in $2N$ cases and discard the remaining cases. Comparing N cases to check if Eve is in action, they find approximately $N/4$ bits disagree, from which they detect there is an eavesdropper in the quantum channel. They may try different quantum channels until they find one whose security is verified.

EXAMPLE 3.1 Suppose the sent and received sequences are

$$
\begin{array}{llllllllllllll}
\text{Alice sends} & 0 & 1 & 0 & 0 & 1 & 1 & 0 & 1 & 0 & 0 & 1 & 0 & \ldots \\
\text{Alice's code} & (1) & (2) & (1) & (2) & (2) & (1) & (2) & (1) & (2) & (2) & (1) & (1) & \ldots \\
\text{Bob's code} & (1) & (2) & (2) & (1) & (2) & (2) & (1) & (2) & (1) & (2) & (2) & (1) & \ldots \\
\text{Bob reads} & 0 & 1 & ? & ? & 1 & ? & ? & ? & ? & 0 & ? & 0 & \ldots
\end{array}
\tag{3.17}
$$

where ? stands for 0 or 1. Alice and Bob keep the sequence $0, 1, 1, 0, 0, \ldots$ to check the security of the channel and to generate a key.

Suppose Eve eavesdrops their communication. Then their readings may, for example, be

$$
\begin{array}{llllllllllllll}
\text{Alice sends} & 0 & 1 & 0 & 0 & 1 & 1 & 0 & 1 & 0 & 0 & 1 & 0 & \ldots \\
\text{Alice's code} & (1) & (2) & (1) & (2) & (2) & (1) & (2) & (1) & (2) & (2) & (1) & (1) & \ldots \\
\text{Eve's code} & (1) & (2) & (1) & (2) & (1) & (2) & (1) & (2) & (1) & (2) & (1) & (2) & \ldots \\
\text{Eve reads} & 0 & 1 & 0 & 0 & ? & ? & ? & ? & ? & 0 & 1 & ? & \ldots \\
\text{Bob's code} & (1) & (2) & (2) & (1) & (2) & (2) & (1) & (2) & (1) & (2) & (2) & (1) & \ldots \\
\text{Bob reads} & 0 & 1 & ? & ? & \underline{?} & ? & ? & ? & ? & 0 & ? & \underline{?} & \ldots
\end{array}
\tag{3.18}
$$

The 5th and 12th bits Bob obtains may not be the correct ones, even though Alice and Bob employ the same coding system.

Other QKD protols are E91 protocol [8] and B92 protocol [9].

References

[1] E. Riefell and W. Polak, ACM Computing Surveys (CSUR) **32**, 300 (2000).

[2] M. A. Neilsen and I. L. Chuang, *Quantum Computation and Quantum Information*, Cambridge University Press (2000).

[3] D. M. Greenberger, M. A. Horne and A. Zeilinger, in *'Bell's Theorem, Quantum Theory, and Conceptions of the Universe*, ed. M. Kafatos, Kluwer, Dordrecht (1989). Also avilable as arXiv:0712.0921 [quant-ph].

[4] W. Dür, G. Vidal and J. I. Cirac, Phys. Rev. A **62**, 062314 (2000).

[5] A. Einstein, B. Podolsky, N. Rosen, Phys. Rev. **41**, 777 (1935).

[6] C. H. Bennett and G. Brassard, in Proc. IEEE Int. Conf. Comp., Systems and Signal Processing **175** (1984).

[7] http://www.magiqtech.com/

[8] A. Ekert, Phys. Rev. Lett. **67**, 661 (1991).

[9] C. H. Bennett, Phys. Rev. Lett. **68**, 3121 (1992).

4

Quantum Gates, Quantum Circuit and Quantum Computation

4.1 Introduction

Now that we have introduced qubits to store information, it is time to consider operations acting on them. If they are simple, these operations are called gates, or more precisely **quantum gates**, in analogy with those in classical logic circuits. More complicated **quantum circuits** are composed of these simple gates. A collection of quantum circuits for executing a complicated algorithm, a **quantum algorithm**, is a part of a quantum computation.

DEFINITION 4.1 (Quantum Computation) A quantum computation is a collection of the following three elements:

(1) A register or a set of registers,

(2) A unitary matrix U, which is taylored to execute a given quantum algorithm, and

(3) Measurements to extract information we need.

More formally, we say a quantum computation is the set $\{\mathcal{H}, U, \{M_m\}\}$, where $\mathcal{H} = \mathbb{C}^{2^n}$ is the Hilbert space of an n-qubit register, $U \in \mathrm{U}(2^n)$ represents the quantum algorithm and $\{M_m\}$ is the set of measurement operators.

The hardware (1) along with equipment to control the qubits is called a quantum computer.

Suppose the register is set to a fiducial initial state, $|\psi_{\mathrm{in}}\rangle = |00\ldots0\rangle$, for example. A unitary matrix U_{alg} is designed to represent an algorithm which we want to execute. Operation of U_{alg} on $|\psi_{\mathrm{in}}\rangle$ yields the output state $|\psi_{\mathrm{out}}\rangle = U_{\mathrm{alg}}|\psi_{\mathrm{in}}\rangle$. Information is extracted from $|\psi_{\mathrm{out}}\rangle$ by appropriate measurements.

Actual quantum computation processes are very different from those of a classical counterpart. In a classical computer, we input the data from a keyboard or other input devices and the signal is sent to the I/O port of the computer, which is then stored in the memory, then fed into the microprocessor, and the result is stored in the memory before it is printed or it is displayed on the screen. Thus information travels around the circuit. In contrast, information in quantum computation is stored in a register, first of all, and then external fields, such as oscillating magnetic fields, electric fields or laser beams are applied to produce gate operations on the register. These external fields are designed so that they produce desired gate operation, i.e., unitary matrix acting on a particular set of qubits. Therefore the information sits in the register and they are updated each time the gate operation acts on the register.

One of the other distinctions between classical computation and quantum computation is that the former is based upon digital processing and the latter upon hybrid (digital + analogue) processing. A qubit may take an arbitrary superposition of $|0\rangle$ and $|1\rangle$, and hence their coefficients are continuous complex numbers. A gate is also an element of a relevant unitary group, which contains continuous parameters. An operation such as "rotate a specified spin around the x-axis by an angle π" is implemented by applying a particular pulse of specified amplitude, angle and duration. These parameters are continuous numbers and always contain errors. These aspects might cause challenging difficulties in a physical realization of a quantum computer.

Parts of this chapter depend on [1, 2] and [3].

4.2 Quantum Gates

We have so far studied the change of a state upon measurements. When measurements are not made, the time evolution of a state is described by the Schrödinger equation. The system preserves the norm of the state vector during time evolution. Thus the time development is unitary. Let U be such a time-evolution operator; $UU^\dagger = U^\dagger U = I$. We will be free from the Schrödinger equation in the following and assume there exist unitary matrices which we need. Physical implementation of these unitary matrices is another important area of quantum information processing and is a subject of the second part of this book.

One of the important conclusions derived from the unitarity of gates is that the computational process is reversible.

4.2.1 Simple Quantum Gates

Examples of quantum gates which transform a one-qubit state are given below. We call them one-qubit gates in the following. Linearity guarantees that the action of a gate is completely specified as soon as its action on the basis $\{|0\rangle, |1\rangle\}$ is given. Let us consider the gate I whose action on the basis vectors are defined by $I : |0\rangle \to |0\rangle$, $|1\rangle \to |1\rangle$. The matrix expression of this gate is easily found as

$$I = |0\rangle\langle 0| + |1\rangle\langle 1| = \begin{pmatrix} 1 & 0 \\ 0 & 1 \end{pmatrix}. \tag{4.1}$$

Similarly we introduce $X : |0\rangle \to |1\rangle, |1\rangle \to |0\rangle, Y : |0\rangle \to -|1\rangle, |1\rangle \to |0\rangle$, and $Z : |0\rangle \to |0\rangle, |1\rangle \to -|1\rangle$, whose matrix representations are

$$X = |1\rangle\langle 0| + |0\rangle\langle 1| = \begin{pmatrix} 0 & 1 \\ 1 & 0 \end{pmatrix} = \sigma_x, \tag{4.2}$$

$$Y = |0\rangle\langle 1| - |1\rangle\langle 0| = \begin{pmatrix} 0 & -1 \\ 1 & 0 \end{pmatrix} = -i\sigma_y, \tag{4.3}$$

$$Z = |0\rangle\langle 0| - |1\rangle\langle 1| = \begin{pmatrix} 1 & 0 \\ 0 & -1 \end{pmatrix} = \sigma_z. \tag{4.4}$$

The transformation I is the trivial (identity) transformation, while X is the negation (NOT), Z the phase shift and $Y = XZ$ the combination of them. It is easily verified that these gates are unitary.

The **CNOT (controlled-NOT)** gate is a two-qubit gate, which plays quite an important role in quantum computation. The gate flips the second qubit (the **target qubit**) when the first qubit (the **control qubit**) is $|1\rangle$, while leaving the second bit unchanged when the first qubit state is $|0\rangle$. Let $\{|00\rangle, |01\rangle, |10\rangle, |11\rangle\}$ be a basis for the two-qubit system. In the following, we use the standard basis vectors with components

$$|00\rangle = (1, 0, 0, 0)^t, \ |01\rangle = (0, 1, 0, 0)^t, \ |10\rangle = (0, 0, 1, 0)^t, \ |11\rangle = (0, 0, 0, 1)^t.$$

The action of the CNOT gate, whose matrix expression will be written as U_{CNOT}, is

$$U_{\text{CNOT}} : |00\rangle \mapsto |00\rangle, \ |01\rangle \mapsto |01\rangle, \ |10\rangle \mapsto |11\rangle, \ |11\rangle \mapsto |10\rangle.$$

It has two equivalent expressions

$$\begin{aligned} U_{\text{CNOT}} &= |00\rangle\langle 00| + |01\rangle\langle 01| + |11\rangle\langle 10| + |10\rangle\langle 11| \\ &= |0\rangle\langle 0| \otimes I + |1\rangle\langle 1| \otimes X, \end{aligned} \tag{4.5}$$

having a matrix form

$$U_{\text{CNOT}} = \begin{pmatrix} 1 & 0 & 0 & 0 \\ 0 & 1 & 0 & 0 \\ 0 & 0 & 0 & 1 \\ 0 & 0 & 1 & 0 \end{pmatrix}. \tag{4.6}$$

The second expression of the RHS in Eq. (4.5) shows that the action of U_{CNOT} on the target qubit is I when the control qubit is in the state $|0\rangle$, while it is σ_x when the control qubit is in $|1\rangle$. Verify that U_{CNOT} is unitary and, moreover, idempotent, i.e., $U_{\text{CNOT}}^2 = I$.

Let $\{|i\rangle\}$ be the basis vectors, where $i \in \{0,1\}$. The action of CNOT on the input state $|i\rangle|j\rangle$ is written as $|i\rangle|i \oplus j\rangle$, where $i \oplus j$ is an addition mod 2, that is, $0 \oplus 0 = 0, 0 \oplus 1 = 1, 1 \oplus 0 = 1$ and $1 \oplus 1 = 0$.

EXERCISE 4.1 Show that the U_{CNOT} cannot be written as a tensor product of two one-qubit gates.

EXERCISE 4.2 Let $(a|0\rangle + b|1\rangle) \otimes |0\rangle$ be an input state to a CNOT gate. What is the output state?

It is convenient to introduce graphical representations of quantum gates. A one-qubit gate whose unitary matrix representation is U is depicted as

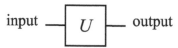

The left horizontal line is the input qubit state, while the right horizontal line is the output qubit state. Therefore the time flows from the left to the right.

A CNOT gate is expressed as

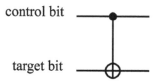

where $\bullet$ denotes the control bit, while $\oplus$ denotes the conditional negation. There may be many control bits (see CCNOT gate below).

More generally, we consider a controlled-U gate,

$$V = |0\rangle\langle 0| \otimes I + |1\rangle\langle 1| \otimes U, \tag{4.7}$$

in which the target bit is acted on by a unitary transformation U only when the control bit is $|1\rangle$. This gate is denoted graphically as

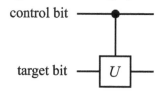

EXERCISE 4.3 (1) Find the matrix representation of the "upside down" CNOT gate (a) in the basis $\{|00\rangle, |01\rangle, |10\rangle, |11\rangle\}$.

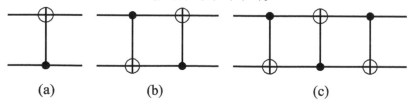

(a)   (b)   (c)

(2) Find the matrix representation of the circuit (b).
(3) Find the matrix representation of the circuit (c). Find the action of the circuit on a tensor product state $|\psi_1\rangle \otimes |\psi_2\rangle$.

The **CCNOT (Controlled-Controlled-NOT)** gate has three inputs, and the third qubit flips when and only when the first two qubits are both in the state $|1\rangle$. The explicit form of the CCNOT gate is

$$U_{\text{CCNOT}} = (|00\rangle\langle00| + |01\rangle\langle01| + |10\rangle\langle10|) \otimes I + |11\rangle\langle11| \otimes X. \qquad (4.8)$$

This gate is graphically expressed as

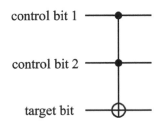

control bit 1

control bit 2

target bit

The CCNOT gate is also known as the **Toffoli gate**.

4.2.2 Walsh-Hadamard Transformation

The **Hadamard gate** or the **Hadamard transformation** H is an important unitary transformation defined by

$$\begin{aligned} U_{\text{H}} : |0\rangle &\to \frac{1}{\sqrt{2}}(|0\rangle + |1\rangle) \\ |1\rangle &\to \frac{1}{\sqrt{2}}(|0\rangle - |1\rangle). \end{aligned} \qquad (4.9)$$

It is used to generate a superposition state from $|0\rangle$ or $|1\rangle$. The matrix representation of H is

$$U_{\text{H}} = \frac{1}{\sqrt{2}}(|0\rangle + |1\rangle)\langle0| + \frac{1}{\sqrt{2}}(|0\rangle - |1\rangle)\langle1| = \frac{1}{\sqrt{2}}\begin{pmatrix} 1 & 1 \\ 1 & -1 \end{pmatrix}. \qquad (4.10)$$

A Hadamard gate is depicted as

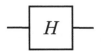

There are numerous important applications of the Hadamard transformation. All possible 2^n states are generated, when U_{H} is applied on each qubit of the state $|00\ldots 0\rangle$:

$$(H \otimes H \otimes \ldots \otimes H)|00\ldots 0\rangle$$

$$= \frac{1}{\sqrt{2}}(|0\rangle + |1\rangle) \otimes \frac{1}{\sqrt{2}}(|0\rangle + |1\rangle) \otimes \ldots \frac{1}{\sqrt{2}}(|0\rangle + |1\rangle)$$

$$= \frac{1}{\sqrt{2^n}} \sum_{x=0}^{2^n-1} |x\rangle. \tag{4.11}$$

Therefore, we produce a superposition of all the states $|x\rangle$ with $0 \le x \le 2^n - 1$ simultaneously. This action of H on an n-qubit system is called the **Walsh transformation**, or **Walsh-Hadamard transformation**, and denoted as W_n. Note that

$$W_1 = U_{\mathrm{H}}, \quad W_{n+1} = U_{\mathrm{H}} \otimes W_n. \tag{4.12}$$

EXERCISE 4.4 Show that W_n is unitary.

EXERCISE 4.5 Show that the two circuits below are equivalent:

This exercise shows that the control bit and the target bit in a CNOT gate are interchangeable by introducing four Hadamard gates.

EXERCISE 4.6 Let us consider the following quantum circuit

$$\tag{4.13}$$

where q_1 denotes the first qubit, while q_2 denotes the second. What are the outputs for the inputs $|00\rangle, |01\rangle, |10\rangle$ and $|11\rangle$?

4.2.3   SWAP Gate and Fredkin Gate

The SWAP gate acts on a tensor product state as

$$U_{\mathrm{SWAP}}|\psi_1, \psi_2\rangle = |\psi_2, \psi_1\rangle. \tag{4.14}$$

The explict form of U_{SWAP} is given by

$$U_{\text{SWAP}} = |00\rangle\langle 00| + |01\rangle\langle 10| + |10\rangle\langle 01| + |11\rangle\langle 11|$$

$$= \begin{pmatrix} 1 & 0 & 0 & 0 \\ 0 & 0 & 1 & 0 \\ 0 & 1 & 0 & 0 \\ 0 & 0 & 0 & 1 \end{pmatrix}. \tag{4.15}$$

Needless to say, it works as a linear operator on a superposition of states. The SWAP gate is expressed as

Note that the SWAP gate is a special gate which maps an arbitrary tensor product state to a tensor product state. In contrast, most two-qubit gates map a tensor product state to an entangled state.

EXERCISE 4.7 Show that the above U_{SWAP} is written as

$$U_{\text{SWAP}} = (|0\rangle\langle 0| \otimes I + |1\rangle\langle 1| \otimes X)(I \otimes |0\rangle\langle 0| + X \otimes |1\rangle\langle 1|)$$
$$(|0\rangle\langle 0| \otimes I + |1\rangle\langle 1| \otimes X). \tag{4.16}$$

This shows that the SWAP gate is implemented with three CNOT gates as given in Exercise 4.3 (3).

The controlled-SWAP gate

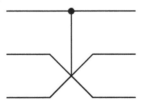

is also called the **Fredkin gate**. It flips the second (middle) and the third (bottom) qubits when and only when the first (top) qubit is in the state $|1\rangle$. Its explicit form is

$$U_{\text{Fredkin}} = |0\rangle\langle 0| \otimes I_4 + |1\rangle\langle 1| \otimes U_{\text{SWAP}}. \tag{4.17}$$

4.3 Correspondence with Classical Logic Gates

Before we proceed further, it is instructive to show that all the elementary logic gates, NOT, AND, XOR, OR and NAND, in classical logic circuits can be implemented with quantum gates. In this sense, quantum information processing contains the classical one.

4.3.1 NOT Gate

Let us consider the **NOT gate** first. It is defined by the following logic function,

$$\text{NOT}(x) = \neg x = \begin{cases} 0 & x = 1 \\ 1 & x = 0 \end{cases} \tag{4.18}$$

where $\neg x$ stands for the **negation** of x. Under the correspondence $0 \leftrightarrow |0\rangle$, $1 \leftrightarrow |1\rangle$, we have already seen in Eq. (4.2) that the gate X negates the basis vectors as

$$X|x\rangle = |\neg x\rangle = |\text{NOT}(x)\rangle, \quad (x = 0, 1). \tag{4.19}$$

Now let us measure the output state. We employ the following measurement operator:

$$M_1 = |1\rangle\langle 1|. \tag{4.20}$$

M_1 has eigenvalues 0 and 1 with the eigenvectors $|0\rangle$ and $|1\rangle$, respectively. When the input is $|0\rangle$, the output is $|1\rangle$ and the measurement gives the value 1 with the probability 1. If, on the other hand, the input is $|1\rangle$, the output is $|0\rangle$ and the measurement yields 1 with probability 0, or in other words, it yields 0 with probability 1. It should be kept in mind that the operator X acts on an arbitrary linear combination $|\psi\rangle = a|0\rangle + b|1\rangle$, which is classically impossible. The output state is then $X|\psi\rangle = a|1\rangle + b|0\rangle$.

We show in the following that the CCNOT gate implements all classical logic gates. The first and the second input qubits are set to $|1\rangle$ to obtain the NOT gate as

$$U_{\text{CCNOT}}|1, 1, x\rangle = |1, 1, \neg x\rangle. \tag{4.21}$$

4.3.2 XOR Gate

Since a quantum gate has to be reversible, we cannot construct a unitary gate corresponding to the classcial **XOR gate** whose function is $x, y \mapsto x \oplus y$ $(x, y \in \{0, 1\})$, where $x \oplus y$ is an addition mod 2; $0 \oplus 0 = 0$, $0 \oplus 1 = 1 \oplus 0 = 1$, $1 \oplus 1 = 0$. Clearly this operation has no inverse. This operation may be made reversible if we keep the first bit x during the gate operation, namely, if we define

$$f(x, y) = (x, x \oplus y), \quad x, y \in \{0, 1\}. \tag{4.22}$$

We call this function f, also the XOR gate. The quantum gate that does this operation is nothing but the CNOT gate defined by Eq. (4.5),

$$U_{\text{XOR}} = U_{\text{CNOT}} = |0\rangle\langle 0| \otimes I + |1\rangle\langle 1| \otimes X. \tag{4.23}$$

Note that the XOR gate may be also obtained from the CCNOT gate. Suppose the first qubit of the CCNOT gate is fixed to $|1\rangle$. Then it is easy to verify that

$$U_{\text{CCNOT}}|1, x, y\rangle = |1, x, x \oplus y\rangle. \qquad (4.24)$$

Thus the CCNOT gate can be used to construct the XOR gate.

4.3.3 AND Gate

The logical **AND gate** is defined by

$$\text{AND}(x, y) \equiv x \wedge y \equiv \begin{cases} 1 & x = y = 1 \\ 0 & \text{otherwise} \end{cases} \quad x, y \in \{0, 1\}. \qquad (4.25)$$

Clearly this operation is not reversible and we have to introduce the same sort of prescription which we employed in the XOR gate.

Let us define the logic function

$$f(x, y, 0) \equiv (x, y, x \wedge y), \qquad (4.26)$$

which we also call AND. Note that we have to keep both x and y for f to be reversible since $x = x \wedge y = 0$ implies *both* $x = y = 0$ and $x = 0, y = 1$. The unitary matrix that computes f is

$$U_{\text{AND}} = (|00\rangle\langle00| + |01\rangle\langle01| + |10\rangle\langle10|) \otimes I$$
$$+|11\rangle\langle11| \otimes X. \qquad (4.27)$$

It is readily verified that

$$U_{\text{AND}}|x, y, 0\rangle = |x, y, x \wedge y\rangle, \quad x, y \in \{0, 1\}. \qquad (4.28)$$

Observe that the third qubit in the RHS is 1 if and only if $x = y = 1$ and 0 otherwise. Thus the CCNOT gate may be employed to implement the AND gate. It follows from Eq. (4.28) that the AND gate is denoted graphically as

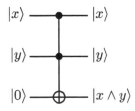

4.3.4 OR Gate

The **OR gate** represents the logical function

$$\text{OR}(x, y) = x \vee y = \begin{cases} 0 & x = y = 0 \\ 1 & \text{otherwise} \end{cases} \quad x, y \in \{0, 1\}. \qquad (4.29)$$

This function OR is not reversible either and special care must be taken.
Let us define

$$f(x, y, 0) \equiv (\neg x, \neg y, x \vee y), \quad x, y \in \{0, 1\}, \tag{4.30}$$

which we also call OR. Although the first and the second bits are negated, it is not essential in the construction of the OR gate. These negations appear due to our construction of the OR gate based on the de Morgan theorem

$$x \vee y = \neg(\neg x \wedge \neg y). \tag{4.31}$$

They may be removed by adding extra NOT gates if necessary.
Let $|x, y, 0\rangle$ be the input state. The unitary matrix that represents f is

$$U_{\text{OR}} = |00\rangle\langle 11| \otimes X + |01\rangle\langle 10| \otimes X + |10\rangle\langle 01| \otimes X + |11\rangle\langle 00| \otimes I. \tag{4.32}$$

EXERCISE 4.8 Verify that the above matrix U_{OR} indeed satisfies

$$U_{\text{OR}}|x, y, 0\rangle = |\neg x, \neg y, x \vee y\rangle, \quad x, y \in \{0, 1\}. \tag{4.33}$$

Now it is obvious why negations in the first and the second qubits appear in the OR gate. Since we have already constructed the NOT gate and AND gate, we take advantage of this in the construction of the OR gate. The equality (4.31) leads us to the following diagram:

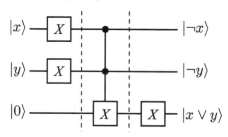

Accordingly, the first and the second qubits are negated. The unitary matrix obtained from this diagram is

$$U_{\text{OR}} = (I \otimes I \otimes X)$$
$$\cdot (|00\rangle\langle 00| \otimes I + |01\rangle\langle 01| \otimes I + |10\rangle\langle 10| \otimes I + |11\rangle\langle 11| \otimes X)$$
$$\cdot (X \otimes X \otimes I). \tag{4.34}$$

The matrix products are readily evaluated to yield

$$U_{\text{OR}} = (|00\rangle\langle 00| \otimes X + |01\rangle\langle 01| \otimes X + |10\rangle\langle 10| \otimes X + |11\rangle\langle 11| \otimes I)$$
$$\cdot (X \otimes X \otimes I)$$
$$= |00\rangle\langle 11| \otimes X + |01\rangle\langle 10| \otimes X + |10\rangle\langle 01| \otimes X + |11\rangle\langle 00| \otimes I,$$

which verifies Eq. (4.32).

Observe that the OR gate is implemented with the X and the CCNOT gates and, moreover, the X gate is obtained from the CCNOT gate by putting the first and the second bits to $|1\rangle$.

If we want to have a gate $V_{OR}|x, y, 0\rangle = |x, y, x \vee y\rangle$, we may multiply $X \otimes X \otimes I$ to U_{OR} from the left so that $V_{OR} = (X \otimes X \otimes I)U_{OR}$.

EXERCISE 4.9 Show that the NAND gate can be obtained from the CC-NOT gate. Here NAND is defined by the function

$$\text{NAND}(x, y) = \neg(x \wedge y) = \begin{cases} 0 & x = y = 1 \\ 1 & \text{otherwise} \end{cases} \quad x, y \in \{0, 1\}. \quad (4.35)$$

In summary, we have shown that all the classical logic gates, NOT, AND, OR, XOR and NAND gates, may be obtained from the CCNOT gate. Thus all the classical computation may be carried out with a quantum computor. Note, however, that these gates belong to a tiny subset of the set of unitary matrices.

In the next section, we show that copying unknown information is impossible in quantum computing. However, it is also shown that this does not restrict the superiority of quantum computing over the classical counterpart.

4.4 No-Cloning Theorem

We copy classical data almost every day. In fact, this is amongst the most common functions with digital media. (Of course we should not copy media that are copyright protected.) This cannot be done in quantum information theory! We cannot clone an unknown quantum state with unitary operations.

THEOREM 4.1 (Wootters and Zurek [4], Dieks [5]) An unknown quantum system cannot be cloned by unitary transformations.

Proof. Suppose there would exist a unitary transformation U that makes a clone of a quantum system. Namely, suppose U acts, for any state $|\varphi\rangle$, as

$$U : |\varphi 0\rangle \rightarrow |\varphi\varphi\rangle. \quad (4.36)$$

Let $|\varphi\rangle$ and $|\phi\rangle$ be two states that are linearly independent. Then we should have $U|\varphi 0\rangle = |\varphi\varphi\rangle$ and $U|\phi 0\rangle = |\phi\phi\rangle$ by definition. Then the action of U on $|\psi\rangle = \frac{1}{\sqrt{2}}(|\varphi\rangle + |\phi\rangle)$ yields

$$U|\psi 0\rangle = \frac{1}{\sqrt{2}}(U|\varphi 0\rangle + U|\phi 0\rangle) = \frac{1}{\sqrt{2}}(|\varphi\varphi\rangle + |\phi\phi\rangle).$$

If U were a cloning transformation, we must also have

$$U|\psi 0\rangle = |\psi\psi\rangle = \frac{1}{2}(|\varphi\varphi\rangle + |\varphi\phi\rangle + |\phi\varphi\rangle + |\phi\phi\rangle),$$

which contradicts the previous result. Therefore, there does not exist a unitary cloning transformation.   ∎

Clearly, there is no way to clone a state by measurements. A measurement is probabilistic and non-unitary, and it gets rid of the component of the state which is in the orthogonal complement of the observed subspace.

EXERCISE 4.10 Suppose U is a cloning unitary transformation, such that

$$|\Psi\rangle \equiv U|\psi\rangle|0\rangle = |\psi\rangle|\psi\rangle$$
$$|\Phi\rangle \equiv U|\phi\rangle|0\rangle = |\phi\rangle|\phi\rangle$$

for arbitrary $|\psi\rangle$ and $|\phi\rangle$.
(1) Write down $\langle\Psi|\Phi\rangle$ in all possible ways.
(2) Show, by inspecting the result of (1), that such U does not exist.

It was mentioned in the end of the previous section that a quantum computer can simulate arbitrary classical logic circuits. Then how about copying data? It should be kept in mind that the no-cloning theorem states that we cannot copy an *arbitrary* state $|\psi\rangle = a|0\rangle + b|1\rangle$. The loophole is that the theorem does not apply if the states to be cloned are limited to $|0\rangle$ and $|1\rangle$. For these cases, the copying operator U should work as

$$U : |00\rangle \mapsto |00\rangle, \quad : |10\rangle \mapsto |11\rangle.$$

We can assign arbitrary action of U on a state whose second input is $|1\rangle$ since this case does not happen. What we have to keep in our mind is only that U be unitary. An example of such U is

$$U = (|00\rangle\langle 00| + |11\rangle\langle 10|) + (|01\rangle\langle 01| + |10\rangle\langle 11|), \qquad (4.37)$$

where the first set of operators renders U the cloning operator and the second set is added just to make U unitary. We immediately notice that U is nothing but the CNOT gate introduced in §4.2.

Therefore, if the data under consideration are limited within $|0\rangle$ and $|1\rangle$, we can copy the qubit states even in a quantum computer. This fact is used to construct quantum error correcting codes.

4.5   Dense Coding and Quantum Teleportation

Now we are ready to introduce two simple applications of qubits and quantum gates: **dense coding** and **quantum teleportation**. The Bell state has been delivered beforehand, and one of the qubits carries two classical bits of information in the dense coding system. In the quantum teleportation, on the other hand, two classical bits are used to transmit a single qubit. At first glance, the quantum teleportation may seem to be in contradiction with the no-cloning theorem. However, this is not the case since the original state is destroyed.

Entanglement is the keyword in both applications. The setting is common for both cases. Suppose Alice wants to send Bob information. Each of them has been sent each of the qubits of the Bell state

$$|\Phi^+\rangle = \frac{1}{\sqrt{2}}(|00\rangle + |11\rangle) \tag{4.38}$$

in advance. Suppose Alice has the first qubit and Bob has the second.

4.5.1 Dense Coding

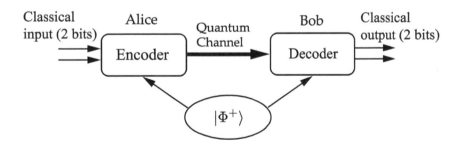

FIGURE 4.1
Communication from Alice to Bob using dense coding. Each qubit of the Bell state $|\Phi^+\rangle$ has been distributed to each of them beforehand. Then two bits of classical information can be transmitted by sending a single qubit through the quantum channel.

<u>Alice</u>: Alice wants to send Bob a binary number $x \in \{00, 01, 10, 11\}$. She picks up one of $\{I, X, Y, Z\}$ according to x she has chosen and applies the transformation on her qubit (the first qubit of the Bell state). Applying the transformation to only her qubit means she applies an identity transformation

to the second qubit which Bob keeps with him. This results in

x	transformation U	state after transformation
$0 = 00$	$I \otimes I$	$\lvert\psi_0\rangle = \dfrac{1}{\sqrt{2}}(\lvert 00\rangle + \lvert 11\rangle)$
$1 = 01$	$X \otimes I$	$\lvert\psi_1\rangle = \dfrac{1}{\sqrt{2}}(\lvert 10\rangle + \lvert 01\rangle)$
$2 = 10$	$Y \otimes I$	$\lvert\psi_2\rangle = \dfrac{1}{\sqrt{2}}(\lvert 10\rangle - \lvert 01\rangle)$
$3 = 11$	$Z \otimes I$	$\lvert\psi_3\rangle = \dfrac{1}{\sqrt{2}}(\lvert 00\rangle - \lvert 11\rangle).$

(4.39)

Alice sends Bob her qubit after the transformation given above is applied. Note that the set of four states in the rightmost column is nothing but the four Bell basis vectors.

Bob: Bob applies CNOT to the entangled pair in which the first qubit, the received qubit, is the control bit, while the second one, which Bob keeps, is the target bit. This results in a tensor-product state:

Received state	Output of CNOT	1st qubit	2nd qubit
$\lvert\psi_0\rangle$	$\dfrac{1}{\sqrt{2}}(\lvert 00\rangle + \lvert 10\rangle)$	$\dfrac{1}{\sqrt{2}}(\lvert 0\rangle + \lvert 1\rangle)$	$\lvert 0\rangle$
$\lvert\psi_1\rangle$	$\dfrac{1}{\sqrt{2}}(\lvert 11\rangle + \lvert 01\rangle)$	$\dfrac{1}{\sqrt{2}}(\lvert 1\rangle + \lvert 0\rangle)$	$\lvert 1\rangle$
$\lvert\psi_2\rangle$	$\dfrac{1}{\sqrt{2}}(\lvert 11\rangle - \lvert 01\rangle)$	$\dfrac{1}{\sqrt{2}}(\lvert 1\rangle - \lvert 0\rangle)$	$\lvert 1\rangle$
$\lvert\psi_3\rangle$	$\dfrac{1}{\sqrt{2}}(\lvert 00\rangle - \lvert 10\rangle)$	$\dfrac{1}{\sqrt{2}}(\lvert 0\rangle - \lvert 1\rangle)$	$\lvert 0\rangle$

(4.40)

Note that Bob can measure the first and second qubits independently since the output is a tensor-product state. The number x is either 00 or 11 if the measurement outcome of the second qubit is $\lvert 0\rangle$, while it is either 01 or 10 if the meansurement outcome is $\lvert 1\rangle$.

Finally, a Hadamard transformation H is applied on the first qubit. Bob obtains

Received state	1st qubit	$U_{\mathrm{H}}\lvert$1st qubit$\rangle$
$\lvert\psi_0\rangle$	$\dfrac{1}{\sqrt{2}}(\lvert 0\rangle + \lvert 1\rangle)$	$\lvert 0\rangle$
$\lvert\psi_1\rangle$	$\dfrac{1}{\sqrt{2}}(\lvert 1\rangle + \lvert 0\rangle)$	$\lvert 0\rangle$
$\lvert\psi_2\rangle$	$\dfrac{1}{\sqrt{2}}(\lvert 1\rangle - \lvert 0\rangle)$	$-\lvert 1\rangle$
$\lvert\psi_3\rangle$	$\dfrac{1}{\sqrt{2}}(\lvert 0\rangle - \lvert 1\rangle)$	$\lvert 1\rangle$

(4.41)

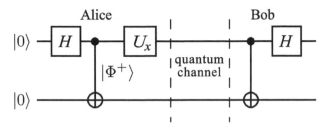

FIGURE 4.2
Quantum circuit implementation of the dense coding system. The leftmost Hadamard gate and the next CNOT gate generate the Bell state. Then a unitary gate U, depending on the bits Alice wants to send, is applied to the first qubit. Bob applies the rightmost CNOT gate and the Hadamard gate to decode Alice's message.

The number x is either 00 or 01 if the measurement of the first qubit results in $|0\rangle$, while it is either 10 or 11 if it is $|1\rangle$. Therefore, Bob can tell what x is in every case.

Quantum circuit implementation for the dense coding is given in Fig. 4.2

4.5.2 Quantum Teleportation

The purpose of **quantum teleportation** is to transmit an unknown quantum *state* of a qubit using two classical bits such that the recipient reproduces exactly the same state as the original qubit state. Note that the qubit itself is not transported but the information required to reproduce the quantum state is transmitted. The original state is destroyed such that quantum teleportation should not be in contradiction with the no-cloning theorem. Quantum teleportation has already been realized under laboratory conditions using photons [6, 7, 8, 9], coherent light field [10], NMR [11], and trapped ions [12, 13]. The teleportation scheme introduced in this section is due to [11]. Figure 4.3 shows the schematic diagram of quantum teleportation, which will be described in detail below.

Alice: Alice has a qubit, whose state she does not know. She wishes to send Bob the quantum state of this qubit through a classical communication channel. Let

$$|\phi\rangle = a|0\rangle + b|1\rangle \tag{4.42}$$

be the state of the qubit. Both of them have been given one of the qubits of the entangled pair

$$|\Phi^+\rangle = \frac{1}{\sqrt{2}}(|00\rangle + |11\rangle)$$

as in the case of the dense coding.

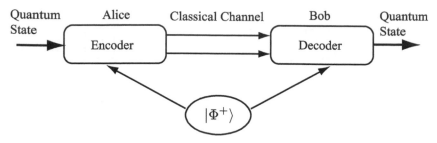

FIGURE 4.3
In quantum teleportation, Alice sends Bob two classical bits so that Bob reproduces a qubit state Alice used to have.

Alice applies the decoding step in the dense coding to the qubit $|\phi\rangle = a|0\rangle + b|1\rangle$ to be sent and her qubit of the entangled pair. They start with the state

$$|\phi\rangle \otimes |\Phi^+\rangle = \frac{1}{\sqrt{2}} [a|0\rangle \otimes (|00\rangle + |11\rangle) + b|1\rangle \otimes (|00\rangle + |11\rangle)]$$
$$= \frac{1}{\sqrt{2}} (a|000\rangle + a|011\rangle + b|100\rangle + b|111\rangle), \qquad (4.43)$$

where Alice has the first two qubits while Bob has the third. Alice applies $U_{\text{CNOT}} \otimes I$ followed by $U_H \otimes I \otimes I$ to this state, which results in

$$(U_H \otimes I \otimes I)(U_{\text{CNOT}} \otimes I)(|\phi\rangle \otimes |\Phi^+\rangle)$$
$$= (U_H \otimes I \otimes I)(U_{\text{CNOT}} \otimes I)\frac{1}{\sqrt{2}} (a|000\rangle + a|011\rangle + b|100\rangle + b|111\rangle)$$
$$= \frac{1}{2} [a(|000\rangle + |011\rangle + |100\rangle + |111\rangle) + b(|010\rangle + |001\rangle - |110\rangle - |101\rangle)]$$
$$= \frac{1}{2} [|00\rangle(a|0\rangle + b|1\rangle) + |01\rangle(a|1\rangle + b|0\rangle)$$
$$+ |10\rangle(a|0\rangle - b|1\rangle) + |11\rangle(a|1\rangle - b|0\rangle)]. \qquad (4.44)$$

If Alice measures the two qubits in her hand, she will obtain one of the states $|00\rangle, |01\rangle, |10\rangle$ or $|11\rangle$ with equal probability $1/4$. Bob's qubit (a qubit from the Bell state initially) collapses to $a|0\rangle + b|1\rangle, a|1\rangle + b|0\rangle, a|0\rangle - b|1\rangle$ or $a|1\rangle - b|0\rangle$, respectively, depending on the result of Alice's measurement. Alice then sends Bob her result of the measurement using two classical bits.

Notice that Alice has totally destroyed the initial qubit $|\phi\rangle$ upon her measurement. This makes quantum teleportation consistent with the no-cloning theorem.

<u>Bob</u>: After receiving two classical bits, Bob knows the state of the qubit in

his hand;

Received bits	Bob's state	Decoding
00	$a\|0\rangle + b\|1\rangle$	I
01	$a\|1\rangle + b\|0\rangle$	X
10	$a\|0\rangle - b\|1\rangle$	Z
11	$a\|1\rangle - b\|0\rangle$	Y

$$(4.45)$$

Bob reconstructs the intial state $|\phi\rangle$ by applying the decoding process shown above. Suppose Alice sends Bob the classical bits 10, for example. Then Bob applies Z to his state to reconstruct $|\phi\rangle$ as follows:

$$Z : (a|0\rangle - b|1\rangle) \mapsto (a|0\rangle + b|1\rangle) = |\phi\rangle.$$

Figure 4.4 shows the actual quantum circuit for quantum teleportation.

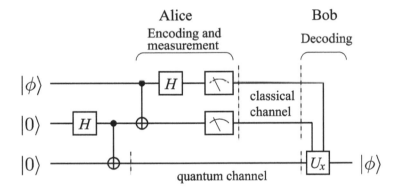

FIGURE 4.4
Quantum circuit implementation of quantum teleportation. Alice operates gates in the left side. The first Hadamard gate and the next CNOT gates generate the Bell state $|\Phi^+\rangle$ from $|00\rangle$. The bottom qubit is sent to Bob through a quantum channel while the first and the second qubits are measured after applying the second set of the CNOT gate and the Hadamard gate on them. The measurement outcome x is sent to Bob through a classical channel. Bob operates a unitary operation U_x, which depends on the received message x, on his qubit.

EXERCISE 4.11 Let $|\psi\rangle = a|00\rangle + b|11\rangle$ be a two-qubit state. Apply a Hadamard gate to the first qubit and then measure the first qubit. Find the second qubit state after the measurement corresponding to the outcome of the first qubit measurement.

4.6   Universal Quantum Gates

It can be shown that any classical logic gate can be constructed by using a small set of gates, AND, NOT and XOR, for example. Such a set of gates is called the *universal* set of classical gates. Since the CCNOT gate can simulate these classical gates, quantum circuits simulate any classical circuits. It should be noted that the set of quantum gates is much larger than those classical gates which can be simulated by quantum gates. Thus we want to find a universal set of *quantum* gates from which any quantum circuits, i.e., any unitary matrix, can be constructed.

In the following, it will be shown that

(1) the set of single qubit gates and

(2) CNOT gate

form a universal set of quantum circuits (**universality theorem**).

We will prove the following Lemma before stating the main theorem. Let us start with a definition. A **two-level unitary matrix** is a unitary matrix which acts non-trivially only on two vector components. Suppose V is a two-level unitary matrix. Then V has the same matrix elements as those of the unit matrix except for certain four elements V_{aa}, V_{ab}, V_{ba} and V_{bb}. An example of a two-level unitary matrix is

$$V = \begin{pmatrix} \alpha^* & 0 & 0 & \beta^* \\ 0 & 1 & 0 & 0 \\ 0 & 0 & 1 & 0 \\ -\beta & 0 & 0 & \alpha \end{pmatrix}, \quad (|\alpha|^2 + |\beta|^2 = 1),$$

where $a = 1$ and $b = 4$.

LEMMA 4.1 Let U be a unitary matrix acting on $\mathbb{C}^d$. Then there are $N \leq d(d-1)/2$ two-level unitary matrices $U_1, U_2, \ldots, U_N$ such that

$$U = U_1 U_2 \ldots U_N. \tag{4.46}$$

Proof. The proof requires several steps. It is instructive to start with the case $d = 3$. Let

$$U = \begin{pmatrix} a & d & g \\ b & e & h \\ c & f & j \end{pmatrix}$$

be a unitary matrix. We want to find two-level unitary matrices U_1, U_2, U_3 such that

$$U_3 U_2 U_1 U = I.$$

Then it follows that

$$U = U_1^\dagger U_2^\dagger U_3^\dagger.$$

(Never mind the daggers! If U_k is two-level unitary, $U_k^\dagger$ is also two-level unitary.) We prove the above decomposition by constructing U_k explicitly.

(i) Let

$$U_1 = \begin{pmatrix} \frac{a^*}{u} & \frac{b^*}{u} & 0 \\ -\frac{b}{u} & \frac{a}{u} & 0 \\ 0 & 0 & 1 \end{pmatrix},$$

where $u = \sqrt{|a|^2 + |b|^2}$. Verify that U_1 is unitary. Then we obtain

$$U_1 U = \begin{pmatrix} a' & d' & g' \\ 0 & e' & h' \\ c' & f' & j' \end{pmatrix},$$

where $a', \ldots, j'$ are some complex numbers, whose details are not necessary. Observe that, with this choice of U_1, the first component of the second row vanishes.

(ii) Let

$$U_2 = \begin{pmatrix} \frac{a'^*}{u'} & 0 & \frac{c'^*}{u'} \\ 0 & 1 & 0 \\ -\frac{c'}{u'} & 0 & \frac{a'}{u'} \end{pmatrix} = \begin{pmatrix} a'^* & 0 & c'^* \\ 0 & 1 & 0 \\ -c' & 0 & a' \end{pmatrix},$$

where $u' = \sqrt{|a'|^2 + |c'|^2} = 1$. Then

$$U_2 U_1 U = \begin{pmatrix} 1 & d'' & g'' \\ 0 & e'' & h'' \\ 0 & f'' & j'' \end{pmatrix} = \begin{pmatrix} 1 & 0 & 0 \\ 0 & e'' & h'' \\ 0 & f'' & j'' \end{pmatrix},$$

where the equality $d'' = g'' = 0$ follows from the fact that $U_2 U_1 U$ is unitary, and hence the first row must be normalized.

(iii) Finally let

$$U_3 = (U_2 U_1 U)^\dagger = \begin{pmatrix} 1 & 0 & 0 \\ 0 & e''^* & f''^* \\ 0 & h''^* & j''^* \end{pmatrix}.$$

Then, by definition, $U_3 U_2 U_1 U = I$ is obvious. This completes the proof for $d = 3$.

Suppose U is a unitary matrix acting on $\mathbb{C}^d$ with a general dimension d. Then by repeating the above arguments, we find two-level unitary matrices $U_1, U_2, \ldots, U_{d-1}$ such that

$$U_{d-1} \ldots U_2 U_1 U = \begin{pmatrix} 1 & 0 & 0 & \ldots & 0 \\ 0 & * & * & \ldots & * \\ 0 & * & * & \ldots & * \\ & & \ldots \ldots \\ 0 & * & * & \ldots & * \end{pmatrix},$$

namely the $(1, 1)$ component is unity and other components of the first row and the first column vanish. The number of matrices $\{U_k\}$ to achieve this form is the same as the number of zeros in the first column, hence $(d-1)$.

We then repeat the same procedure to the $(d-1) \times (d-1)$ block unitary matrix using $(d-2)$ two-level unitary matrices. After repeating this, we finally decompose U into a product of two-level unitary matrices

$$U = V_1 V_2 \ldots V_N,$$

where $N \le (d-1) + (d-2) + \ldots + 1 = d(d-1)/2$.   ∎

EXERCISE 4.12 Let U be a general 4×4 unitary matrix. Find two-level unitary matrices U_1, U_2 and U_3 such that

$$U_3 U_2 U_1 U = \begin{pmatrix} 1 & 0 & 0 & 0 \\ 0 & * & * & * \\ 0 & * & * & * \\ 0 & * & * & * \end{pmatrix}.$$

EXERCISE 4.13 Let

$$U = \frac{1}{2} \begin{pmatrix} 1 & 1 & 1 & 1 \\ 1 & i & -1 & -i \\ 1 & -1 & 1 & -1 \\ 1 & -i & -1 & i \end{pmatrix}. \tag{4.47}$$

Decompose U into a product of two-level unitary matrices.

Let us consider a unitary matrix acting on an n-qubit system. Then this unitary matrix is decomposed into a product of at most $2^n(2^n - 1)/2 = 2^{n-1}(2^n - 1)$ two-level unitary matrices. Now we are in a position to state the main theorem.

THEOREM 4.2 (Barenco *et al.*)[14] The set of single qubit gates and CNOT gate are universal. Namely, any unitary gate acting on an n-qubit register can be implemented with single qubit gates and CNOT gates.

Proof. We closely follow [1] for the proof here. Thanks to the previous Lemma, it suffices to prove the theorem for a two-level unitary matrix. Let U be a two-level unitary matrix acting nontrivially only on $|s\rangle$ and $|t\rangle$ basis vectors of an n-qubit system, where $s = s_{n-1}2^{n-1} + \ldots + s_1 2 + s_0$ and $t = t_{n-1}2^{n-1} + \ldots + t_1 2 + t_0$ are binary expressions for decimal numbers s and t. This means that matrix elements U_{ss}, U_{st}, U_{ts} and U_{tt} are different from those of the unit matrix, while all the others are the same, where $|s\rangle$ stands for $|s_{n-1}\rangle |s_{n-2}\rangle \ldots |s_0\rangle$, for example. We can construct $\tilde{U}$, the non-trivial 2×2 unitary submatrix of U. $\tilde{U}$ may be thought of as a unitary matrix acting on a single qubit, whose basis is $\{|s\rangle, |t\rangle\}$.

STEP 1: U is reduced to $\tilde{U} \in \mathrm{U}(2)$.

The basis vectors $|s\rangle$ and $|t\rangle$ may be put together to form a basis for a single qubit using the following trick. This is done by introducing **Gray codes**. For two binary numbers $s = s_{n-1} \ldots s_1 s_0$ and $t = t_{n-1} \ldots t_1 t_0$, a Gray code connecting s and t is a sequence of binary numbers $\{g_1, \ldots, g_m\}$ where the adjacent numbers, g_k and g_{k+1}, differ in exactly one bit. Moreover, the sequence satisfies the boundary conditions $g_1 = s$ and $g_m = t$.

Suppose $s = 100101$ and $t = 110110$, for example. An example of a Gray code connecting s and t is

$$s = g_1 = 100101$$
$$g_2 = 1\hat{1}0101$$
$$g_3 = 1101\hat{1}1$$
$$g_4 = 11011\hat{0} = t,$$

where the digit with ˆ has been renewed. It is clear from this construction that if s and t differ in p bits, the shortest Gray code is made of $p+1$ elements. It should be also clear that if s and t are of n digits, then $m \leq (n+1)$ since s and t differ at most in n bits.

With these preparations, we consider the implementation of U. The strategy is to find gates providing the sequence of state changes

$$|s\rangle = |g_1\rangle \rightarrow |g_2\rangle \rightarrow \ldots \rightarrow |g_{m-1}\rangle. \tag{4.48}$$

Then g_{m-1} and g_m differ only in one bit, which is identified with the single qubit on which $\tilde{U}$ acts. In the example above, we have $|g_3\rangle = |11011\rangle \otimes |1\rangle$ and $|t\rangle = |g_4\rangle = |11011\rangle \otimes |0\rangle$. Now the operator $\tilde{U}$ may be introduced so that it acts on a two-dimensional subspace of the total Hilbert space, in which the first five qubits are in the state $|11011\rangle$. Then we undo the sequence (4.48) so that $|g_{m-1}\rangle \rightarrow |g_{m-2}\rangle \rightarrow \ldots \rightarrow |g_1\rangle = |s\rangle$. Each of these steps can be easily implemented using simple gates that have been introduced previously (see below).

Let us consider the following example of a three-qubit system, whose basis is $\{|000\rangle, |001\rangle, \ldots, |111\rangle\}$. Let

$$U = \begin{pmatrix} a & 0 & 0 & 0 & 0 & 0 & 0 & c \\ 0 & 1 & 0 & 0 & 0 & 0 & 0 & 0 \\ 0 & 0 & 1 & 0 & 0 & 0 & 0 & 0 \\ 0 & 0 & 0 & 1 & 0 & 0 & 0 & 0 \\ 0 & 0 & 0 & 0 & 1 & 0 & 0 & 0 \\ 0 & 0 & 0 & 0 & 0 & 1 & 0 & 0 \\ 0 & 0 & 0 & 0 & 0 & 0 & 1 & 0 \\ b & 0 & 0 & 0 & 0 & 0 & 0 & d \end{pmatrix}, \quad (a, b, c, d \in \mathbb{C}) \tag{4.49}$$

be a two-level unitary matrix which we wish to implement. Note that U acts non-trivially only in the subspace spanned by $|000\rangle$ and $|111\rangle$. The unitarity

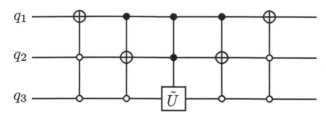

FIGURE 4.5
Example of circuit implementing the gate U.

of U ensures that the matrix

$$\tilde{U} = \begin{pmatrix} a & c \\ b & d \end{pmatrix}$$ (4.50)

is also unitary. An example of a Gray code connecting 000 and 111 is

$$\begin{array}{cccc} & q_1 & q_2 & q_3 \\ g_1 = & 0 & 0 & 0 \\ g_2 = & 1 & 0 & 0 \\ g_3 = & 1 & 1 & 0 \\ g_4 = & 1 & 1 & 1 \end{array}$$ (4.51)

Since g_3 and g_4 differ only in the third qubit, which we call q_3, we have to bring g_1 to g_3 and then operate $\tilde{U}$ on the qubit q_3 provided that the first and the second qubits are in the state $|11\rangle$. (Namely we have a controlled-$\tilde{U}$ gate with the target bit q_3 and the control bits q_1 and q_2.) After this controlled operation is done, we have to put $|g_3\rangle = |110\rangle$ back to the state $|000\rangle$ as

$$|110\rangle \rightarrow |100\rangle \rightarrow |000\rangle.$$

This operation is graphically shown in Fig. 4.5. Here ○ denotes the negated control node. This means that the unitary gate acts on the target bit only when the control bit is in the state $|0\rangle$. This is easily implemented by adding two X gates as

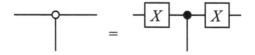

It is easy to see that this gate indeed implements U. Suppose the input is $|101\rangle$, for example. Figure 4.6 shows that the gate has no effect on this basis vector since U should act as a unit matrix on this vector. The operation of U

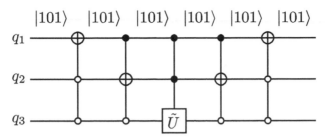

FIGURE 4.6
U-gate has no effect on the vector $|101\rangle$.

on the input $\alpha|000\rangle + \beta|111\rangle$ is

$$U(\alpha|000\rangle + \beta|111\rangle) = \begin{pmatrix} a & 0 & 0 & 0 & 0 & 0 & 0 & c \\ 0 & 1 & 0 & 0 & 0 & 0 & 0 & 0 \\ 0 & 0 & 1 & 0 & 0 & 0 & 0 & 0 \\ 0 & 0 & 0 & 1 & 0 & 0 & 0 & 0 \\ 0 & 0 & 0 & 0 & 1 & 0 & 0 & 0 \\ 0 & 0 & 0 & 0 & 0 & 1 & 0 & 0 \\ 0 & 0 & 0 & 0 & 0 & 0 & 1 & 0 \\ b & 0 & 0 & 0 & 0 & 0 & 0 & d \end{pmatrix} \begin{pmatrix} \alpha \\ 0 \\ 0 \\ 0 \\ 0 \\ 0 \\ 0 \\ \beta \end{pmatrix} = \begin{pmatrix} \alpha a + \beta c \\ 0 \\ 0 \\ 0 \\ 0 \\ 0 \\ 0 \\ \alpha b + \beta d \end{pmatrix}. \quad (4.52)$$

If we use the circuit shown in Fig. 4.5, we produce the same result as shown in Fig. 4.7

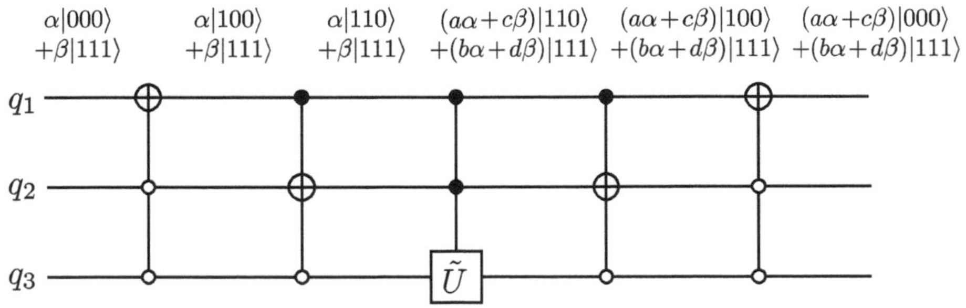

$\alpha|000\rangle$ $\alpha|100\rangle$ $\alpha|110\rangle$ $(a\alpha+c\beta)|110\rangle$ $(a\alpha+c\beta)|100\rangle$ $(a\alpha+c\beta)|000\rangle$
$+\beta|111\rangle$ $+\beta|111\rangle$ $+\beta|111\rangle$ $+(b\alpha+d\beta)|111\rangle$ $+(b\alpha+d\beta)|111\rangle$ $+(b\alpha+d\beta)|111\rangle$

FIGURE 4.7
U-gate acting on $\alpha|000\rangle + \beta|111\rangle$ yields the desired output $(a\alpha + c\beta)|000\rangle + (b\alpha + d\beta)|111\rangle$.

This construction is easily generalized to any two-level unitary matrix $U \in U(2^n)$. It will be shown below that all the gates in the above circuit can be implemented with single-qubit gates and CNOT gates, which proves the universality of these gates.

EXERCISE 4.14 (1) Find the shortest Gray code which connects 000 with 110.
(2) Use this result to find a quantum circuit, such as Fig. 4.5, implementing a two-level unitary gate

$$U = \begin{pmatrix} a\,0\,0\,0\,0\,0\,c\,0 \\ 0\,1\,0\,0\,0\,0\,0\,0 \\ 0\,0\,1\,0\,0\,0\,0\,0 \\ 0\,0\,0\,1\,0\,0\,0\,0 \\ 0\,0\,0\,0\,1\,0\,0\,0 \\ 0\,0\,0\,0\,0\,1\,0\,0 \\ b\,0\,0\,0\,0\,0\,d\,0 \\ 0\,0\,0\,0\,0\,0\,0\,1 \end{pmatrix}, \quad \tilde{U} \equiv \begin{pmatrix} a & c \\ b & d \end{pmatrix} \in U(2).$$

You may use various controlled-NOT gates and controlled-$\tilde{U}$ gates.

STEP 2: Two-level unitary gates are decomposed into single-qubit gates and CNOT gates.

A controlled-U gate can be constructed from at most four single-qubit gates and two CNOT gates for any single-qubit unitary $U \in U(2)$. Let us prove several Lemmas before we prove this statement.

LEMMA 4.2 Let $U \in SU(2)$. Then there exist $\alpha, \beta, \gamma \in \mathbb{R}$ such that $U = R_z(\alpha)R_y(\beta)R_z(\gamma)$, where

$$R_z(\alpha) = \exp(i\alpha\sigma_z/2) = \begin{pmatrix} e^{i\alpha/2} & 0 \\ 0 & e^{-i\alpha/2} \end{pmatrix},$$

$$R_y(\beta) = \exp(i\beta\sigma_y/2) = \begin{pmatrix} \cos(\beta/2) & \sin(\beta/2) \\ -\sin(\beta/2) & \cos(\beta/2) \end{pmatrix}.$$

Proof. After some calculation, we obtain

$$R_z(\alpha)R_y(\beta)R_z(\gamma) = \begin{pmatrix} e^{i(\alpha+\gamma)/2}\cos(\beta/2) & e^{i(\alpha-\gamma)/2}\sin(\beta/2) \\ -e^{i(-\alpha+\gamma)/2}\sin(\beta/2) & e^{-i(\alpha+\gamma)/2}\cos(\beta/2) \end{pmatrix}. \quad (4.53)$$

Any $U \in SU(2)$ may be written in the form

$$U = \begin{pmatrix} a & b \\ -b^* & a^* \end{pmatrix} = \begin{pmatrix} \cos\theta e^{i\lambda} & \sin\theta e^{i\mu} \\ -\sin\theta e^{-i\mu} & \cos\theta e^{-i\lambda} \end{pmatrix}, \quad (4.54)$$

where we used the fact that $\det U = |a|^2 + |b|^2 = 1$. Now we obtain $U = R_z(\alpha)R_y(\beta)R_z(\gamma)$ by making identifications

$$\theta = \frac{\beta}{2}, \lambda = \frac{\alpha+\gamma}{2}, \mu = \frac{\alpha-\gamma}{2}. \quad (4.55)$$

∎

LEMMA 4.3 Let $U \in SU(2)$. Then there exist $A, B, C \in SU(2)$ such that $U = AXBXC$ and $ABC = I$, where $X = \sigma_x$.

Proof. Lemma 4.2 states that $U = R_z(\alpha)R_y(\beta)R_z(\gamma)$ for some $\alpha, \beta, \gamma \in \mathbb{R}$. Let

$$A = R_z(\alpha)R_y\left(\frac{\beta}{2}\right), B = R_y\left(-\frac{\beta}{2}\right)R_z\left(-\frac{\alpha+\gamma}{2}\right), C = R_z\left(-\frac{\alpha-\gamma}{2}\right).$$

Then

$$\begin{aligned}
AXBXC &= R_z(\alpha)R_y\left(\frac{\beta}{2}\right)XR_y\left(-\frac{\beta}{2}\right)R_z\left(-\frac{\alpha+\gamma}{2}\right)XR_z\left(-\frac{\alpha-\gamma}{2}\right)\\
&= R_z(\alpha)R_y\left(\frac{\beta}{2}\right)\left[XR_y\left(-\frac{\beta}{2}\right)X\right]\left[XR_z\left(-\frac{\alpha+\gamma}{2}\right)X\right]R_z\left(-\frac{\alpha-\gamma}{2}\right)\\
&= R_z(\alpha)R_y\left(\frac{\beta}{2}\right)R_y\left(\frac{\beta}{2}\right)R_z\left(\frac{\alpha+\gamma}{2}\right)R_z\left(-\frac{\alpha-\gamma}{2}\right)\\
&= R_z(\alpha)R_y(\beta)R_z(\gamma) = U,
\end{aligned}$$

where use has been made of the identities $X^2 = I$ and $X\sigma_{y,z}X = -\sigma_{y,z}$.

It is also verified that

$$\begin{aligned}
ABC &= R_z(\alpha)R_y\left(\frac{\beta}{2}\right)R_y\left(-\frac{\beta}{2}\right)R_z\left(-\frac{\alpha+\gamma}{2}\right)R_z\left(-\frac{\alpha-\gamma}{2}\right)\\
&= R_z(\alpha)R_y(0)R_z(-\alpha) = I.
\end{aligned}$$

This proves the Lemma. ∎

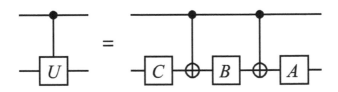

FIGURE 4.8
Controlled-U gate is made of at most three single-qubit gates and two CNOT gates for any $U \in SU(2)$.

LEMMA 4.4 Let $U \in SU(2)$ be factorized as $U = AXBXC$ as in the previous Lemma. Then the controlled-U gate can be implemented with at most three single-qubit gates and two CNOT gates (see Fig. 4.8).

Proof. The proof is almost obvious. When the control bit is 0, the target bit $|\psi\rangle$ is operated by C, B and A in this order so that

$$|\psi\rangle \mapsto ABC|\psi\rangle = |\psi\rangle,$$

while when the control bit is 1, we have

$$|\psi\rangle \mapsto AXBXC|\psi\rangle = U|\psi\rangle.$$

∎

So far, we have worked with $U \in \mathrm{SU}(2)$. To implement a general U-gate with $U \in \mathrm{U}(2)$, we have to deal with the phase. Let us first recall that any $U \in \mathrm{U}(2)$ is decomposed as $U = e^{i\alpha}V$, $V \in \mathrm{SU}(2), \alpha \in \mathbb{R}$.

LEMMA 4.5 Let

$$\Phi(\phi) = e^{i\phi}I = \begin{pmatrix} e^{i\phi} & 0 \\ 0 & e^{i\phi} \end{pmatrix}$$

and

$$D = R_z(-\phi)\Phi\left(\frac{\phi}{2}\right) = \begin{pmatrix} e^{-i\phi/2} & 0 \\ 0 & e^{i\phi/2} \end{pmatrix}\begin{pmatrix} e^{i\phi/2} & 0 \\ 0 & e^{i\phi/2} \end{pmatrix} = \begin{pmatrix} 1 & 0 \\ 0 & e^{i\phi} \end{pmatrix}.$$

Then the controlled-$\Phi(\phi)$ gate is expressed as a tensor product of single qubit gates as

$$U_{\mathrm{C}\Phi(\phi)} = D \otimes I. \tag{4.56}$$

Proof. The LHS is

$$U_{\mathrm{C}\Phi(\phi)} = |0\rangle\langle 0| \otimes I + |1\rangle\langle 1| \otimes \Phi(\phi) = |0\rangle\langle 0| \otimes I + |1\rangle\langle 1| \otimes e^{i\phi}I$$
$$= |0\rangle\langle 0| \otimes I + e^{i\phi}|1\rangle\langle 1| \otimes I,$$

while the RHS is

$$D \otimes I = \begin{pmatrix} 1 & 0 \\ 0 & e^{i\phi} \end{pmatrix} \otimes I$$
$$= [|0\rangle\langle 0| + e^{i\phi}|1\rangle\langle 1|] \otimes I = U_{\mathrm{C}\Phi(\phi)},$$

which proves the lemma.   ∎

Figure 4.9 shows the statement of the above lemma.

EXERCISE 4.15 Let us consider the controlled-V_1 gate $U_{\mathrm{C}V_1}$ and the controlled-V_2 gate $U_{\mathrm{C}V_2}$. Show that the controlled-V_1 gate followed by the controlled-V_2 gate is the controlled-V_2V_1 gate $U_{\mathrm{C}(V_2V_1)}$ as shown in Fig. 4.10.

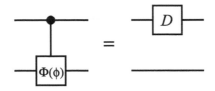

FIGURE 4.9
Equality $U_{C\Phi(\phi)} = D \otimes I$.

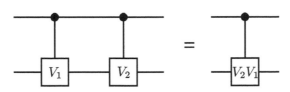

FIGURE 4.10
Equality $U_{CV_2}U_{CV_1} = U_{C(V_2V_1)}$.

Now we are ready to prove the main proposition.

PROPOSITION 4.1 Let $U \in U(2)$. Then the controlled-U gate U_{CU} can be constructed by at most four single-qubit gates and two CNOT gates.

Proof. Let $U = \Phi(\phi)AXBXC$. According to the exercise above, the controlled-U gate is written as a product of the controlled-$\Phi(\phi)$ gate and the controlled-$AXBXC$ gate. Moreover, Lemma 4.5 states that the controlled-$\Phi(\phi)$ gate may be replaced by a single-qubit phase gate acting on the first qubit. The rest of the gate, the controlled-$AXBXC$ gate is implemented with three SU(2) gates and two CNOT gates as proved in Lemma 4.3. Therefore we have the following decomposition:

$$U_{CU} = (D \otimes A)U_{CNOT}(I \otimes B)U_{CNOT}(I \otimes C), \qquad (4.57)$$

where

$$D = R_z(-\phi)\Phi(\phi/2)$$

and use has been made of the identity $(D \otimes I)(I \otimes A) = D \otimes A$. ∎

Figure 4.11 shows the statement of the proposition.

STEP 3: CCNOT gate and its variants are implemented with CNOT gates and their variants.

Now our final task is to prove that controlled-U gates with $n - 1$ control bits are also constructed using single-qubit gates and CNOT gates. Let us start with the simplest case, in which $n = 3$.

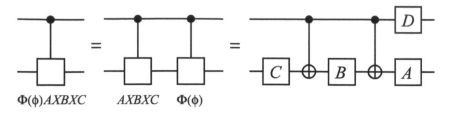

$\Phi(\phi)AXBXC$ $AXBXC$ $\Phi(\phi)$

FIGURE 4.11
Controlled-U gate is implemented with at most four single-qubit gates and
two CNOT gates.

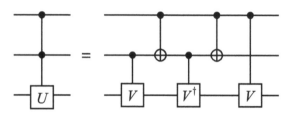

FIGURE 4.12
Controlled-controlled-U gate is equivalent to the gate made of controlled-V
gates with $U = V^2$ and CNOT gates.

LEMMA 4.6 The two quantum circuits in Fig. 4.12 are equivalent, where
$U = V^2$.

Proof. If both the first and the second qubits are 0 in the RHS, all the gates
are ineffective and the third qubit is unchanged; the gate in this subspace
acts as $|00\rangle\langle00| \otimes I$. In case the first qubit is 0 and the second is 1, the
third qubit is mapped as $|\psi\rangle \mapsto V^\dagger V|\psi\rangle = |\psi\rangle$; the gate is then $|01\rangle\langle01| \otimes I$.
When the first qubit is 1 and the second is 0, the third qubit is mapped as
$|\psi\rangle \mapsto VV^\dagger|\psi\rangle = |\psi\rangle$; hence the gate in this subspace is $|10\rangle\langle10| \otimes I$. Finally
let both the first and the second qubits be 1. Then the action of the gate on
the third qubit is $|\psi\rangle \mapsto VV|\psi\rangle = U|\psi\rangle$; namely the gate in this subspace is
$|11\rangle\langle11| \otimes U$. Thus it has been proved that the RHS of Fig. 4.12 is

$$(|00\rangle\langle00| + |01\rangle\langle01| + |10\rangle\langle10|) \otimes I + |11\rangle\langle11| \otimes U, \qquad (4.58)$$

namely the controlled-controlled-U gate.   ∎

This decomposition is explained intuitively as follows. The first V operates
on the third qubit $|\psi\rangle$ if and only if the second qubit is 1. $V^\dagger$ is in action
if and only if $x_1 \oplus x_2 = 1$, where x_k is the input bit of the kth qubit. The
second V operation is applied if and only if the first qubit is 1. Thus the
action of this gate on the third qubit is $V^2 = U$ only when $x_1 \wedge x_2 = 1$ and

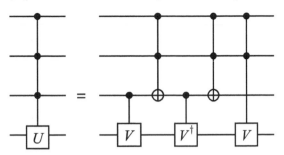

FIGURE 4.13
Decomposition of the $C^3 U$ gate.

I otherwise. This intuitive picture is of help when we implement the U gate with more control qubits.

EXERCISE 4.16 Prove Lemma 4.6 by writing down the action of each gate in the RHS of Fig. 4.12 explicitly using bras, kets and $I, U, V, V^\dagger$. (For example, $U_{\mathrm{CNOT}} = |0\rangle\langle 0| \otimes I + |1\rangle\langle 1| \otimes X$ for a two-qubit system.)

A simple generalization of the above construction is applied to a controlled-U gate with three control bits as the following exercise shows.

EXERCISE 4.17 Show that the circuit in Fig. 4.13 is a controlled-U gate with three control bits, where $U = V^2$.

Now it should be clear how these examples are generalized to gates with more control bits.

PROPOSITION 4.2 The quantum circuit in Fig. 4.14 with $U = V^2$ is a decomposition of the controlled-U gate with $n - 1$ control bits.

The proof of the above proposition is very similar to that of Lemma 4.6 and Exercise 4.17 and is left as an exercise to the readers.

Theorem 4.2 has been now proved. ∎

Other types of gates are also implemented with single-qubit gates and the CNOT gates. See Barenco *et al.* [14] for further details. A few remarks are in order. The above controlled-U gate with $(n - 1)$ control bits requires $\Theta(n^2)$ elementary gates.*† Let us write the number of the elementary gates required

---

*We call single-qubit gates and the CNOT gates elementary gates from now on.
†We will be less strict in the definition of "the order of." In the theory of computational complexity, people use three types of "order of magnitude." One writes "$f(n)$ is $O(g(n))$" if there exist $n_0 \in \mathbb{N}$ and $c \in \mathbb{R}$ such that $f(n) \leq cg(n)$ for $n \geq n_0$. In other words, O sets the asymptotic upper bound of $f(n)$. A function $f(n)$ is said to be $\Omega(g(n))$ if there exist

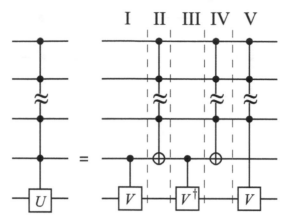

FIGURE 4.14

Decomposition of the $\mathrm{C}^{(n-1)}U$ gate. The number on the top denotes the layer refered to in the text.

to construct the gate in Fig. 4.14 by $C(n)$. Construction of layers I and III requires elementary gates whose number is independent of n. It can be shown that the number of the elementary gates required to construct the controlled NOT gate with $(n-2)$ control bits is $\Theta(n)$ [14]. Therefore layers II and IV require $\Theta(n)$ elementary gates. Finally the layer V, a controlled-V gate with $(n-2)$ control bits, requires $C(n-1)$ basic gates by definition. Thus we obtain a recursion relation

$$C(n) - C(n-1) = \Theta(n). \tag{4.59}$$

The solution to this recursion relation is

$$C(n) = \Theta(n^2). \tag{4.60}$$

Therefore, implementation of a controlled-U gate with $U \in \mathrm{U}(2)$ and $(n-1)$ control bits requires $\Theta(n^2)$ elementary gates.

$n_0 \in \mathbb{N}$ and $c \in \mathbb{R}$ such that $f(n) \geq cg(n)$ for $n \geq n_0$. In other words, Ω sets the asymptotic lower bound of $f(n)$. Finally $f(n)$ is said to be $\Theta(f(n))$ if $f(n)$ behaves asymptotically as $g(n)$, namely if $f(n)$ is both $O(g(n))$ and $\Omega(g(n))$.

4.7   Quantum Parallelism and Entanglement

Given an input x, a typical quantum computer computes $f(x)$ in such a way
as

$$U_f : |x\rangle|0\rangle \mapsto |x\rangle|f(x)\rangle, \tag{4.61}$$

where U_f is a unitary matrix that implements the function f.

Suppose U_f acts on the input which is a superposition of many states. Since
U_f is a linear operator, it acts simultaneously on all the vectors that constitute
the superposition. Thus the output is also a superposition of all the results;

$$U_f : \sum_x |x\rangle|0\rangle \mapsto \sum_x |x\rangle|f(x)\rangle. \tag{4.62}$$

Namely, when the input is a superposition of n states, U_f computes n values
$f(x_k)$ $(1 \leq k \leq n)$ simultaneously. This feature, called the *quantum parallelism*, gives a quantum computer an enormous power. A quantum computer
is advantageous compared to a classical counterpart in that it makes use of
this quantum parallelism and also entanglement.

A unitary transformation acts on a superposition of all possible states in
most quantum algorithms. This superposition is prepared by the action
of the Walsh-Hadamard transformation on an n-qubit register in the state
$|00\ldots0\rangle = |0\rangle \otimes |0\rangle \otimes \ldots \otimes |0\rangle$ resulting in

$$\frac{1}{\sqrt{2^n}} (|00\ldots0\rangle + |00\ldots1\rangle + \ldots |11\ldots1\rangle) = \frac{1}{\sqrt{2^n}} \sum_{x=0}^{2^n-1} |x\rangle. \tag{4.63}$$

This state is a superposition of vectors encoding all the integers between 0
and $2^n - 1$. Then the linearity of U_f leads to

$$U_f \left(\frac{1}{\sqrt{2^n}} \sum_{x=0}^{2^n-1} |x\rangle|0\rangle \right) = \frac{1}{\sqrt{2^n}} \sum_{x=0}^{2^n-1} U_f|x\rangle|0\rangle = \frac{1}{\sqrt{2^n}} \sum_{x=0}^{2^n-1} |x\rangle|f(x)\rangle. \tag{4.64}$$

Note that the superposition is made of $2^n = e^{n \ln 2}$ states, which makes quantum computation exponentially faster than the classical counterpart in a certain kind of computation.

What about the limitation of a quantum computer? Let us consider the
CCNOT gate, for example. This gate flips the third qubit if and only if the
first and the second qubits are both in the state $|1\rangle$, while it leaves the third
qubit unchanged otherwise. Let us fix the third input qubit to $|0\rangle$. It was
shown in §4.3.3 that the third output is $|x \wedge y\rangle$, where $|x\rangle$ and $|y\rangle$ are the first
and the second input qubit states, respectively. Suppose the input state is a
superposition of all possible states while the third qubit is fixed to $|0\rangle$. This

can be achieved by the Walsh-Hadamard transformation as

$$U_H|0\rangle \otimes U_H|0\rangle \otimes |0\rangle = \frac{1}{\sqrt{2}}(|0\rangle + |1\rangle) \otimes \frac{1}{\sqrt{2}}(|0\rangle + |1\rangle) \otimes |0\rangle$$

$$= \frac{1}{2}(|000\rangle + |010\rangle + |100\rangle + |110\rangle). \qquad (4.65)$$

By operating CCNOT on this state, we obtain

$$U_{\text{CCNOT}}(U_H|0\rangle \otimes U_H|0\rangle \otimes |0\rangle) = \frac{1}{2}(|000\rangle + |010\rangle + |100\rangle + |111\rangle). \qquad (4.66)$$

This output may be thought of as the truth table of AND: $|x, y, x \wedge y\rangle$. It is extremely important to note that the output is an entangled state and the measurement projects the state to *one line* of the truth table, i.e., a single term in the RHS of Eq. (4.66). The order of the measurements of the three qubits does not matter at all. The measurement of the third qubit projects the state to the superposition of the states with the given value of the third qubit. Repeating the measurements on the rest of the qubits leads to the collapse of the output state to one of $|x, y, x \wedge y\rangle$.

There is no advantage of quantum computation over classical at this stage. This is because only *one* result may be obtained by a single set of measurements. What is worse, we cannot choose a specific vector $|x, y, x \wedge y\rangle$ at our will! Thus any quantum algorithm should be programmed so that the particular vector we want to observe should have larger probability to be measured compared to other vectors. This step has no classical analogy and is very special in quantum computation. The programming strategies to deal with this feature are [2]

1. to amplify the amplitude, and hence the probability, of the vector that we want to observe. This strategy is employed in the Grover's database search algorithm.

2. to find a common property of all the $f(x)$. This idea was employed in the quantum Fourier transform to find the order[‡] of f in the Shor's factoring algorithm.

Now we consider the power of entanglement. Suppose we have an n-qubit register, whose Hilbert space is 2^n-dimensional. Since each qubit has two basis vectors $|0\rangle$ and $|1\rangle$, there are $2n$ basis vectors (n $|0\rangle$'s and n $|1\rangle$'s) involved to span this 2^n-dimensional Hilbert space. Imagine that we have a single quantum system, instead, which has the same Hilbert space. One might think that the system may do the same quantum computation as the n-qubit register does. One possible problem is that one cannot measure the "kth digit"

[‡]Let $m, N \in \mathbb{N}$ ($m < N$) be numbers coprime to each other. Then there exists $P \in \mathbb{N}$ such that $m^P \equiv 1 \pmod{N}$. The smallest such number P is called the **period** or the **order**. It is easily seen that $m^{x+P} \equiv m^x \pmod{N}$, $\forall x \in \mathbb{N}$.

leaving other digits unaffected. Even worse, consider how many different basis vectors are required for this system. This single system must have an enormous number, 2^n, of basis vectors! Let us consider 20 spin-1/2 particles in a magnetic field. We can employ the spin-up and spin-down energy eigenstates of each particle as the qubit basis vectors. Then there are merely 40 energy eigenvectors involved. Suppose we use energy eigenstates of a certain molecule to replace this register. Then we have to use $2^{20} \sim 10^6$ eigenstates. Separation and control of so many eigenstates are certainly beyond current technology. These simple consideration shows that multipartite implementation of a quantum algorithm requires an exponentially smaller number of basis vectors than monopartite implementation since the former makes use of entanglement as a computational resource.

References

[1] M. A. Neilsen and I. L. Chuang, *Quantum Computation and Quantum Information*, Cambridge University Press (2000).

[2] E. Riefell and W. Polak, ACM Computing Surveys (CSUR) **32**, 300 (2000).

[3] Y. Uesaka, *Mathematical Principle of Quantum Computation*, Corona Publishing, Tokyo, in Japanese (2000).

[4] W. K. Wootters and W. H. Zurek, Nature **299**, 802 (1982).

[5] D. Dieks, Phys. Lett. A **92**, 271 (1982).

[6] D. Bouwmeester *et al.*, Nature **390**, 575 (1997).

[7] D. Boschi *et al.*, Phys. Rev. Lett. **80**, 1121 (1998).

[8] I. Marcikic *et al.*, Nature **421**, 509 (2003).

[9] R. Ursin *et al.*, Nature **430**, 849 (2004).

[10] A. Furusawa *et al.*, Science **282**, 706 (1998).

[11] M. A. Nielsen *et al.*, Nature **396**, 52 (1998).

[12] M. Riebe *et al.*, Nature **429**, 734 (2004).

[13] M. D. Barret *et al.*, Nature **429**, 737 (2004).

[14] A. Barenco *et al.*, Phys. Rev. A **52**, 3457 (1995).

5

Simple Quantum Algorithms

Before we start presenting "useful" but rather complicated quantum algorithms, we introduce a few simple quantum algorithms which will be of help for readers to understand how quantum algorithms are different from and superior to classical algorithms. We follow closely Meglicki [1].

5.1 Deutsch Algorithm

The **Deutsch algorithm** is one of the first quantum algorithms which showed quantum algorithms may be more efficient than their classical counterparts. In spite of its simplicty, full use of the superposition principle has been made here.

Let $f : \{0, 1\} \to \{0, 1\}$ be a binary function. Note that there are only four possible f, namely

$$f_1 : 0 \mapsto 0, \ 1 \mapsto 0, \quad f_2 : 0 \mapsto 1, \ 1 \mapsto 1,$$
$$f_3 : 0 \mapsto 0, \ 1 \mapsto 1, \quad f_4 : 0 \mapsto 1, \ 1 \mapsto 0.$$

The first two cases, f_1 and f_2, are called *constant*, while the rest, f_3 and f_4, are *balanced*. If we only have classical resources, we need to evaluate f twice to tell if f is constant or balanced. There is a quantum algorithm, however, with which it is possible to tell if f is constant or balanced with a single evaluation of f, as was shown by Deutsch [2].

Let $|0\rangle$ and $|1\rangle$ correspond to classical bits 0 and 1, respectively, and consider the state $|\psi_0\rangle = \frac{1}{2}(|00\rangle - |01\rangle + |10\rangle - |11\rangle)$. We apply f on this state in terms of the unitary operator $U_f : |x, y\rangle \mapsto |x, y \oplus f(x)\rangle$, where $\oplus$ is an addition mod 2. To be explicit, we obtain

$$|\psi_1\rangle = U_f|\psi_0\rangle$$
$$= \frac{1}{2}(|0, f(0)\rangle - |0, 1 \oplus f(0)\rangle + |1, f(1)\rangle - |1, 1 \oplus f(1)\rangle)$$
$$= \frac{1}{2}(|0, f(0)\rangle - |0, \neg f(0)\rangle + |1, f(1)\rangle - |1, \neg f(1)\rangle),$$

where $\neg$ stands for negation. Therefore this operation is nothing but the CNOT gate with the control bit $f(x)$; the target bit y is flipped if and only if

$f(x) = 1$ and left unchanged otherwise. Subsequently we apply a Hadamard gate on the first qubit to obtain

$$|\psi_2\rangle = (U_H \otimes I)|\psi_1\rangle$$
$$= \frac{1}{2\sqrt{2}} [(|0\rangle + |1\rangle)(|f(0)\rangle - |\neg f(0)\rangle) + (|0\rangle - |1\rangle)(|f(1)\rangle - |\neg f(1)\rangle)].$$

The wave function reduces to

$$|\psi_2\rangle = \frac{1}{\sqrt{2}}|0\rangle(|f(0)\rangle - |\neg f(0)\rangle) \tag{5.1}$$

in case f is constant, for which $|f(0)\rangle = |f(1)\rangle$, and

$$|\psi_2\rangle = \frac{1}{\sqrt{2}}|1\rangle(|f(0)\rangle - |\neg f(0)\rangle) \tag{5.2}$$

if f is balanced, for which $|\neg f(0)\rangle = |f(1)\rangle$. Therefore the measurement of the first qubit tells us whether f is constant or balanced.

Let us consider a quantum circuit which implements the Deutsch algorithm. We first apply Walsh-Hadamard transformation $W_2 = U_H \otimes U_H$ on $|01\rangle$ to obtain $|\psi_0\rangle$. We need to introduce a conditional gate U_f, i.e., the controlled-NOT gate with the control bit $f(x)$, whose action is $U_f : |x, y\rangle \rightarrow |x, y \oplus f(x)\rangle$. Then a Hadamard gate is applied on the first qubit before it is measured. Figure 5.1 depicts this implementation.

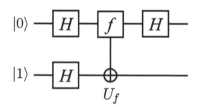

FIGURE 5.1
Implementation of the Deutsch algorithm.

In the quauntum circuit, we assume the gate U_f is a black box for which we do not ask the explicit implementation. We might think it is a kind of subroutine. Such a black box is often called an **oracle**. The gate U_f is called the Deutsch oracle. Its implementation is given only after f is specified.

Then what is the merit of the Deutsch algorithm? Suppose your friend gives you a unitary matrix U_f and asks you to tell if f is constant or balanced. Instead of applying $|0\rangle$ and $|1\rangle$ separately, you may construct the circuit in Fig. 5.1 with the given matrix U_f and apply the circuit on the input state $|01\rangle$. Then you can tell your friend whether f is constant or balanced with a single use of U_f.

5.2 Deutsch-Jozsa Algorithm and Bernstein-Vazirani Algorithm

The Deutsch algorithm introduced in the previous section may be generalized to the **Deutsch-Jozsa algorithm** [3].

Let us first define the **Deutsch-Jozsa problem**. Suppose there is a binary function

$$f : S_n \equiv \{0, 1, \ldots, 2^n - 1\} \to \{0, 1\}. \tag{5.3}$$

We require that f be either *constant* or *balanced* as before. When f is constant, it takes a constant value 0 or 1 irrespetive of the input value x. When it is balanaced the value $f(x)$ for the half of $x \in S_n$ is 0, while it is 1 for the rest of x. In other words, $|f^{-1}(0)| = |f^{-1}(1)| = 2^{n-1}$, where $|A|$ denotes the number of elements in a set A, known as the cardinality of A. Although there are functions which are neither constant nor balanced, we will not consider such cases here. Our task is to find an algorithm which tells if f is constant or balanced with the least possible number of evaluations of f.

It is clear that we need at least $2^{n-1} + 1$ steps, in the worst case with classical manipulations, to make sure if $f(x)$ is constant or balanced with 100% confidence. It will be shown below that the number of steps reduces to a single step if we are allowed to use a quantum algorithm.

The algorithm is divided into the following steps:

1. Prepare an $(n + 1)$-qubit register in the state $|\psi_0\rangle = |0\rangle^{\otimes n} \otimes |1\rangle$. First n qubits work as input qubits, while the $(n + 1)$st qubit serves as a "scratch pad." Such qubits, which are neither input qubits nor output qubits, but work as a scratch pad to store temporary information are called **ancillas** or **ancillary qubits**.

2. Apply the Walsh-Hadamard transforamtion to the register. Then we have the state

$$|\psi_1\rangle = U_H^{\otimes n+1} |\psi_0\rangle = \frac{1}{\sqrt{2^n}} (|0\rangle + |1\rangle)^{\otimes n} \otimes \frac{1}{\sqrt{2}} (|0\rangle - |1\rangle)$$

$$= \frac{1}{\sqrt{2^n}} \sum_{x=0}^{2^n-1} |x\rangle \otimes \frac{1}{\sqrt{2}} (|0\rangle - |1\rangle). \tag{5.4}$$

3. Apply the $f(x)$-controlled-NOT gate on the register, which flips the $(n + 1)$st qubit if and only if $f(x) = 1$ for the input x. Therefore we need a U_f gate which evaluates $f(x)$ and acts on the register as $U_f|x\rangle|c\rangle = |x\rangle|c \oplus f(x)\rangle$, where $|c\rangle$ is the one-qubit state of the $(n+1)$st qubit. Observe that $|c\rangle$ is flipped if and only if $f(x) = 1$ and left

unchanged otherwise. We then obtain a state

$$|\psi_2\rangle = U_f|\psi_1\rangle$$

$$= \frac{1}{\sqrt{2^n}} \sum_{x=0}^{2^n-1} |x\rangle \frac{1}{\sqrt{2}}(|f(x)\rangle - |\neg f(x)\rangle)$$

$$= \frac{1}{\sqrt{2^n}} \sum_{x}(-1)^{f(x)}|x\rangle \frac{1}{\sqrt{2}}(|0\rangle - |1\rangle). \qquad (5.5)$$

Although the gate U_f is applied once for all, it is applied to *all* the n-qubit states $|x\rangle$ simultaneously.

4. The Walsh-Hadamard transformation (4.11) is applied on the first n qubits next. We obtain

$$|\psi_3\rangle = (W_n \otimes I)|\psi_2\rangle = \frac{1}{\sqrt{2^n}} \sum_{x=0}^{2^n-1} (-1)^{f(x)} U_{\mathrm{H}}^{\otimes n}|x\rangle \frac{1}{\sqrt{2}}(|0\rangle - |1\rangle). \quad (5.6)$$

It is instructive to write the action of the one-qubit Hadamard gate in the following form,

$$U_{\mathrm{H}}|x\rangle = \frac{1}{\sqrt{2}}(|0\rangle + (-1)^x|1\rangle) = \frac{1}{\sqrt{2}} \sum_{y\in\{0,1\}} (-1)^{xy}|y\rangle,$$

where $x \in \{0,1\}$, to find the resulting state. The action of the Walsh-Hadamard transformation on $|x\rangle = |x_{n-1} \ldots x_1 x_0\rangle$ yields

$$W_n|x\rangle = (U_{\mathrm{H}}|x_{n-1}\rangle)(U_{\mathrm{H}}|x_{n-2}\rangle)\ldots(U_{\mathrm{H}}|x_0\rangle)$$

$$= \frac{1}{\sqrt{2^n}} \sum_{y_{n-1},y_{n-2},\ldots,y_0\in\{0,1\}} (-1)^{x_{n-1}y_{n-1}+x_{n-2}y_{n-2}+\ldots+x_0y_0}$$

$$\times |y_{n-1}y_{n-2}\ldots y_0\rangle$$

$$= \frac{1}{\sqrt{2^n}} \sum_{y=0}^{2^n-1} (-1)^{x\cdot y}|y\rangle, \qquad (5.7)$$

where $x\cdot y = x_{n-1}y_{n-1} \oplus x_{n-2}y_{n-2} \oplus \ldots \oplus x_0y_0$. Substituting this result into Eq. (5.6), we obtain

$$|\psi_3\rangle = \frac{1}{2^n} \left(\sum_{x,y=0}^{2^n-1} (-1)^{f(x)}(-1)^{x\cdot y}|y\rangle \right) \frac{1}{\sqrt{2}}(|0\rangle - |1\rangle). \qquad (5.8)$$

5. The first n qubits are measured. Suppose $f(x)$ is constant. Then $|\psi_3\rangle$ is put in the form

$$|\psi_3\rangle = \frac{1}{2^n} \sum_{x,y}(-1)^{x\cdot y}|y\rangle \frac{1}{\sqrt{2}}(|0\rangle - |1\rangle)$$

up to an overall phase. Now let us consider the summation

$$\frac{1}{2^n} \sum_{x=0}^{2^n-1} (-1)^{x \cdot y}$$

with a fixed $y \in S_n$. Clearly it vanishes since $x \cdot y$ is 0 for half of x and 1 for the other half of x unless $y = 0$. Therefore the summation yields δ_{y0}. Now the state reduces to

$$|\psi_3\rangle = |0\rangle^{\otimes n} \frac{1}{\sqrt{2}}(|0\rangle - |1\rangle),$$

and the measurement outcome of the first n qubits is always $00\ldots0$. Suppose $f(x)$ is balanced next. The probability amplitude of $|y = 0\rangle$ in $|\psi_3\rangle$ is proportional to

$$\sum_{x=0}^{2^n-1} (-1)^{f(x)} (-1)^{x \cdot 0} = \sum_{x=0}^{2^n-1} (-1)^{f(x)} = 0.$$

Therefore the probability of obtaining measurement outcome $00\ldots0$ for the first n qubits vanishes. In conclusion, the function f is constant if we obtain $00\ldots0$ upon the meaurement of the first n qubits in the state $|\psi_3\rangle$, and it is balanced otherwise.

EXERCISE 5.1 Let us take $n = 2$ for definiteness. Consider the following cases and find the final wave function $|\psi_3\rangle$ and evaluate the measurement outcomes and their probabilities for each case.
(1) $f(x) = 1 \ \forall x \in S_2$.
(2) $f(00) = f(01) = 1, f(10) = f(11) = 0$.
(3) $f(00) = 0, f(01) = f(10) = f(11) = 1$. (This function is neither constant nor balanced.)

The above exercise shows that the measurement gives $|00\rangle$ with probability 1 if f is constant and with probability 0 if balanced. If f is neither constant nor balanced $|\psi_3\rangle$ is a superposition of several states including $|00\rangle$, which is attributed to "incomplete" interference.

A quantum circuit which implements the Deutsch-Jozsa algorithm is given in Fig. 5.2. The gate U_f is called the **Deutsch-Jozsa oracle**.

The **Bernstein-Vazirani algorithm** is a special case of the Deutsch-Jozsa algorithm, in which $f(x)$ is given by $f(x) = c \cdot x$, where $c = c_{n-1}c_{n-2}\ldots c_0$ is an n-bit binary number [4]. Our aim is to find c with the smallest number of evaluations of f. If we apply the Deutsch-Jozsa algorithm with this f, we obtain

$$|\psi_3\rangle = \frac{1}{2^n} \left[\sum_{x,y=0}^{2^n-1} (-1)^{c \cdot x} (-1)^{x \cdot y} |y\rangle \right] \frac{1}{\sqrt{2}}(|0\rangle - |1\rangle).$$

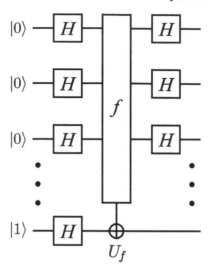

FIGURE 5.2

Quantum circuit implementing the Deutsch-Jozsa algorithm. The gate U_f is the Deutsch-Jozsa oracle.

Let us fix y first. If we take $y = c$, we obtain

$$\sum_x (-1)^{c \cdot x}(-1)^{x \cdot c} = \sum_x (-1)^{2c \cdot x} = 2^n.$$

If $y \neq c$, on the other hand, there will be the same number of x such that $c \cdot x = 0$ and x such that $c \cdot x = 1$ in the summation over x and, as a result, the probability amplitude of $|y \neq c\rangle$ vanishes. By using these results, we end up with

$$|\psi_3\rangle = |c\rangle \frac{1}{\sqrt{2}}(|0\rangle - |1\rangle). \tag{5.9}$$

We are able to tell what c is by measuring the first n qubits. Note that this is done by a single application of the circuit in Fig. 5.2.

EXERCISE 5.2 Consider the Bernstein-Vazirani algorithm with $n = 3$ and $c = 101$. Work out the quantum circuit depicted in Fig. 5.2 to show that the measurement outcome of the first three qubits is $c = 101$.

5.3 Simon's Algorithm

The final example of simple quantum algorithms is **Simon's algorithm**. Let us consider a function (oracle) $f : \{0,1\}^n \to \{0,1\}^n$ such that

1. f is 2 to 1; namely, for any x_1, there is one and only one $x_2 \neq x_1$ such that $f(x_1) = f(x_2)$.

2. f is periodic; namely, there exists $p \in \{0,1\}^n$ such that $f(x \oplus p) = f(x)$, $\forall x \in \{0,1\}^n$, where $\oplus$ is a bitwise addition mod 2.

The function f is made of n component functions $f_k : \{0,1\}^n \to \{0,1\}$ as $f = (f_1, f_2, \ldots, f_n)$.

Suppose we want to find the period p, given an unknown oracle f. Since p can be any number between $00\ldots0$ and $11\ldots1$, we have to try $\sim 2^n$ possibilities classically before we hit the right number. It is shown below that the number of trials required to find p is reduced to $O(n)$ if Simon's algorithm is employed.

The algorithm is decomposed into the following steps:

1. Prepare two sets of n-qubit regiters in the state $|\psi_0\rangle = |0\rangle|0\rangle$. Then the Walsh-Hadamard transformation W_n is applied on the first register to yield

$$|\psi_1\rangle = (W_n \otimes I)|\psi_0\rangle = \frac{1}{\sqrt{2^n}} \sum_{x=0}^{2^n-1} |x\rangle|0\rangle.$$

2. Introduce n controlled-NOT gates with control qubits $f_k(x)$ $(1 \leq k \leq n)$ and the target bit is the kth qubit of the second register. We write

$$U_f : |x\rangle|0\rangle \mapsto |x\rangle|f(x)\rangle,$$

where $|0\rangle$ is an n-qubit register state and $|f(x)\rangle = |f_1(x)\rangle|f_2(x)\rangle \ldots |f_n(x)\rangle$. Linearity implies the state $|\psi_2\rangle$ after the U_f gate operation on $|\psi_1\rangle$ is

$$|\psi_2\rangle = \frac{1}{\sqrt{2^n}} \sum_{x=0}^{2^n-1} |x\rangle|f(x)\rangle. \tag{5.10}$$

3. Now we measure the second register. In fact, we do not need to know the measurement outcome. What we have to do is to project the second register to a certain state $|f(x_0)\rangle$, for example. After one of these operations, the state is now projected to

$$|\psi_3\rangle = \frac{1}{\sqrt{2}}(|x_0\rangle + |x_0 \oplus p\rangle)|f(x_0)\rangle, \tag{5.11}$$

where we noted that there are exactly two states $|x_0\rangle$ and $|x_0 \oplus p\rangle$ that give the second register state $|f(x_0)\rangle$ in step 2.

4. Now we apply W_n again on the first register to obtain

$$|\psi_4\rangle = \frac{1}{\sqrt{2}} \frac{1}{\sqrt{2^n}} \sum_{y=0}^{2^n-1} \left[(-1)^{x_0 \cdot y} + (-1)^{(x_0 \oplus p) \cdot y} \right] |y\rangle \otimes |f(x_0)\rangle$$

$$= \frac{1}{\sqrt{2}} \frac{1}{\sqrt{2^n}} \sum_{y=0}^{2^n-1} (-1)^{x_0 \cdot y} \left[1 + (-1)^{p \cdot y} \right] |y\rangle \otimes |f(x_0)\rangle. \quad (5.12)$$

The inner product $p \cdot y$ takes two values 0 and 1. We immediately notice that such y which satisfies $1 + (-1)^{p \cdot y} = 0$, namely, $p \cdot y = 1$ does not contribute to the summation in Eq. (5.12). Now we are left with

$$|\psi_4\rangle = \frac{2}{\sqrt{2}} \frac{1}{\sqrt{2^n}} \left[\sum_{p \cdot y = 0} (-1)^{x_0 \cdot y} |y\rangle \right] \otimes |f(x_0)\rangle. \quad (5.13)$$

5. Finally we measure the first register. Upon this measurement, we obtain $|y\rangle$ such that $p \cdot y = 0$. Of course, this equation is not enough to identify the period p. Now we repeat the algorithm many times to obtain

$$p \cdot y_1 = p \cdot y_2 = \ldots = p \cdot y_m = 0. \quad (5.14)$$

It should be clear that we need at least n trials since not all equations are linearly independent. For a sufficiently large number of trials m, we are able to solve Eq. (5.14) for p classically. The number of trials necessary for this is $O(n)$ with a good probability.

Figure 5.3 shows the quantum circuit to implement Simon's algorithm for the case $n = 3$.

Simon's algorithm has been improved so that it may be executed in deterministic polynomial time [6].

References

[1] Z. Meglicki, http://beige.ucs.indiana.edu/M743/index.html

[2] D. Deutsch, Proc. Roy. Soc. Lond. A, **400**, 97 (1985).

[3] D. Deutsch and R. Jozsa, Proc. Roy. Soc. Lond. A, **439**, 553 (1992).

[4] E. Bernstein and U. Vazirani, SIAM J. Comput., **26**, 1411 (1997).

[5] D. R. Simon, Proc. 35th Annual Sympo. Found. Comput. Science, IEEE Comput. Soc. Press, Los Alamitos, 116 (1994).

[6] T. Mihara and S. C. Sung, Comput. Complex. **12**, 162 (2003).

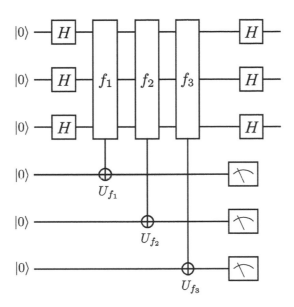

FIGURE 5.3
Quantum circuit implementing Simon's algorithm. The second register is measured only for projection purposes, and reading the outcome is not necessary.

6

Quantum Integral Transforms

We demonstrated in the previous chapter that there are some quantum algorithms superior to their classical counterparts. It is, however, rather difficult to find any practical use of these algorithms. There are two quantum algorithms, known to date, which are potentially useful: Grover's search algorithm and Shor's prime number factorization algorithm. Both of them depend on quantum integral transforms, which will be introduced in the present chapter. We mainly follow [1] in our presentation.

6.1 Quantum Integral Transforms

DEFINITION 6.1 (Discrete Integral Transform) Let $n \in \mathbb{N}$ and $S_n = \{0, 1, \ldots, 2^n - 1\}$ be a set of integers. Consider a map

$$K : S_n \times S_n \to \mathbb{C}. \tag{6.1}$$

For any function $f : S_n \to \mathbb{C}$, its **discrete integral transform** (DIT) $\tilde{f} : S_n \to \mathbb{C}$ with the **kernel** K is defined as:

$$\tilde{f}(y) = \sum_{x=0}^{2^n-1} K(y, x) f(x). \tag{6.2}$$

The transformation $f \to \tilde{f}$ is also called the discrete integral transform.

We define $N \equiv 2^n$ to simplify our notations. The kernel K is expressed as a matrix,

$$K = \begin{pmatrix} K(0,0) & \ldots & K(0, N-1) \\ K(1,0) & \ldots & K(1, N-1) \\ \ldots & \ldots & \ldots \\ K(N-1, 0) & \ldots & K(N-1, N-1), \end{pmatrix} \tag{6.3}$$

and the function f as a vector,

$$f = (f(0), f(1), \ldots f(N-1))^t.$$

The definition of DIT then reduces to the ordinary multiplication of a matrix on a vector as

$$\tilde{f} = Kf.$$

PROPOSITION 6.1 Suppose the kernel K is unitary: $K^\dagger = K^{-1}$. Then the inverse transform $\tilde{f} \to f$ of a DIT exists and is given by

$$f(x) = \sum_{y=0}^{N-1} K^\dagger(x,y)\tilde{f}(y). \tag{6.4}$$

Proof. By substituting Eq. (6.2) into Eq. (6.4), we prove

$$\sum_{y=0}^{N-1} K^\dagger(x,y)\tilde{f}(y) = \sum_{y=0}^{N-1} K^\dagger(x,y)\left[\sum_{z=0}^{N-1} K(y,z)f(z)\right]$$

$$= \sum_{z=0}^{N-1}\left[\sum_{y=0}^{N-1} K^\dagger(x,y)K(y,z)\right]f(z)$$

$$= \sum_{z=0}^{N-1} \delta_{xz}f(z) = f(x).$$

∎

Let U be an $N \times N$ unitary matrix which acts on the n-qubit space $\mathcal{H} = (\mathbb{C}^2)^{\otimes n}$. Let $\{|x\rangle = |x_{n-1}, x_{n-2} \ldots, x_0\rangle\}$ $(x_k \in \{0,1\})$ be the standard binary basis of $\mathcal{H}$, where $x = x_{n-1}2^{n-1} + x_{n-2}2^{n-2} + \ldots + x_0 2^0$. Then

$$U|x\rangle = \sum_{y=0}^{N-1} |y\rangle\langle y|U|x\rangle = \sum_{y=0}^{N-1} U(y,x)|y\rangle. \tag{6.5}$$

The complex number $U(x,y) = \langle x|U|y\rangle$ is the (x,y)-component of U in this basis.

PROPOSITION 6.2 Let U be a unitary transformation, acting on $\mathcal{H} = (\mathbb{C}^2)^{\otimes n}$. Suppose U acts on a basis vector $|x\rangle$ as

$$U|x\rangle = \sum_{y=0}^{N-1} K(y,x)|y\rangle. \tag{6.6}$$

Then U *computes** the DIT $\tilde{f}(y) = \sum_{x=0}^{N-1} K(y,x)f(x)$ for any $y \in S_n$, in the sense that

$$U\left[\sum_{x=0}^{N-1} f(x)|x\rangle\right] = \sum_{y=0}^{N-1} \tilde{f}(y)|y\rangle. \tag{6.7}$$

---

*The proposition claims that U maps a state with the probability amplitude $f(x)$ to another state with the probability amplitude $\tilde{f}(y)$ that is related with $f(x)$ through the kernel K.

Here $|x\rangle$ and $|y\rangle$ are basis vectors of $\mathcal{H}$.

Proof. In fact,

$$
U\left[\sum_{x=0}^{N-1} f(x)|x\rangle\right] = \sum_{x=0}^{N-1} f(x)U|x\rangle
$$

$$
= \sum_{x=0}^{N-1} f(x)\left[\sum_{y=0}^{N-1} K(y,x)|y\rangle\right] = \sum_{y=0}^{N-1}\left[\sum_{x=0}^{N-1} K(y,x)f(x)\right]|y\rangle
$$

$$
= \sum_{y=0}^{N-1} \tilde{f}(y)|y\rangle. \tag{6.8}
$$

∎

Note that the unitary matrix U computes the discrete integral transform $\tilde{f}(y)$ for *all* variables y by a single operation if it acts on the superposition state $\sum_x f(x)|x\rangle$. There are exponentially large numbers 2^n of y for an n-qubit register, and this fact provides a quantum computer with exponentially fast computing power for a certain kind of computations compared to classical alternatives.

The unitary matrix U implementing a discrete integral transform as in Eq. (6.7) is called the **quantum integral transform (QIT)**.

EXERCISE 6.1 Let $f \to \tilde{f}$ be a DFT with a unitary kernel K. Prove **Parseval's theorem**

$$
\sum_{x=0}^{N-1} |f(x)|^2 = \sum_{y=0}^{N-1} |\tilde{f}(y)|^2. \tag{6.9}
$$

6.2 Quantum Fourier Transform (QFT)

One of the most important quantum integral transforms is the quantum Fourier transform. Let ω_n be the Nth primitive root of 1;

$$
\omega_n = e^{2\pi i/N}, \tag{6.10}
$$

where $N = 2^n$ as before. The complex number ω_n defines a kernel K by

$$
K(x,y) = \frac{1}{\sqrt{N}}\omega_n^{-xy}. \tag{6.11}
$$

The discrete integral transform with the kernel K,

$$
\tilde{f}(y) = \frac{1}{\sqrt{N}}\sum_{x=0}^{N-1}\omega_n^{-xy}f(x), \tag{6.12}
$$

is called the **discrete Fourier transform (DFT)**.

The kernel K is unitary since

$$(KK^\dagger)(x,y) = \langle x|K \sum_z |z\rangle\langle z|K^\dagger|y\rangle = \sum_z K(x,z)K^\dagger(z,y)$$

$$= \frac{1}{N} \sum_z \omega_n^{-xz} \omega_n^{yz} = \frac{1}{N} \sum_z \omega^{-(x-y)z} = \delta_{xy}.$$

The quantum integral transform defined with this kernel is called the **quantum Fourier transform (QFT)**.

The kernel for $n = 1$ is

$$K_1 = \frac{1}{\sqrt{2}} \begin{pmatrix} 1 & 1 \\ 1 & e^{2\pi i/2} \end{pmatrix} = \frac{1}{\sqrt{2}} \begin{pmatrix} 1 & 1 \\ 1 & -1 \end{pmatrix}, \tag{6.13}$$

which is nothing but our familiar Hadamard gate. For $n = 2$, we have $\omega_2 = e^{2\pi i/4} = i$ and

$$K_2 = \frac{1}{2} \begin{pmatrix} 1 & 1 & 1 & 1 \\ 1 & \omega_2^{-1} & \omega_2^{-2} & \omega_2^{-3} \\ 1 & \omega_2^{-2} & \omega_2^{-4} & \omega_2^{-6} \\ 1 & \omega_2^{-3} & \omega_2^{-6} & \omega_2^{-9} \end{pmatrix} = \frac{1}{2} \begin{pmatrix} 1 & 1 & 1 & 1 \\ 1 & -i & -1 & i \\ 1 & -1 & 1 & -1 \\ 1 & i & -1 & -i \end{pmatrix}. \tag{6.14}$$

The inverse DFT is given by

$$f(x) = \frac{1}{\sqrt{N}} \sum_{y=0}^{N-1} \omega_n^{xy} \tilde{f}(y). \tag{6.15}$$

It is important to note that

$$U_{\text{QFT}n}|0\rangle = \frac{1}{\sqrt{2^n}} \sum_{y=0}^{2^n-1} |y\rangle, \tag{6.16}$$

where $U_{\text{QFT}n}$ is the n-qubit QFT gate. This equality shows that the QFT of $f(x) = \delta_{x0}$ is $\tilde{f}(y) = 1/\sqrt{2^n}$, which is similar to the FT of the Dirac delta function $\delta(x)$. Observe that a single application of $U_{\text{QFT}n}$ on the state $|0\rangle$ has produced the superposition of all the basis vectors of $\mathcal{H}$.

EXERCISE 6.2 Let

$$|\psi\rangle = \mathcal{N} \sum_{x=0}^{N-1} \cos\left(\frac{2\pi x}{N}\right) |x\rangle$$

be an n-qubit state.
(1) Normalize $|\psi\rangle$.
(2) Find $U_{\text{QFT}n}|\psi\rangle$.

TABLE 6.1
Coefficient of a vector $|x\rangle|f(y)\rangle$. Only the diagonal combination $|x\rangle|f(x)\rangle$ has nonvanishing amplitude in the initial state $|\Psi\rangle$. Moreover all the non-vanishing coefficients have vanishing phase.

	$|0\rangle$	$|1\rangle$	$|2\rangle$	$|3\rangle$	$|4\rangle$	$|5\rangle$	$|6\rangle$	$|7\rangle$
$|f(7)\rangle$	0	0	0	0	0	0	0	$\frac{1}{\sqrt{8}}$
$|f(6)\rangle$	0	0	0	0	0	0	$\frac{1}{\sqrt{8}}$	0
$|f(5)\rangle$	0	0	0	0	0	$\frac{1}{\sqrt{8}}$	0	0
$|f(4)\rangle$	0	0	0	0	$\frac{1}{\sqrt{8}}$	0	0	0
$|f(3)\rangle$	0	0	0	$\frac{1}{\sqrt{8}}$	0	0	0	0
$|f(2)\rangle$	0	0	$\frac{1}{\sqrt{8}}$	0	0	0	0	0
$|f(1)\rangle$	0	$\frac{1}{\sqrt{8}}$	0	0	0	0	0	0
$|f(0)\rangle$	$\frac{1}{\sqrt{8}}$	0	0	0	0	0	0	0
	$|0\rangle$	$|1\rangle$	$|2\rangle$	$|3\rangle$	$|4\rangle$	$|5\rangle$	$|6\rangle$	$|7\rangle$

6.3 Application of QFT: Period-Finding

There is a cool application of QFT, which is essential in Shor's factorization algorithm. The following example is taken from [2]. Let $|REG_1\rangle \in \mathcal{H}_1$ be the input register and $|REG_2\rangle \in \mathcal{H}_2$ be the output register. Each register is a 3-qubit system, to make our argument concrete, and the total system Hilbert space is $\mathcal{H}_1 \otimes \mathcal{H}_2$. Let the initial state of $|REG_1\rangle$ be

$$\frac{1}{\sqrt{2^3}}(|000\rangle + |001\rangle + \ldots + |111\rangle) = \frac{1}{\sqrt{8}}(|0\rangle + |1\rangle + \ldots + |7\rangle). \qquad (6.17)$$

Let $S_3 = \{0, 1, \ldots, 2^3 - 1 = 7\}$ and let $f : S_3 \to S_3$ be a function. Apply f on the initial state to obtain

$$|\Psi\rangle = U_f \frac{1}{\sqrt{8}} \sum_x |x\rangle|0\rangle = \frac{1}{\sqrt{8}} \sum_x |x, f(x)\rangle = \frac{1}{\sqrt{8}}(|0, f(0)\rangle + \ldots + |7, f(7)\rangle).$$

$$(6.18)$$

It is interesting to visualize the coefficient of each vector as in Table. 6.1.

TABLE 6.2

Coefficient of a vector $|y\rangle|f(x)\rangle$ in $|\psi'\rangle$. The amplitude is $1/8$ for all the coefficients. The arrow denotes the phase associated with each coefficient. ↗ denotes $e^{i\pi/4}$, for example.

$\lvert f(7)\rangle$	→	↘	↓	↙	←	↖	↑	↗
$\lvert f(6)\rangle$	→	↓	←	↑	→	↓	←	↑
$\lvert f(5)\rangle$	→	↙	↑	↘	←	↗	↓	↖
$\lvert f(4)\rangle$	→	←	→	←	→	←	→	←
$\lvert f(3)\rangle$	→	↖	↓	↗	←	↘	↑	↙
$\lvert f(2)\rangle$	→	↑	←	↓	→	↑	←	↓
$\lvert f(1)\rangle$	→	↗	↑	↖	←	↙	↓	↘
$\lvert f(0)\rangle$	→	→	→	→	→	→	→	→
	$\lvert 0\rangle$	$\lvert 1\rangle$	$\lvert 2\rangle$	$\lvert 3\rangle$	$\lvert 4\rangle$	$\lvert 5\rangle$	$\lvert 6\rangle$	$\lvert 7\rangle$

Let us apply the following QFT,

$$|x\rangle \to \frac{1}{\sqrt{8}}\sum_{y=0}^{7} e^{-2\pi i x y/8}|y\rangle,\tag{6.19}$$

on the first register. Then we obtain

$$|\Psi'\rangle = \frac{1}{8}\sum_{x,y} e^{-2\pi i x y/8}|y, f(x)\rangle$$

$$= \frac{1}{8}|0\rangle \otimes [|f(0)\rangle + |f(1)\rangle + \ldots + |f(7)\rangle] \quad (y=0)$$

$$+ \frac{1}{8}|1\rangle \otimes \left[|f(0)\rangle + e^{-2\pi i/8}|f(1)\rangle + \ldots + e^{-2\pi i 7/8}|f(7)\rangle\right] \quad (y=1)$$

$$\ldots$$

$$+ \frac{1}{8}|7\rangle \otimes \left[|f(0)\rangle + e^{-14\pi i/8}|f(1)\rangle + \ldots + e^{-14\pi i 7/8}|f(7)\rangle\right]. \quad (y=7)$$

$$\tag{6.20}$$

The coefficient of each component $|y\rangle|f(x)\rangle$ is shown in Table 6.2.

Suppose $f(x)$ is a periodic function satisfying $f(x + P) = f(x)$, $P \in \mathbb{N}$. This period is found from the measurement outcome of $|\mathrm{REG}_1\rangle$. Let $P = 2$, for example. Then it follows that

$$f(0) = f(2) = f(4) = f(6) = a, \quad f(1) = f(3) = f(5) = f(7) = b,$$

where $a, b \in S_3$ and $a \neq b$. The state $|\Psi'\rangle$ now reduces to

$$|\Psi'\rangle = \frac{1}{8}\sum_{x,y\in S_3} e^{-2\pi i x y/8}|y, f(x)\rangle$$

TABLE 6.3
Coefficient of a vector $|y\rangle|f(x)\rangle$ in the state
$|\Psi'\rangle$ in which $f(0) = f(2) = f(4) = f(6) = a$
and $f(1) = f(3) = f(5) = f(7) = b$. The
amplitude of all the non-vanishing
coefficients is $1/2$.

$|b\rangle$	$\rightarrow$	0	0	0	$\leftarrow$	0	0	0
$|a\rangle$	$\rightarrow$	0	0	0	$\rightarrow$	0	0	0
		$|0\rangle$	$|1\rangle$	$|2\rangle$	$|3\rangle$	$|4\rangle$	$|5\rangle$	$|6\rangle$	$|7\rangle$

$$= \frac{1}{2}|0\rangle \otimes [|a\rangle + |b\rangle]$$
$$+ \frac{1}{8}|1\rangle \otimes \left[|a\rangle \left(1 + e^{-1\cdot2\cdot2\pi i/8} + e^{-1\cdot4\cdot2\pi i/8} + e^{-1\cdot6\cdot2\pi i/8} \right) \right.$$
$$\left. + |b\rangle \left(e^{-1\cdot1\cdot2\pi i/8} + e^{-1\cdot3\cdot2\pi i/8} + e^{-1\cdot5\cdot2\pi i/8} + e^{-1\cdot7\cdot2\pi i/8} \right) \right]$$
$$\cdots \tag{6.21}$$

As a result, all the vectors but

$$|0, a\rangle, |0, b\rangle, |4, a\rangle, |4, b\rangle$$

cancel out to vanish, and we are left with

$$|\Psi'\rangle = \frac{1}{2} \left(|0, a\rangle + |0, b\rangle + |4, a\rangle + e^{-i\pi}|4, b\rangle \right) \tag{6.22}$$

(see Table 6.3).

If we measure the first register $|\text{REG}_1\rangle$, the result is either 0 or 4, which is the direct consequence of the periodicity $P = 2$ of $f(x)$.

EXERCISE 6.3 Suppose each register above is an n-qubit system. Let $f(x)$ be a periodic function with the period P. Show that the observed value of the first register is one of

$$0, \frac{1 \cdot 2^n}{P}, \frac{2 \cdot 2^n}{P}, \frac{3 \cdot 2^n}{P}, \cdots \frac{(P-1)2^n}{P}, \tag{6.23}$$

where it is assumed that $2^n/P \in \mathbb{N}$.

The cancellation observed above is extensively made use of in Shor's factorization algorithm.

6.4 Implementation of QFT

We now consider a quantum circuit $U_{\text{QFT}n}$ which implements the n-qubit QFT. The circuit $U_{\text{QFT}n}$ maps a state $\sum_x f(x)|x\rangle$ to a state $\sum_y \tilde{f}(y)|y\rangle$, where

$$\tilde{f}(y) = \frac{1}{\sqrt{N}} \sum_x \omega_n^{-xy} f(x), \quad \omega_n = e^{2\pi i/N}.$$

Thus for $f(x') = \delta_{x'x}$, we obtain $\tilde{f}(y) = \omega_n^{-xy}/\sqrt{N}$, namely

$$U_{\text{QFT}n}|x\rangle = \frac{1}{\sqrt{N}} \sum_{y=0}^{N-1} e^{-2\pi i x y/N} |y\rangle.$$

Let us start our implmentation of QFT with $n = 1, 2$ and 3 to familiarize ourselves with the problem.

<u>$n = 1$</u>
Eq. (6.13) shows that the kernel for $n = 1$ QFT is the Hadamard gate H, whose action on $|x\rangle$, $x \in \{0, 1\}$, is concisely written as

$$U_{\text{H}}|x\rangle = \frac{1}{\sqrt{2}}(|0\rangle + (-1)^x|1\rangle) = \frac{1}{\sqrt{2}} \sum_{y=0}^{1} (-1)^{xy}|y\rangle. \tag{6.24}$$

In fact, this is the defining equation for $n = 1$ QFT as

$$U_{\text{QFT}1}|x\rangle = \frac{1}{\sqrt{2}} \sum_{y=0}^{1} \omega_1^{-xy}|y\rangle = \frac{1}{\sqrt{2}} \sum_{y=0}^{1} (-1)^{xy}|y\rangle. \tag{6.25}$$

It is instructive to demonstrate Eq. (6.7) explicitly here. Let $|\psi\rangle = f(0)|0\rangle + f(1)|1\rangle$ be any one-qubit state. Then

$$U_{\text{QFT}1}|\psi\rangle = f(0)\frac{1}{\sqrt{2}}(|0\rangle + |1\rangle) + f(1)\frac{1}{\sqrt{2}}(|0\rangle - |1\rangle)$$

$$= \frac{1}{\sqrt{2}}(f(0) + f(1))|0\rangle + \frac{1}{\sqrt{2}}(f(0) - f(1))|1\rangle = \sum_{y=0}^{1} \tilde{f}(y)|y\rangle.$$

<u>$n = 2$</u>
This case is considerably more complicated than the case $n = 1$. It also gives important insights into implementing QFT with $n \geq 3$. Let us introduce an important gate, the **controlled-B_{jk}** gate. The B_{jk} gate is defined by the matrix

$$B_{jk} = \begin{pmatrix} 1 & 0 \\ 0 & e^{-i\theta_{jk}} \end{pmatrix}, \quad \theta_{jk} = \frac{2\pi}{2^{k-j+1}}, \tag{6.26}$$

where $j, k \in \{0, 1, 2, \ldots\}$ and $k \geq j$.

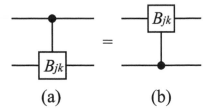

FIGURE 6.1
(a) Controlled-B_{jk} gate. The inverted controlled-B_{jk} gate (b) and the controlled-B_{jk} gate are equivalent (see Lemma 6.1).

LEMMA 6.1 The controlled-B_{jk} gate U_{jk} in Fig. 6.1 (a) acts on $|x\rangle|y\rangle$, $x, y \in \{0, 1\}$, as

$$U_{jk}|x, y\rangle = e^{-i\theta_{jk}xy}|x, y\rangle = \exp\left(-\frac{2\pi i}{2^{k-j+1}}xy\right)|x, y\rangle. \qquad (6.27)$$

Proof. The controlled-B_{jk} gate is written as

$$U_{jk} = |0\rangle\langle 0| \otimes I + |1\rangle\langle 1| \otimes B_{jk}, \qquad (6.28)$$

and its action on $|x, y\rangle$ is

$$U_{jk}|x, y\rangle = |0\rangle\langle 0|x\rangle \otimes |y\rangle + |1\rangle\langle 1|x\rangle \otimes B_{jk}|y\rangle$$
$$= \begin{cases} |x\rangle \otimes |y\rangle & x = 0 \\ |x\rangle \otimes B_{jk}|y\rangle & x = 1. \end{cases} \qquad (6.29)$$

Moreover, when $x = 1$ we have

$$B_{jk}|y\rangle = \begin{cases} |y\rangle & y = 0 \\ e^{-i\theta_{jk}}|y\rangle & y = 1. \end{cases} \qquad (6.30)$$

Thus the action of U_{jk} on $|y\rangle$ is trivial if $xy = 0$ and nontrivial if and only if $x = y = 1$. These results may be summarized as Eq. (6.27). ∎

The action of the controlled-B_{jk} gate on a basis vector $|x\rangle|y\rangle$ is detemined by the combination xy and not by x and y independently. Therefore the controlled-B_{jk} gate and the "inverted" controlled-B_{jk} gate are equivalent; see Fig. 6.1.

The DFT for $n = 2$ is defined as

$$\tilde{f}(y) = \frac{1}{2}\sum_{x=0}^{3} \omega_2^{-xy} f(x), \quad \omega_2 = e^{2\pi i/4} = i, \ y \in S_2. \qquad (6.31)$$

Equation (6.6) in Proposition 6.2 states that our task is to find a unitary matrix U_{QFT2} such that

$$U_{\text{QFT2}}|x\rangle = \frac{1}{2}\sum_{y=0}^{3} \omega_2^{-xy}|y\rangle. \tag{6.32}$$

Let us write x and y in the binary form as $x = 2x_1 + x_0$ and $y = 2y_1 + y_0$, respectively. The action of U_{QFT2} on $|x\rangle$ is

$$U_{\text{QFT2}}|x_1x_0\rangle = \frac{1}{2}\sum_{y=0}^{3} e^{-2\pi ixy/2^2}|y\rangle = \frac{1}{2}\sum_{y_0,y_1=0}^{1} e^{-2\pi ix(2y_1+y_0)/2^2}|y_1y_0\rangle$$

$$= \frac{1}{2}\sum_{y_1} e^{-2\pi ixy_1/2}|y_1\rangle \otimes \sum_{y_0} e^{-2\pi ixy_0/2^2}|y_0\rangle$$

$$= \frac{1}{2}\left(|0\rangle + e^{-2\pi ix/2}|1\rangle\right) \otimes \left(|0\rangle + e^{-2\pi ix/2^2}|1\rangle\right)$$

$$= \frac{1}{2}\left(|0\rangle + e^{-2\pi i(2x_1+x_0)/2}|1\rangle\right) \otimes \left(|0\rangle + e^{-2\pi i(2x_1+x_0)/2^2}|1\rangle\right)$$

$$= \frac{1}{2}\left(|0\rangle + e^{-\pi ix_0}|1\rangle\right) \otimes \left(|0\rangle + e^{-\pi ix_1}e^{-i(\pi/2)x_0}|1\rangle\right)$$

$$= \frac{1}{2}\left(|0\rangle + (-1)^{x_0}|1\rangle\right) \otimes B_{12}^{x_0}\left(|0\rangle + (-1)^{x_1}|1\rangle\right), \tag{6.33}$$

where use has been made of the fact $\theta_{12} = 2\pi/2^{2-1+1} = \pi/2$ to obtain the last expression. Note that $B_{12}^{x_0}$ is the controlled-B_{12} gate with the control bit x_0 and the target bit x_1; $B_{12}^0 = I$ while $B_{12}^1 = B_{12}$. Note also that, in spite of its tensor product looking appearance, the last line of Eq. (6.33) is entangled due to this conditional operation. Equation (6.33) suggests that the $n = 2$ QFT are implemented with the Hadamard and the U_{12} gates. Before writing down the quantum circuit realizing Eq. (6.33), we should note that the first qubit has a power $(-1)^{x_0}$, while the second one has $(-1)^{x_1}$, when the input state is $|x_1x_0\rangle$. If we naively applied the Hadamard gate to the second qubit, we would obtain

$$(I \otimes U_{\text{H}})|x_1x_0\rangle = |x_1\rangle \otimes \frac{1}{\sqrt{2}}(|0\rangle + (-1)^{x_0}|1\rangle).$$

These facts suggest that we need to swap the first and second qubits at the beginning of the implementation so that

$$U_{\text{QFT2}}|x_1x_0\rangle = \frac{1}{\sqrt{2^2}}\left(|0\rangle + (-1)^{x_0}|1\rangle\right) \otimes B_{12}^{x_0}\left(|0\rangle + (-1)^{x_1}|1\rangle\right)$$

$$= (U_{\text{H}} \otimes I)U_{12}(I \otimes U_{\text{H}})|x_0,x_1\rangle$$

$$= (U_{\text{H}} \otimes I)U_{12}(I \otimes U_{\text{H}})U_{\text{SWAP}}|x_1x_0\rangle. \tag{6.34}$$

Since Eq. (6.34) is true for any $|x_1x_0\rangle$, we should have $U_{\text{QFT2}} = (U_{\text{H}} \otimes I)U_{12}(I \otimes U_{\text{H}})U_{\text{SWAP}}$, which proves the following proposition.

PROPOSITION 6.3 The $n = 2$ QFT gate is implemented as

$$U_{\mathrm{QFT2}} = (U_{\mathrm{H}} \otimes I)U_{12}(I \otimes U_{\mathrm{H}})U_{\mathrm{SWAP}} \qquad (6.35)$$

(see Fig. 6.2).

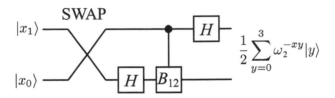

FIGURE 6.2
Implementation of the $n = 2$ QFT, U_{QFT2}.

The reader should verify the above implementation by explicitly writing down the gates as matrices.

EXERCISE 6.4 It is also possible to have the SWAP gate in the very end of the implementation. Design such an $n = 2$ QFT gate.

This construction is easily generalized to $n \geq 3$ as we see next.

$n = 3$ and beyond

It is instructive to rewrite the construction of $n = 2$ QFT in a more generalizable form. The state $|x_1 x_0\rangle$ has been transformed as in Eq. (6.33):

$$|x_1 x_0\rangle \rightarrow \frac{1}{\sqrt{2^2}} \sum_{y=0}^{2^2-1} e^{-2\pi i x y/2^2} |y\rangle$$

$$= \frac{1}{\sqrt{2^2}} (|0\rangle + e^{-2\pi i x_0/2}|1\rangle) \otimes (|0\rangle + e^{-2\pi i(x_1/2 + x_0/2^2)}|1\rangle).$$

This observation suggests the following construction of $n = 3$ QFT:

$$U_{\mathrm{QFT3}}|x_2 x_1 x_0\rangle$$

$$= \frac{1}{\sqrt{2^3}}(|0\rangle + e^{-2\pi i x_0/2}|1\rangle) \otimes (|0\rangle + e^{-2\pi i(x_1/2 + x_0/2^2)}|1\rangle)$$

$$\otimes(|0\rangle + e^{-2\pi i(x_2/2 + x_1/2^2 + x_0/2^3)}|1\rangle)$$

$$= \frac{1}{\sqrt{2^3}}(|0\rangle + (-1)^{x_0}|1\rangle) \otimes B_{01}^{x_0}(|0\rangle + (-1)^{x_1}|1\rangle)$$

$$\otimes B_{02}^{x_0} B_{12}^{x_1}(|0\rangle + (-1)^{x_2}|1\rangle)$$

$$= (U_H \otimes I \otimes I)U_{01}(I \otimes U_H \otimes I)U_{02}U_{12}(I \otimes I \otimes U_H)|x_0x_1x_2\rangle$$
$$= (U_H \otimes I \otimes I)U_{01}(I \otimes U_H \otimes I)U_{02}U_{12}(I \otimes I \otimes U_H)P|x_2x_1x_0\rangle, \quad (6.36)$$

where U_{jk} is the controlled-B_{jk} gate with the control qubit x_j, and the gate P reverses the order of the qubits as $P|x_2x_1x_0\rangle = |x_0x_1x_2\rangle$. For a three-qubit QFT, P is a SWAP gate between the first qubit (x_2) and the third qubit (x_0). Again note here that we should be careful in ordering the gates so that the control bit x_j acts in U_{jk} before it is acted by a Hadamard gate.

EXERCISE 6.5 Let $x = 2^2x_2 + 2x_1 + x_0$ and $y = 2^2y_2 + 2y_1 + y_0$.
(1) Write down the RHS of

$$U_{\text{QFT3}}|x_2x_1x_0\rangle = \frac{1}{\sqrt{2^3}}\sum_{y=0}^{2^3-1} e^{-2\pi ixy/2^3}|y\rangle \quad (6.37)$$

explicitly in terms of x_i and y_i.
(2) Show that the RHS of Eq. (6.37) agrees with the first line of the RHS of Eq. (6.36).

Since Eq. (6.36) is true for any $|x_2x_1x_0\rangle$, we have found

$$U_{\text{QFT3}} = (U_H \otimes I \otimes I)U_{01}(I \otimes U_H \otimes I)U_{02}U_{12}(I \otimes I \otimes U_H)P. \quad (6.38)$$

Equation (6.38) readily leads us to the quantum circuit in Fig. 6.3.

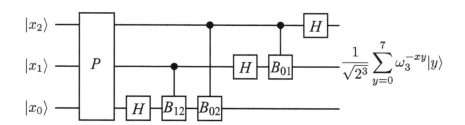

FIGURE 6.3
Implementation of the $n = 3$ QFT.

EXERCISE 6.6 Design a quantum circuit U_{QFT3} in which the permutation gate P is at the very end of the circuit.

Now the generalization of the present construction to $n \geq 4$ should be easy. The equation that generalizes Eq. (6.36) is

$$U_{\text{QFT}n}|x_{n-1}\ldots x_1x_0\rangle$$

$$= \frac{1}{\sqrt{N}} (|0\rangle + e^{-2\pi i x_0/2}|1\rangle) \otimes (|0\rangle + e^{-2\pi i(x_1/2 + x_0/2^2)}|1\rangle)$$

$$\otimes (|0\rangle + e^{-2\pi i(x_2/2 + x_1/2^2 + x_0/2^3)}|1\rangle) \otimes \ldots$$

$$\ldots \otimes (|0\rangle + e^{-2\pi i(x_{n-1}/2 + x_{n-2}/2^2 + \ldots x_1/2^{n-1} + x_0/2^n)}|1\rangle)$$

$$= (U_{\mathrm{H}} \otimes I \otimes \ldots \otimes I) U_{01}(I \otimes U_{\mathrm{H}} \otimes I \otimes \ldots \otimes I) U_{02} U_{12}$$

$$\times (I \otimes I \otimes U_{\mathrm{H}} \otimes \ldots \otimes I) \ldots$$

$$\times U_{0,n-1} U_{1,n-1} \ldots U_{n-2,n-1}(I \otimes \ldots \otimes I \otimes U_{\mathrm{H}})|x_0 x_1 \ldots x_{n-1}\rangle$$

$$= (U_{\mathrm{H}} \otimes I \otimes \ldots \otimes I) U_{01}(I \otimes U_{\mathrm{H}} \otimes I \otimes \ldots \otimes I) U_{02} U_{12}$$

$$\times (I \otimes I \otimes U_{\mathrm{H}} \otimes \ldots \otimes I) \ldots U_{0,n-1} U_{1,n-1} \ldots U_{n-2,n-1}$$

$$\times (I \otimes \ldots \otimes I \otimes U_{\mathrm{H}}) P |x_{n-1} \ldots x_1 x_0\rangle, \tag{6.39}$$

where P reverses the order of x_k as $P|x_{n-1} \ldots x_1 x_0\rangle = |x_0 x_1 \ldots x_{n-1}\rangle$.
We finally find the following decompostion of $U_{\mathrm{QFT}n}$:

$$U_{\mathrm{QFT}n} = (U_{\mathrm{H}} \otimes I \otimes \ldots \otimes I) U_{01}(I \otimes U_{\mathrm{H}} \otimes I \otimes \ldots \otimes I) U_{02} U_{12}$$

$$\times (I \otimes I \otimes U_{\mathrm{H}} \otimes \ldots \otimes I) \ldots$$

$$\times U_{0,n-1} U_{1,n-1} \ldots U_{n-2,n-1}(I \otimes \ldots \otimes I \otimes U_{\mathrm{H}}) P. \tag{6.40}$$

A quantum circuit which implements $U_{\mathrm{QFT}n}$ is found from Eq. (6.40) as in
Fig. 6.4. It may be proved, by induction, for example, that the circuit in

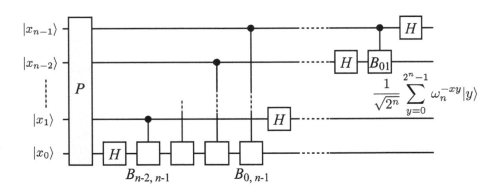

FIGURE 6.4
Implementation of the n-qubit QFT.

Fig. 6.4 really implements the n-qubit QFT.

PROPOSITION 6.4 The n-qubit QFT may be constructed with $\Theta(n^2)$
elementary gates.

Proof. The n-qubit QFT is made of a P gate, n Hadamard gates and $(n -$

$1) + (n - 2) + \ldots + 2 + 1 = n(n - 1)/2$ controlled-B_{jk} gates (see Fig. 6.4). It has been shown in §4.2.3 that it requres three CNOT gates to construct a SWAP gate. Furthermore, a P gate for n qubits requires $\lfloor n/2 \rfloor$ SWAP gates,[†] assuming that there exists a SWAP gate for any pair of qubits. Thus a P gate requires $3 \times \lfloor n/2 \rfloor = \Theta(n)$ elementary gates. Proposition 4.1 states that a controlled-B_{ij} gate is constructed with at most six elementary gates. Thus it has been proved that the n-qubit QFT is made of $\Theta(n^2)$ elementary gates.   ∎

The above proposition is quite important in estimating the efficiency of quantum algorithms. If we look at the definition

$$\tilde{f}(y) = \frac{1}{\sqrt{N}} \sum_{x=0}^{2^n - 1} \omega^{-xy} f(x),$$

we naively expect that $N = 2^n$ steps (including the evaluation of exponential functions followed by multiplication) are required for *each* y and $N \times N$ steps for all y's. In other words, it takes exponentially large steps ($\sim N^2 = e^{2n \ln 2}$) to carry out the QFT. The above proposition states that this is done in $\Theta(n^2)$ steps with the QFT gate if the initial state is a superposition of all x's.

6.5   Walsh-Hadamard Transform

There are two other quantum integral transforms, the Walsh-Hadamard transform and the selective phase rotation transform, which are often employed in quantum computing.

We have already encountered the Walsh-Hadamard transform in §4.2.2 and §5.2. Let $x, y \in S_n = \{0, 1, \ldots, N-1\}$ with binary expressions $x_{n-1} x_{n-2} \ldots x_0$ and $y_{n-1} y_{n-2} \ldots y_0$, where $N = 2^n$. The Walsh-Hadamard transform, written in the form of Eq. (5.7), shows that it is a quantum integral transform with a kernel $W_n : S_n \times S_n \to \mathbb{C}$ defined by

$$W_n(x, y) = \frac{1}{\sqrt{N}} (-1)^{x \cdot y} \quad (x, y \in S_n), \tag{6.41}$$

where $x \cdot y = x_{n-1} y_{n-1} \oplus x_{n-2} y_{n-2} \oplus \ldots \oplus x_0 y_0$. This kernel defines a discrete integral transform

$$\tilde{f}(y) = \frac{1}{\sqrt{N}} \sum_{x=0}^{N-1} (-1)^{x \cdot y} f(x). \tag{6.42}$$

[†] $\lfloor x \rfloor$ is the largest integer which is less than or equal to $x \in \mathbb{R}$ and called the floor of x.

6.6 Selective Phase Rotation Transform

DEFINITION 6.2 (Selective Phase Rotation Transform) Let us define a kernel

$$K_n(x,y) = e^{i\theta_x}\delta_{xy}, \quad \forall x, y \in S_n, \tag{6.43}$$

where $\theta_x \in \mathbb{R}$. The discrete integral transform

$$\tilde{f}(y) = \sum_{x=0}^{N-1} K(x,y)f(x) = \sum_{x=0}^{N-1} e^{i\theta_x}\delta_{xy}f(x) = e^{i\theta_y}f(y) \tag{6.44}$$

with the kernel K_n is called the **selective phase rotation transform**.

EXERCISE 6.7 Show that K_n defined above is unitary. Write down the inverse transformation K_n^{-1}.

The matrix representations for K_1 and K_2 are

$$K_1 = \begin{pmatrix} e^{i\theta_0} & 0 \\ 0 & e^{i\theta_1} \end{pmatrix}, \quad K_2 = \begin{pmatrix} e^{i\theta_0} & 0 & 0 & 0 \\ 0 & e^{i\theta_1} & 0 & 0 \\ 0 & 0 & e^{i\theta_2} & 0 \\ 0 & 0 & 0 & e^{i\theta_3} \end{pmatrix}.$$

The implementation of K_n is achieved with the universal set of gates as follows. Take $n = 2$, for example. The kernel K_2 has been given above. This is decomposed as a product of two two-level unitary matrices as

$$K_2 = A_0 A_1, \tag{6.45}$$

where

$$A_0 = \begin{pmatrix} e^{i\theta_0} & 0 & 0 & 0 \\ 0 & e^{i\theta_1} & 0 & 0 \\ 0 & 0 & 1 & 0 \\ 0 & 0 & 0 & 1 \end{pmatrix}, A_1 = \begin{pmatrix} 1 & 0 & 0 & 0 \\ 0 & 1 & 0 & 0 \\ 0 & 0 & e^{i\theta_2} & 0 \\ 0 & 0 & 0 & e^{i\theta_3} \end{pmatrix}. \tag{6.46}$$

Note that

$$A_0 = |0\rangle\langle 0| \otimes U_0 + |1\rangle\langle 1| \otimes I, \quad U_0 = \begin{pmatrix} e^{i\theta_0} & 0 \\ 0 & e^{i\theta_1} \end{pmatrix},$$

$$A_1 = |0\rangle\langle 0| \otimes I + |1\rangle\langle 1| \otimes U_1, \quad U_1 = \begin{pmatrix} e^{i\theta_2} & 0 \\ 0 & e^{i\theta_3} \end{pmatrix}.$$

Thus A_1 is realized as an ordinary controlled-U_1 gate while the control bit is negated in A_0. Then what we have to do for A_0 is to negate the control bit first and then to apply ordinary controlled-U_0 gate and finally to negate the control bit back to its input state. In summary, A_0 is implemented as in

Fig. 6.5. In fact, it can be readily verified that the gate in Fig. 6.5 is written as

$$(X \otimes I)(|0\rangle\langle 0| \otimes I + |1\rangle\langle 1| \otimes U_0)(X \otimes I)$$
$$= X|0\rangle\langle 0|X \otimes I + X|1\rangle\langle 1|X \otimes U_0 = |1\rangle\langle 1| \otimes I + |0\rangle\langle 0| \otimes U_0 = A_0.$$

Thus these gates are implemented with the set of universal gates. In fact, the order of A_i does not matter since $[A_0, A_1] = 0$.

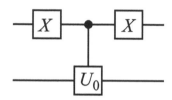

FIGURE 6.5
Implementation of A_0.

EXERCISE 6.8 Repeat the above arguments for $n = 3$. In this case K_3 is written as a product of four two-level unitary matrices. Write down these matrices and find the quantum circuits similar to that in Fig. 6.5 that implements these two-level unitary matrices.

References

[1] Y. Uesaka, *Mathematical Principle of Quantum Computation*, Corona Publishing, Tokyo, in Japanese (2000).

[2] G. P. Berman, G. D. Doolen, R. Mainieri and V. I. Tsifrinovich, *Introduction to Quantum Computers*, World Scientific, Singapore (1998).

7

Grover's Search Algorithm

Suppose there is a stack of N unstructured files and we want to find a particular file (or files) out of the N files. To find someone's phone number in a telephone directory is an easy structured datebase search problem, while to find a person's name who has a particular phone number is a more difficult unstructured database seach problem, with which we are concerned in this chapter. It is required to take $O(N)$ steps on average if a classical algorithm is employed. If we check the files one by one, we will hit the right file with probability $1/2$ after $N/2$ files are examined. It turns out that this takes only $O(\sqrt{N})$ steps with a quantum algorithm, first discovered by Grover [1, 2, 3]. Our presentation in this chapter closely follows [4] and [5].

7.1 Searching for a Single File

Suppose there is a stack of $N = 2^n$ files, randomly placed, that are numbered by $x \in S_n \equiv \{0, 1, \ldots, N-1\}$. Our task is to find an algorithm which picks out a particular file which satisfies a certain condition.

In mathematical language, this is expressed as follows. Let $f : S_n \to \{0, 1\}$ be a function defined by

$$f(x) = \begin{cases} 1 & (x = z) \\ 0 & (x \neq z), \end{cases} \tag{7.1}$$

where z is the address of the file we are looking for. It is assumed that $f(x)$ is *instantaneously* calculable, such that this process does not require any computational steps. A function of this sort is often called an oracle as noted in Chapter 5. Thus, the problem is to find z such that $f(z) = 1$, given a function $f : S_n \to \{0, 1\}$ which assumes the value 1 only at a single point.

Clearly we have to check one file after another in a classical algorithm, and it will take $O(N)$ steps on average. It is shown below that it takes only $O(\sqrt{N})$ steps with Grover's algorithm. This is accomplished by *amplifying* the amplitude of the vector $|z\rangle$ while cancelling that of the vectors $|x\rangle$ $(x \neq z)$.

We describe the algorithm in several steps.

STEP 1 (Selective phase rotation transform; see §6.6.)

Define the kernel of the selective phase rotation transform R_f by

$$K_f(x, y) = e^{i\pi f(x)}\delta_{xy} = (-1)^{f(x)}\delta_{xy}, \tag{7.2}$$

where $x, y \in S_n$. Since R_f maps $|z\rangle \mapsto -|z\rangle$, while leaving all the other vectors unchanged, it can be expressed as

$$R_f = I - 2|z\rangle\langle z|. \tag{7.3}$$

Let us consider a state

$$|\varphi\rangle = \sum_{x=0}^{N-1} w_x|x\rangle, \quad \sum_x |w_x|^2 = 1. \tag{7.4}$$

Then it is easy to verify

$$R_f|\varphi\rangle = w_0|0\rangle + \ldots + (-1)w_z|z\rangle + \ldots + w_{N-1}|N-1\rangle. \tag{7.5}$$

(In other words, R_f changes the sign of w_z while leaving all other coefficients unchanged.)

STEP 2 Define a unitary matrix

$$D = W_n R_0 W_n, \tag{7.6}$$

where W_n is the Walsh-Hadamard transform,

$$W_n(x, y) = \frac{1}{\sqrt{N}}(-1)^{x \cdot y}, \quad (x, y \in S_n) \tag{7.7}$$

and R_0 is the selective phase rotation transform defined by

$$R_0(x, y) = e^{i\pi(1-\delta_{x0})}\delta_{xy} = (-1)^{1-\delta_{x0}}\delta_{xy}. \tag{7.8}$$

PROPOSITION 7.1 Let

$$|\varphi_0\rangle = \frac{1}{\sqrt{N}}\sum_{x=0}^{N-1}|x\rangle. \tag{7.9}$$

Then

$$D = -I + 2|\varphi_0\rangle\langle\varphi_0|. \tag{7.10}$$

Moreover

$$D|\varphi\rangle = \sum_{x=0}^{N-1}(\bar{w} - (w_x - \bar{w}))|x\rangle, \tag{7.11}$$

where $|\varphi\rangle$ is given in Eq. (7.4) and

$$\bar{w} = \frac{1}{N}\sum_{x=0}^{N-1}w_x \tag{7.12}$$

is the avarage of w_x over S_n.

Proof. Let us evaluate the matrix elements of the RHS of Eq. (7.10). We obtain from

$$-I + 2|\varphi_0\rangle\langle\varphi_0| = -I + \frac{2}{N}\sum_x |x\rangle\sum_y\langle y| = -I + \frac{2}{N}\sum_{x,y}|x\rangle\langle y|$$

that the (x, y)-component of the RHS is

$$\langle x|\text{RHS}|y\rangle = -\delta_{xy} + \frac{2}{N}. \tag{7.13}$$

Let us turn to the LHS next. The (x, y)-component of $D = W_n R_0 W_n$ is

$$\langle x|W_n R_0 W_n|y\rangle = \sum_{u,v}\langle x|W_n|u\rangle\langle u|R_0|v\rangle\langle v|W_n|y\rangle$$

$$= \frac{1}{N}\sum_{u,v}(-1)^{x\cdot u}$$

$$\times (-1)^{1-\delta_{u0}}\delta_{uv}(-1)^{v\cdot y}.$$

The summation over u is evaluated as

$$\sum_{u=0}^{N-1}(-1)^{x\cdot u}(-1)^{1-\delta_{u0}}\delta_{uv}$$

$$= (-1)^0(-1)^0\delta_{0v} - \sum_{u=1}^{N-1}(-1)^{x\cdot u}\delta_{uv}$$

$$= 2\delta_{0v} - \sum_{u=0}^{N-1}(-1)^{x_{n-1}u_{n-1}+...+x_1 u_1 + x_0 u_0}\delta_{u_{n-1}v_{n-1}}\cdots\delta_{u_1 v_1}\delta_{u_0 v_0}$$

$$= 2\delta_{0v} - \left[\sum_{u_{n-1}=0}^{1}(-1)^{x_{n-1}u_{n-1}}\delta_{u_{n-1}v_{n-1}}\right]\cdots\left[\sum_{u_1=0}^{1}(-1)^{x_1 u_1}\delta_{u_1 v_1}\right]$$

$$\times\left[\sum_{u_0=0}^{1}(-1)^{x_0 u_0}\delta_{u_0 v_0}\right].$$

Then the LHS becomes

$$\langle x|D|y\rangle = \frac{1}{N}\sum_{v=0}^{N-1}\left[2\delta_{0v} - \left(\sum_{u_{n-1}=0}^{1}(-1)^{x_{n-1}u_{n-1}}\delta_{u_{n-1}v_{n-1}}\right)\right.$$

$$\left.\cdots\left(\sum_{u_0=0}^{1}(-1)^{x_0 u_0}\delta_{u_0 v_0}\right)\right](-1)^{v\cdot y}$$

$$= \frac{2}{N} - \frac{1}{N} \left[\sum_{u_{n-1},v_{n-1}=0}^{1} (-1)^{x_{n-1}u_{n-1}+v_{n-1}y_{n-1}} \delta_{u_{n-1}v_{n-1}} \right]$$

$$\cdots \left[\sum_{u_0,v_0=0}^{1} (-1)^{x_0 u_0 + v_0 y_0} \delta_{u_0 v_0} \right]$$

$$= \frac{2}{N} - \frac{1}{N} \left[1 + (-1)^{x_{n-1}+y_{n-1}} \right] \cdots \left[1 + (-1)^{x_0+y_0} \right]$$

$$= \frac{2}{N} - \frac{2^n}{N} \delta_{x_{n-1}y_{n-1}} \cdots \delta_{x_0 y_0}$$

$$= \frac{2}{N} - \delta_{xy},$$

which proves Eq. (7.10).

Equation (7.11) is proved as

$$D|\varphi\rangle = (-I + 2|\varphi_0\rangle\langle\varphi_0|)|\varphi\rangle = \left(-I + \frac{2}{N} \sum_{y,z} |y\rangle\langle z| \right) \sum_x w_x |x\rangle$$

$$= -\sum_x w_x |x\rangle + \frac{2}{N} \sum_{x,y,z} w_x |y\rangle \delta_{xz} = -\sum_x w_x |x\rangle + \frac{2}{N} \sum_{y,z} w_z |y\rangle$$

$$= -\sum_x w_x |x\rangle + 2\sum_y \bar{w}|y\rangle = \sum_{x=0}^{N-1} [\bar{w} - (w_x - \bar{w})] |x\rangle.$$

∎

Equation (7.11) shows that D is an operator that produces "inversion about the average" since the quantity $\bar{w}-(w_x-\bar{w}) = 2\bar{w}-w_x$ is obtained by reflecting w_x about $\bar{w}$.

STEP 3 Now let us consider the unitary transformation U_f defined by

$$U_f = DR_f = (-I + 2|\varphi_0\rangle\langle\varphi_0|)(I - 2|z\rangle\langle z|) \qquad (7.14)$$

and consider its action on $|\varphi\rangle = \sum_x w_x |x\rangle$. Direct application of the results in step 1 and step 2 yields

$$U_f|\varphi\rangle = D\left(\sum_{x \neq z} w_x |x\rangle - w_z |z\rangle \right) = \sum_{x \neq z} [\bar{w} - (w_x - \bar{w})]|x\rangle + [\bar{w} + (w_z + \bar{w})]|z\rangle$$

$$= \sum_{\substack{x=0 \\ x \neq z}}^{N-1} (2\bar{w} - w_x)|x\rangle + (2\bar{w} + w_z)|z\rangle, \qquad (7.15)$$

where $\sum_{x=0}^{N-1} |w_x|^2 = 1$ and

$$\bar{w} = \frac{1}{N} \left(\sum_{x=0, x \neq z}^{N-1} w_x - w_z \right) \tag{7.16}$$

is the average value of the coefficients of the state $R_f |\varphi\rangle$.

This result shows that the amplitude of $|z\rangle$ has increased upon the operation of U_f while that of $|x\rangle$ $(x \neq z)$ has decreased, assuming that all the weights w_x are positive. Thus repeated applications of U_f increase the amplitude of $|z\rangle$ so that this particular state is observed with probability close to 1 when the system is measured. Let us find the state obtained after U_f is applied k times on the initial state $|\varphi_0\rangle$.

PROPOSITION 7.2 Let us write

$$U_f^k |\varphi_0\rangle = a_k |z\rangle + b_k \sum_{x \neq z} |x\rangle \tag{7.17}$$

with the initial condition

$$a_0 = b_0 = \frac{1}{\sqrt{N}}.$$

Then the coefficients $\{a_k, b_k\}$ for $k \geq 1$ satisfy the recursion relations

$$a_k = \frac{N-2}{N} a_{k-1} + \frac{2(N-1)}{N} b_{k-1}, \tag{7.18}$$

$$b_k = -\frac{2}{N} a_{k-1} + \frac{N-2}{N} b_{k-1} \tag{7.19}$$

for $k = 1, 2, \ldots$.

Proof. It is easy to see the recursion relations are satified for $k = 1$ by making use of Eqs. (7.15) and (7.16). Let $U_f^{k-1} |\varphi_0\rangle = a_{k-1} |z\rangle + b_{k-1} \sum_{x \neq z} |x\rangle$. Then

$$U_f^k |\varphi_0\rangle = U_f \left(a_{k-1} |z\rangle + b_{k-1} \sum_{x \neq z} |x\rangle \right)$$

$$= (-I + 2|\varphi_0\rangle\langle\varphi_0|) \left(-a_{k-1} |z\rangle + b_{k-1} \sum_{x \neq z} |x\rangle \right)$$

$$= -b_{k-1} \sum_{x \neq z} |x\rangle + a_{k-1} |z\rangle + \frac{2}{\sqrt{N}}(N-1) b_{k-1} |\varphi_0\rangle - \frac{2a_{k-1}}{\sqrt{N}} |\varphi_0\rangle$$

$$= -b_{k-1} \sum_{x \neq z} |x\rangle + a_{k-1} |z\rangle + \frac{2}{N}(N-1) b_{k-1} \sum_x |x\rangle - \frac{2a_{k-1}}{N} \sum_x |x\rangle$$

$$= \left[\frac{N-2}{N} a_{k-1} + \frac{2(N-1)}{N} b_{k-1} \right] |z\rangle$$

$$+ \left[-\frac{2}{N} a_{k-1} + \frac{N-2}{N} b_{k-1} \right] \sum_{x \neq z} |x\rangle,$$

and proposition is proved.   ∎

PROPOSITION 7.3 The solutions of the recursion relations in Proposition 7.2 are explicitly given by

$$a_k = \sin[(2k+1)\theta], \quad b_k = \frac{1}{\sqrt{N-1}} \cos[(2k+1)\theta], \tag{7.20}$$

for $k = 0, 1, 2, \ldots$, where

$$\sin \theta = \sqrt{\frac{1}{N}}, \quad \cos \theta = \sqrt{1 - \frac{1}{N}}. \tag{7.21}$$

Proof. Let $c_k = \sqrt{N-1} b_k$. The recursion relations (7.18) and (7.19) are written in a matrix form,

$$\begin{pmatrix} a_k \\ c_k \end{pmatrix} = M \begin{pmatrix} a_{k-1} \\ c_{k-1} \end{pmatrix}, \quad M = \begin{pmatrix} (N-2)/N & 2\sqrt{N-1}/N \\ -2\sqrt{N-1}/N & (N-2)/N \end{pmatrix} = \begin{pmatrix} \cos 2\theta & \sin 2\theta \\ -\sin 2\theta & \cos 2\theta \end{pmatrix}.$$

Note that M is a rotation matrix in $\mathbb{R}^2$, and its kth power is another rotation matrix corresponding to a rotation angle $2k\theta$. Thus the above recursion relation is easily solved to yield

$$\begin{pmatrix} a_k \\ c_k \end{pmatrix} = M^k \begin{pmatrix} a_0 \\ c_0 \end{pmatrix} = \begin{pmatrix} \cos 2k\theta & \sin 2k\theta \\ -\sin 2k\theta & \cos 2k\theta \end{pmatrix} \begin{pmatrix} \sin \theta \\ \cos \theta \end{pmatrix} = \begin{pmatrix} \sin[(2k+1)\theta] \\ \cos[(2k+1)\theta] \end{pmatrix}.$$

Replacing c_k by b_k proves the proposition.   ∎

We have proved that the application of U_f k times on $|\varphi_0\rangle$ results in the state

$$U_f^k |\varphi_0\rangle = \sin[(2k+1)\theta]|z\rangle + \frac{1}{\sqrt{N-1}} \cos[(2k+1)\theta] \sum_{x \neq z} |x\rangle. \tag{7.22}$$

Measurement of the state $U_f^k |\varphi_0\rangle$ yields $|z\rangle$ with the probability

$$P_{z,k} = \sin^2[(2k+1)\theta]. \tag{7.23}$$

It is instructive to visualize what is going on with a simple example. Let us take $n = 4$, for which $N = 2^n = 16$. The probabilities (a_k, b_k) are given by

$$a_0^2 = b_0^2 = 1/16, \quad a_k^2 = \sin^2[(2k+1)\theta], b_k^2 = \frac{\cos^2[(2k+1)\theta]}{16-1},$$

where $\theta = \sin^{-1}(1/4)$. Figure 7.1 shows the probability distributions for $k = 1, 2, 3$ and 4 where we have chosen $z = 10$.

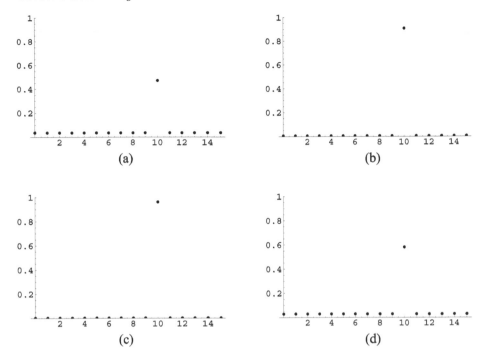

FIGURE 7.1
Probability distribution of $U_f^k|\varphi_0\rangle$, where z is chosen as 10. The number k of iteration is (a) 1, (b) 2, (c) 3 and (d) 4. Observe that $P_{z,k}$ takes its maximum value $\simeq 1$ when $k = 3$.

It should be noted that a_k does not increase monotonically with k, but there is a k ($= 3$ in the present case) that maximizes $P_{z,k} = a_k^2$.

STEP 4 Our final task is to find the k that maximizes $P_{z,k}$. A rough estimate for the maximizing k is obtained by putting

$$(2k + 1)\theta = \frac{\pi}{2} \rightarrow k = \frac{1}{2}\left(\frac{\pi}{2\theta} - 1\right). \tag{7.24}$$

The previous example gave $k = 3$, which is consistent with this estimate:

$$\theta = \sin^{-1}(1/4) \simeq 0.25268 \rightarrow k \simeq 2.6.$$

This can be refined as the following proposition.

PROPOSITION 7.4 Let $N \gg 1$ and let

$$m = \left\lfloor \frac{\pi}{4\theta} \right\rfloor, \tag{7.25}$$

where $\lfloor x \rfloor$ stands for the floor of x. The file we are searching for will be obtained in $U_f^m |\varphi_0\rangle$ with the probability

$$P_{z,m} \geq 1 - \frac{1}{N} \tag{7.26}$$

and

$$m = O(\sqrt{N}). \tag{7.27}$$

Proof. Equation (7.25) leads to the inequality $\pi/4\theta - 1 < m \leq \pi/4\theta$. Let us define $\tilde{m}$ by

$$(2\tilde{m} + 1)\theta = \frac{\pi}{2} \rightarrow \tilde{m} = \frac{\pi}{4\theta} - \frac{1}{2}.$$

Observe that m and $\tilde{m}$ satisfy

$$|m - \tilde{m}| \leq \frac{1}{2}, \tag{7.28}$$

from which it follows that

$$|(2m + 1)\theta - (2\tilde{m} + 1)\theta| = \left|(2m + 1)\theta - \frac{\pi}{2}\right| \leq \theta. \tag{7.29}$$

Considering that $\theta \sim 1/\sqrt{N}$ is a small number when $N \gg 1$ and $\sin x$ is monotonically increasing in the neighborhood of $x = 0$, we obtain

$$0 < \sin|(2m + 1)\theta - \pi/2| < \sin\theta$$

or

$$\cos^2[(2m + 1)\theta] \leq \sin^2\theta = \frac{1}{N}. \tag{7.30}$$

Thus it has been shown that

$$P_{m,z} = \sin^2[(2m + 1)\theta] = 1 - \cos^2[(2m + 1)\theta] \geq 1 - \frac{1}{N}. \tag{7.31}$$

It also follows from $\theta > \sin\theta = 1/\sqrt{N}$ that

$$m = \left\lfloor\frac{\pi}{4\theta}\right\rfloor \leq \frac{\pi}{4\theta} \leq \frac{\pi}{4}\sqrt{N}. \tag{7.32}$$

∎

It is important to note that this quantum algorithm takes only $O(\sqrt{N})$ steps and this is much faster than the classical counterpart which requires $O(N)$ steps.

Figure 7.2 shows the quantum circuit which implements Grover's search algorithm. We gave working space for oracles explicitly.

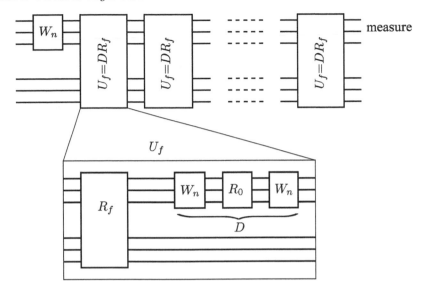

FIGURE 7.2
Implementation of Grover's search algorithm. Details of the box denoted by $U_f = DR_f$ are given in the lower diagram. The box U_f is repeated m times to maximize $P_{z,k}$. The gate R_f is the oracle, and working qubits to implement the oracle are given explicitly.

7.2   Searching for d Files

Suppose there are d $(1 < d \le N)$ files that satisfy a given condition and we are asked to find all of them. This problem is formulated with a help of an oracle

$$f(x) = \begin{cases} 1 \ (x \in A) \\ 0 \ (x \notin A). \end{cases} \tag{7.33}$$

where A is the subset of S_n, whose elements satisfy the given condition. The subset A is of course unknown to us beforehand.

This problem is solved similarly to the single-file searching problem. Let us define

$$R_f = I - 2 \sum_{z \in A} |z\rangle\langle z|. \tag{7.34}$$

Then an application of R_f on $|\varphi\rangle = \sum_{x=0}^{N-1} w_x |x\rangle$ $(\sum_x |w_x|^2 = 1)$ yields

$$R_f|\varphi\rangle = \sum_{x \notin A} w_x |x\rangle - \sum_{z \in A} w_z |z\rangle. \tag{7.35}$$

Consider

$$U_f = DR_f = (-I + 2|\varphi_0\rangle\langle\varphi_0|)\left(I - 2\sum_{z \in A}|z\rangle\langle z|\right), \qquad (7.36)$$

where $D = W_n R_0 W_n$ has been defined in Eq. (7.6). Application of U_f on $|\varphi\rangle$ produces the state

$$U_f|\varphi\rangle = \sum_{x \notin A}(2\bar{w} - w_x)|x\rangle + \sum_{z \in A}(2\bar{w} + w_z)|z\rangle, \qquad (7.37)$$

where

$$\bar{w} = \frac{1}{N}\left(\sum_{x \notin A}w_x - \sum_{z \in A}w_z\right). \qquad (7.38)$$

EXERCISE 7.1 Prove Eq. (7.37).

EXERCISE 7.2 Let $|\varphi_0\rangle = (1/\sqrt{N})\sum_{x=0}^{N-1}|x\rangle$. Show that

$$U_f^k|\varphi_0\rangle = a_k \sum_{z \in A}|z\rangle + b_k \sum_{x \notin A}|x\rangle, \qquad (7.39)$$

where $a_0 = b_0 = 1/\sqrt{N}$ and

$$a_k = \frac{N - d}{N}a_{k-1} + \frac{2(N - d)}{N}b_{k-1} \qquad (7.40)$$

$$b_k = -\frac{2d}{N}a_{k-1} + \frac{N - 2d}{N}b_{k-1}, \qquad (7.41)$$

where $d = |A|$.

The above recursion relations are easily solved to yield

$$a_k = \frac{1}{\sqrt{d}}\sin[(2k + 1)\theta], \quad b_k = \frac{1}{\sqrt{N - d}}\cos[(2k + 1)\theta], \qquad (7.42)$$

where

$$\sin\theta = \sqrt{\frac{d}{N}}, \quad \cos\theta = \sqrt{1 - \frac{d}{N}}. \qquad (7.43)$$

EXERCISE 7.3 Prove Eq. (7.42).

The above solution shows that the application of U_f on $|\varphi_0\rangle$ k times yields the state

$$U_f^k|\varphi_0\rangle = \frac{1}{\sqrt{d}}\sin[(2k + 1)\theta]\sum_{z \in A}|z\rangle + \frac{1}{\sqrt{N - d}}\cos[(2k + 1)\theta]\sum_{x \notin A}|x\rangle. \quad (7.44)$$

In order to maximize the probability with which the desired files are observed upon measurements, we have to maximize

$$P_{A,k} = \sum_{z \in A} \left(\frac{1}{\sqrt{d}} \sin[(2k+1)\theta] \right)^2 = \sin^2[(2k+1)\theta]. \qquad (7.45)$$

By repeating the arguments in the previous section, we arrive at the following conclusions. Suppose $d \ll N$ and define

$$m = \left\lfloor \frac{\pi}{4\theta} \right\rfloor. \qquad (7.46)$$

Then the probability $P_{A,m}$ with which one of the files in A is observed in the state $U_f^m |\varphi_0\rangle$ satisfies

$$P_{A,m} \geq 1 - \frac{d}{N} \qquad (7.47)$$

and, moreover,

$$m = O(\sqrt{N/d}). \qquad (7.48)$$

EXERCISE 7.4 Prove Eqs. (7.47) and (7.48).

Implementation of the Grover's algorithm with many search files is also given by the quantum circuit in Fig. 7.2 provided that the oracle R_f is properly modified.

References

[1] L. K. Grover, *Proc. 28th Annual ACM Symposium on the Theory of Computing*, ACM, New York, 212 (1996).

[2] L. K. Grover, Phys. Rev. Lett. **79**, 325 (1997).

[3] L. K. Grover, Amer. J. Phys. **69**, 769 (2001).

[4] Y. Uesaka, *Mathematical Principle of Quantum Computation*, Corona Publishing, Tokyo, in Japanese (2000).

[5] M. A. Neilsen and I. L. Chuang, *Quantum Computation and Quantum Information*, Cambridge University Press (2000).

8

Shor's Factorization Algorithm

Shor's factorization algorithm is one of the prime examples in which a quantum computer demonstrates enormous power surpassing its classical counterpart [1, 2]. Although the factorization algorithm may be carried out with a classical computer, it takes an exponentially longer time (i.e., practically impossible) compared to Shor's quantum algorithm. Shor's algorithm is almost identical with the classical one; it consists of a sequence of classical steps with one exception, which is replaced by a quantum algorithm. Our presentation here closely follows [3] and [4].

Let us first consider why factorization of a large number is important.

8.1 The RSA Cryptosystem

The RSA public-key cryptosystem and its variations are widely used to transmit messages over public lines, such as Internet communications, securing the privacy. It is based on the assumption that

"It takes an enormous time to factor a large integer."

The current chapter is written with a PC with a 1.4 GHz Pentium M CPU. The table below shows how long it takes for this PC to factor a large integer N by using `FactorInteger` commmand of Mathematica;

N	time[s]
45878443254366745	0.02
7536576836238936804738907362515346578697687343	3.084
7536576836287436733893680473896754073625189021153465786976687	98.88

As the number of digits grows up, it takes more and more time to factorize an integer. Readers who are interested in challenging large number factorization should visit the RSA Security website

http://www.rsasecurity.com/rsalabs/node.asp?id=2092

Factorization of a 617-digit decimal number is worth a $200,000 reward!

It should be noted that it is easier to verify whether a number is prime or not [5], but it is very difficult to find the factors of a big number. The **RSA cryptography** [6] makes use of this fact to encode and decode messages.

Let us start with an example. Bob wants to send Alice a message through a public communication channel. He encrypts his message with a key Alice publicizes. Although the key is publicly available, Alice is the only person who can decode the message.

1. Alice prepares two large prime numbers p and q, which she keeps secret and publishes the product N of them. We use the following example here:

$$p = 928101320540413151847590244727697333896 9$$

and

$$q = 9591715349237194999547050068718930514279,$$

for which

$$N = 89020836818747907956831989272091600303613264603794247$$
$$03263764762563155496163835 1.$$

It takes quite long (in practical situations with more digits for N) to factor N into p and q. Alice also prepares a number called an **exponent** e ($< N$), which is relatively prime to $(p-1)(q-1)$. She can easily find such a number:

$$e = 1234567, \ \gcd[e, (p-1)(q-1)] = 1,$$

for example. This number e is also published along with N. She then calculates the modular inverse d of e mod $(p-1)(q-1)$:

$$de \equiv 1 \ \mathrm{mod} \ (p-1)(q-1) \rightarrow$$
$$d = 37853991457169688722835964472412302649896709869911699 3$$
$$5543701913266864573727079 9.$$

Alice keeps d secret.

2. Bob wants to send Alice a message, "hello," for example. This message is transformed into a sequence of decimal numbers less than N under a certain scheme (ASCII etc.). Suppose his scheme transforms the message as

$$\mathrm{hello} \rightarrow 123000456000789000123,$$

for example. He encodes his message as hello^e mod N and sends Alice the result through an open channel:

$$\mathrm{encrypted} = \mathrm{hello}^e \ \mathrm{mod} \ N$$
$$= 37853991457169688722835964472412302649896709869 9$$
$$1169935543701913266864573727079 9.$$

3. Alice now decodes the message she received, using d, as

$$\text{encrypted}^d \bmod N = 123000456000789000123.$$

We outline how the RSA system works below. Readers who are eager to proceed to Shor's algorithm may skip this part and jump to the next section. Let us start with **Fermat's little theorem**.

THEOREM 8.1 Let p be an odd prime number and a be any positive integer which is not a multiple of p. Then

$$a^{p-1} \equiv 1 \bmod p. \tag{8.1}$$

Proof. First we prove the congruence

$$m^p - m \equiv 0 \bmod p \tag{8.2}$$

for any $m \in \mathbb{N}$ by induction. Equation (8.2) is true for $m = 1$. Let (8.2) be satisfied with $m = k$. Then we obtain for $m = k + 1$

$$(k+1)^p - (k+1) = k^p + \binom{p}{1} k^{p-1} + \binom{p}{2} k^{p-2} + \ldots + \binom{p}{p-1} k - k$$
$$\equiv k^p - k \bmod p,$$

where we noted that $\binom{p}{k}$ is a multiple of p for any k satisfying $1 \le k \le p - 1$. Since we assumed $k^p - k \equiv 0 \bmod p$, we obtain $(k+1)^p - (k+1) \equiv 0 \bmod p$.

Now let $m = a$ and write Eq. (8.2) as $a^p - a = a(a^{p-1} - 1) \equiv 0 \bmod p$. Since a is not divisible by p by assumption, $a^{p-1} - 1$ must be divisible by p. In other words, $a^{p-1} - 1 \equiv 0 \bmod p$, and Eq. (8.1) has been proved. ∎

Suppose $N = pq$, p and q being primes, and e $(1 < e < (p-1)(q-1))$ is the encryption exponent which is coprime to $(p-1)(q-1)$ as was assumed previously. The modular inverse d of e satisfies $de \equiv 1 \bmod (p-1)(q-1)$ and $1 < d < (p-1)(q-1)$. Let m $(< N)$ be a message to be encrypted using the public key e as $m_{\text{encrypted}} \equiv m^e \bmod N$. Decryption is possible only with the secret key d since

$$m_{\text{encrypted}}^d = m^{de} \equiv m \bmod N.$$

In fact, the congruence $de \equiv 1 \bmod (p-1)(q-1)$ leads to $de = s(p-1)(q-1) + 1$ $(s \in \mathbb{N})$ and

$$m^{de} = m^{s(p-1)(q-1)+1} = m \left[m^{s(q-1)} \right]^{p-1}.$$

Now suppose m is not a multiple of p. Then Fermat's little theorem asserts that $\left[m^{s(q-1)} \right]^{p-1} \equiv 1 \bmod p$. If m is a multiple of p, then $m^{de} \equiv 0 \bmod p$. By making a trade of p for q, we obtain $[m^{s(p-1)}]^{q-1} \equiv 1 \bmod q$ if q does not divide m, while $m^{de} \equiv 0 \bmod q$ if q divides m. Since p and q are prime, these equalities imply $m^{de} \equiv m \bmod N$.

The RSA cryptosystem depends heavily on the *belief* that factorization of a large number into its prime factors is practically impossible. Shor's algorithm would demolish this myth, as we will see below.

8.2 Factorization Algorithm

Let p and q be prime numbers and let $N = pq$. We want to factor N into a product of p and q. A naive method for the factorization takes $\sqrt{N}$ trials, in the worst case, before p and q are found. Since $\sqrt{N} = e^{(r/2)\ln 2}$ for $N = 2^r$, this method is inefficient. It turns out that the following algorithm is best suited for our purpose.

STEP 1 Take a positive integer m less than N randomly. Calculate the greatest common divisor $\gcd(m, N)$ by the Euclidean algorithm. If $\gcd(m, N) \neq 1$, we are extremely lucky: m is either p or q, and we are done. Suppose $\gcd(m, N) = 1$.

STEP 2 Define $f_N : \mathbb{N} \to \mathbb{N}$ by $a \mapsto m^a \bmod N$. Find the smallest $P \in \mathbb{N}$, such that $m^P \equiv 1 \bmod N$. The number P is called the **order** or **period**. It is known that this takes exponentially large steps in any classical algorithm, but it takes only polynomial steps in Shor's algorithm. A quantum computer is required only in this step, and the rest may be executed in polynomial steps even with a classical computer.

STEP 3 If P is odd, it cannot be used in the following steps. Go back to step 1 and repeat the above steps with different m until an even P is obtained. If P is even, proceed to step 4.

STEP 4 Since P is even, it holds that

$$(m^{P/2} - 1)(m^{P/2} + 1) = m^P - 1 \equiv 0 \bmod N. \tag{8.3}$$

If $m^{P/2} + 1 \equiv 0 \bmod N$, then $\gcd(m^{P/2} - 1, N) = 1$; go back to step 1 and try with different m. If $m^{P/2} + 1 \not\equiv 0 \bmod N$, $m^{P/2} - 1$ contains either p or q, and we proceed to step 5. Note that the number $m^{P/2} - 1$ cannot be a multiple of N in the latter case. If this is the case, it leads to $m^{P/2} \equiv 1 \bmod N$, which contradicts the assumption that P is the smallest number which satisifes $m^P \equiv 1 \bmod N$.

STEP 5 The number

$$d = \gcd(m^{P/2} - 1, N) \tag{8.4}$$

is either p or q, and factorization is done.

EXAMPLE 8.1 An example will clarify the above steps. Let $N = 799 = 17 \cdot 47$.*

STEP 1: The choice $m = 7$ leads to $\gcd(799, 7) = 1$. So this is OK.

STEP 2: It follows from Fig. 8.1 that $7^{368} \equiv 1 \bmod 799$. Thus $P = 368$. Of course we have cheated here and a quantum computer must be used for a large N.

STEP 3: The order thus found is even: $P/2 = 184$. Let us proceed to step 4.

STEP 4: $(7^{184} - 1)(7^{184} + 1) \equiv 0 \bmod 799$. It is easy to see that

$$\gcd(7^{184} + 1, 799) = 17 \neq 1,$$

and we proceed to step 5.

STEP 5: $7^{184} - 1$ and $N = 799$ must have a common prime factor. Indeed, it is found that $d = \gcd(7^{184} - 1, 799) = 47$. It is also found that $799/47 = 17$, which leads to $799 = 47 \cdot 17$.

EXERCISE 8.1 Let $N = 35$. Repeat the above steps to find the factors of N. (There are m whose orders are less than 10. If your m does not give $P < 10$, try another m. Good luck!)

It should be emphasized again that a quantum computation is required only in step 2, where the order P of the function $f : \mathbb{N} \to \mathbb{Z}/N\mathbb{Z}$ ($a \mapsto m^a \bmod N$) must be found. Here $\mathbb{Z}/N\mathbb{Z}$ stands for the set of equivalence classes in which x and $x + kN$ ($k \in \mathbb{Z}$) are identified. Clearly, we may take x satisfying $0 \leq x \leq N - 1$ as a representative of each equivalence class.[†]

8.3 Quantum Part of Shor's Algorithm

8.3.1 Settings for STEP 2

Let $N = pq \in \mathbb{N}$ be a number to be factored, where p and q are primes. Find $n \in \mathbb{N}$, such that

$$N^2 \leq 2^n < 2N^2. \tag{8.5}$$

Let us write $Q = 2^n$ hereafter. Denote $f : a \mapsto m^a \bmod N$ restricted on

$$S_n = \{0, 1, \ldots, Q - 1\} \tag{8.6}$$

---

[*]This example is repeatedly studied in due course.
[†]In fact $x = 0$ is omitted since m and N are coprime to each other.

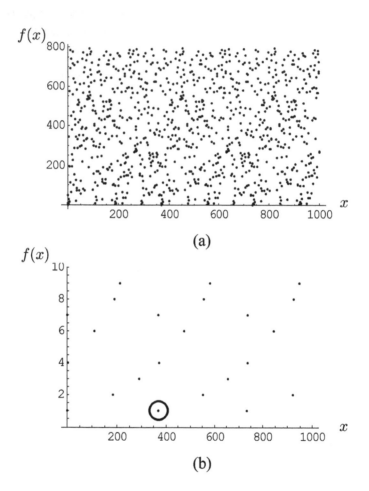

FIGURE 8.1
(a) Graph of 7^x mod 799. (b) The same graph as (a) with the range $0 \sim 10$.
The point encircled is at $x = 368$, which shows that $P = 368$ is the smallest
positive integer satisfying $7^P \equiv 1$ mod 799.

by the same symbol $f : S_n \to \mathbb{Z}/N\mathbb{Z}$.[‡]

Our quantum computer has two n-qubit registers which we call $|\text{REG1}\rangle$ and $|\text{REG2}\rangle$:

$$|\text{REG1}\rangle|\text{REG2}\rangle = |a\rangle|b\rangle = |a_{n-1}\ldots a_1 a_0\rangle|b_{n-1}\ldots b_1 b_0\rangle, \qquad (8.7)$$

where decimal numbers $a, b \in S_n$ are expressed in binary numbers in the RHS;

$$a = \sum_{j=0}^{n-1} a_j 2^j, \quad b = \sum_{j=0}^{n-1} b_j 2^j.$$

In the following, extensive use of QFT will be made on an n-qubit system, which is given by

$$|x\rangle \to U_{\text{QFT}n}|x\rangle = \frac{1}{\sqrt{Q}} \sum_{y=0}^{Q-1} \omega_n^{-xy}|y\rangle, \qquad (8.8)$$

where $x, y \in S_n$ and $\omega_n = \exp(2\pi i/Q)$. We will denote $U_{\text{QFT}n}$ by $\mathcal{F}$ hereafter.

8.3.2 STEP 2

Let us have a closer look at step 2.

STEP 2.0: Set the registers to the initial state

$$|\psi_0\rangle = |\text{REG1}\rangle|\text{REG2}\rangle = |\underbrace{00\ldots0}_{n\text{ qubits}}\rangle|\underbrace{00\ldots0}_{n\text{ qubits}}\rangle. \qquad (8.9)$$

STEP 2.1: The QFT $\mathcal{F}$ is applied on the first register;

$$|\psi_0\rangle = |0\rangle|0\rangle \overset{\mathcal{F}\otimes I}{\mapsto} |\psi_1\rangle = \frac{1}{\sqrt{Q}} \sum_{x=0}^{Q-1} |x\rangle|0\rangle. \qquad (8.10)$$

The first register is in a superposition of all the states $|x\rangle$ $(0 \le x \le Q-1)$, as remarked in Chapter 6.

STEP 2.2: Let us define a function $f : S_n \to \mathbb{Z}/N\mathbb{Z}$ by

$$f(x) = m^x \bmod N, \quad x \in S_n. \qquad (8.11)$$

Suppose that the unitary gate U_f realizes the action of f on x in such a way that $U_f|x\rangle|0\rangle = |x\rangle|f(x)\rangle$. Apply U_f on the state prepared in step 2.1 to yield

$$U_f|\psi_1\rangle = |\psi_2\rangle \equiv \frac{1}{\sqrt{Q}} \sum_{x=0}^{Q-1} |x\rangle|f(x)\rangle. \qquad (8.12)$$

[‡]It is clear that the range of f is $\mathbb{Z}/N\mathbb{Z}$ since $0 \le f(x) \le N-1 \le \sqrt{Q}-1 < Q-1, \forall x \in S_n$.

This shows that the two registers are entangled in general.

STEP 2.3: Apply QFT on $|\text{REG1}\rangle$ again to yield

$$|\psi_3\rangle = (\mathcal{F} \otimes I)|\psi_2\rangle = \frac{1}{Q} \sum_{x=0}^{Q-1} \sum_{y=0}^{Q-1} \omega_n^{-xy} |y\rangle |f(x)\rangle$$

$$= \frac{1}{Q} \sum_{y=0}^{Q-1} |y\rangle |\Upsilon(y)\rangle = \frac{1}{Q} \sum_{y=0}^{Q-1} \||\Upsilon(y)\rangle\| \cdot |y\rangle \frac{|\Upsilon(y)\rangle}{\||\Upsilon(y)\rangle\|}, \qquad (8.13)$$

where

$$|\Upsilon(y)\rangle = \sum_{x=0}^{Q-1} \omega_n^{-xy} |f(x)\rangle. \qquad (8.14)$$

STEP 2.4: $|\text{REG1}\rangle$ is measured. The result $y \in S_n$ is obtained with the probability

$$\text{Prob}(y) = \frac{\||\Upsilon(y)\rangle\|^2}{Q^2}, \qquad (8.15)$$

and, at the same time, the state collapses to

$$|y\rangle \frac{|\Upsilon(y)\rangle}{\||\Upsilon(y)\rangle\|}.$$

The measurement process generates a random variable following a classical probability distribution $\mathcal{S}$ over S_n, in which "symbols" $y \in S_n$ are generated with the probability (8.15).

STEP 2.5: Extract the order P from the measurement outcome.

EXERCISE 8.2 Let $N = 21$ and $m = 11$. Find n which satisifes Eq. (8.5). Find also the order P.[§] Repeat the above steps to find the wave function $|\psi_3\rangle$ and $\text{Prob}(y)$ ($y \in S_n$).

8.4   Probability Distribution

Let us study the probability distribution $\text{Prob}(y)$ in detail.

[§]The order is less than 10 in this case.

PROPOSITION 8.1 Let $Q = 2^n = Pq + r$, $(0 \leq r < P)$, where q and r are uniquely determined non-negative integers. Let $Q_0 = Pq$. Then

$$
\text{Prob}(y) = \begin{cases} \dfrac{r \sin^2\left(\frac{\pi P y}{Q}\left(\frac{Q_0}{P} + 1\right)\right) + (P - r)\sin^2\left(\frac{\pi P y}{Q} \cdot \frac{Q_0}{P}\right)}{Q^2 \sin^2\left(\frac{\pi P y}{Q}\right)} & (Py \not\equiv 0 \bmod Q) \\[4mm] \dfrac{r(Q_0 + P)^2 + (P - r)Q_0^2}{Q^2 P^2} & (Py \equiv 0 \bmod Q). \end{cases}
$$

Proof. It is found from the definition that[¶]

$$
|\Upsilon(y)\rangle = \sum_{x=0}^{Q-1} \omega^{-xy}|f(x)\rangle = \sum_{x=0}^{Q_0-1} \omega^{-xy}|f(x)\rangle + \sum_{x=Q_0}^{Q-1} \omega^{-xy}|f(x)\rangle
$$

$$
= \sum_{x_0=0}^{P-1} \sum_{x_1=0}^{Q_0/P-1} \omega^{-(Px_1+x_0)y}|f(Px_1 + x_0)\rangle
$$

$$
+ \sum_{x_0=0}^{r-1} \omega^{-[P(Q_0/P)+x_0]y}|f(P(Q_0/P) + x_0)\rangle
$$

$$
= \sum_{x_0=0}^{P-1} \omega^{-x_0 y}\left(\sum_{x_1=0}^{Q_0/P-1} \omega^{-Px_1 y}\right)|f(x_0)\rangle + \sum_{x_0=0}^{r-1} \omega^{-x_0 y}\omega^{-Py(Q_0/P)}|f(x_0)\rangle
$$

$$
= \sum_{x_0=0}^{r-1} \omega^{-x_0 y} \sum_{x_1=0}^{Q_0/P-1} \omega^{-Pyx_1}|f(x_0)\rangle
$$

$$
+ \sum_{x_0=r}^{P-1} \omega^{-x_0 y} \sum_{x_1=0}^{Q_0/P-1} \omega^{-Pyx_1}|f(x_0)\rangle + \sum_{x_0=0}^{r-1} \omega^{-x_0 y}\omega^{-Py(Q_0/P)}|f(x_0)\rangle
$$

$$
= \sum_{x_0=0}^{r-1} \omega^{-x_0 y}\left(\sum_{x_1=0}^{Q_0/P} \omega^{-Pyx_1}\right)|f(x_0)\rangle
$$

$$
+ \sum_{x_0=r}^{P-1} \omega^{-x_0 y}\left(\sum_{x_1=0}^{Q_0/P-1} \omega^{-Pyx_1}\right)|f(x_0)\rangle.
$$

Note that the map $f : a \mapsto m^a \bmod N$ is $1 : 1$ on $\{0, 1, 2, \ldots, P-1\}$, which we prove now. Suppose $m^a \equiv m^b \bmod N$ $(a > b)$; then $m^b(m^{a-b}-1) \equiv 0 \bmod N$. Since m and N are coprime, so are m^b and N. Then $m^P > m^{a-b} \equiv 1 \bmod N$, which contradicts the assumption that P is the smallest natural number such that $m^P \equiv 1 \bmod N$. This implies that $|f(0)\rangle, |f(1)\rangle, \ldots, |f(P-1)\rangle$ are

---

[¶]We drop n from ω_n to simplify our notation.

mutually orthogonal. Accordingly

$$\langle \Upsilon(y)|\Upsilon(y)\rangle = r \left| \sum_{x_1=0}^{Q_0/P} \omega^{-Pyx_1} \right|^2 + (P-r) \left| \sum_{x_1=0}^{Q_0/P-1} \omega^{-Pyx_1} \right|^2.$$

In case $Py \equiv 0 \bmod Q$, we put $Py = aQ$, $a \in \mathbb{N}$ and obtain

$$\omega^{-Pyx_1} = e^{-2\pi i(Py/Q)x_1} = e^{-2\pi iax_1} = 1.$$

Therefore

$$\langle \Upsilon(y)|\Upsilon(y)\rangle = r \cdot \left(\frac{Q_0}{P} + 1 \right)^2 + (P-r) \left(\frac{Q_0}{P} \right)^2,$$

which leads to the result independent of y,

$$\text{Prob}(y) = \frac{r(Q_0 + P)^2 + (P-r)Q_0^2}{P^2Q^2} = \frac{r(q+1)^2 + (P-r)q^2}{Q^2}. \qquad (8.16)$$

If $Py \not\equiv 0 \bmod Q$, on the other hand, we obtain

$$\langle \Upsilon(y)|\Upsilon(y)\rangle = r \left| \frac{\omega^{-Py(Q_0/P+1)} - 1}{\omega^{-Py} - 1} \right|^2 + (P-r) \left| \frac{\omega^{-Py(Q_0/P)} - 1}{\omega^{-Py} - 1} \right|^2$$

$$= r \left| \frac{e^{-(2\pi i/Q)Py(Q_0/P+1)} - 1}{e^{-(2\pi i/Q)Py} - 1} \right|^2 + (P-r) \left| \frac{e^{-(2\pi i/Q)Py(Q_0/P)} - 1}{e^{-(2\pi i/Q)Py} - 1} \right|^2.$$

Here we find from

$$|e^{i\theta} - 1|^2 = 2(1 - \cos\theta) = 4\sin^2\frac{\theta}{2}$$

that

$$\langle \Upsilon(y)|\Upsilon(y)\rangle = r\frac{\sin^2\frac{\pi}{Q}Py\left(\frac{Q_0}{P}+1\right)}{\sin^2\frac{\pi}{Q}Py} + (P-r)\frac{\sin^2\frac{\pi}{Q}Py\frac{Q_0}{P}}{\sin^2\frac{\pi}{Q}Py}.$$

Therefore, the probability distribution is given by

$$\text{Prob}(y) = \frac{\||\Upsilon(y)\rangle\|^2}{Q^2} = \frac{r\sin^2\left[\frac{\pi}{Q}Py\left(\frac{Q_0}{P}+1\right)\right] + (P-r)\sin^2\left[\frac{\pi}{Q}Py\frac{Q_0}{P}\right]}{Q^2\sin^2\frac{\pi}{Q}Py},$$

$$(8.17)$$

which proves the proposition.   ∎

COROLLARY 8.1 Suppose $Q/P \in \mathbb{Z}$ (namely $Q_0 = Q$). Then the probability of obtaining a measurement outcome y is

$$\text{Prob}(y) = \begin{cases} 0 & (Py \not\equiv 0 \bmod Q) \\ \dfrac{1}{P} & (Py \equiv 0 \bmod Q) \end{cases}$$

Proof. When $Py \not\equiv 0 \bmod Q$, $r = 0$ implies $Q = Pq$. Therefore

$$\text{Prob}(y) = \frac{P \sin^2 \pi y}{Q^2 \sin^2 \frac{\pi y}{q}} = 0.$$

In case $Py \equiv 0 \bmod Q$, we obtain

$$\text{Prob}(y) = \frac{PQ^2}{Q^2 P^2} = \frac{1}{P}.$$

∎

Figure 8.2 shows the probability distribution $\text{Prob}(y)$ with the parameters $N = 799 = 17 \cdot 47, P = 368, Q = 2^{20} = 1,048,576$. Note that $N^2 = 638,401$ and $2N^2 = 1,276,802$ so that they satisfy $N^2 \leq Q < 2N^2$. Then $Q \equiv 144 \bmod 368 \to r = 144, Q_0 = Q - r = 1,048,432, \to q = Q_0/P = 2849$. Accordingly, $\text{Prob}(y)$ exhibits sharp peaks at integral multiples of $q = 2849$. Figure 8.3 shows $\text{Prob}(y)$ for $8,520 < y < 8,580$. Observe that it has a sharp peak at $y = 8548$, for which $8548/2849 = 3.00035$. Compare the numbers

$$\text{Prob}(8547) = 0.00005393, \text{Prob}(8548) = 0.00245753, \text{Prob}(8549) = 0.00010892.$$

For neighboring numbers, we obtain $8547/2849 = 3$ and $8549/2849 = 3.0007$. Note that there are $P = 368$ sharp peaks, and $\text{Prob}(y)$ at each peak is roughly on the order of $1/386 \sim 0.00272$.

Since y is restricted within the range $0 \leq y \leq Q-1$, repeated measurements reveal that the minimal distance between the peaks is ~ 2849, which yields the approximate order $P = Q/2849 \sim 368.0505$. The order thus obtained is probabilistic, and its plausibility must be checked by carrying out step 3 $\sim$ step 5. Needless to say, this strategy is not practical when N is considerably large. There is a powerful method of continued fraction expansion by which we find the order P with a single measurement of the first register, which is the subject of §8.5.

It will be shown that factorization of a number $N = pq$ is carried out efficiently by a quantum computer. A quantum algorithm is employed to find the order of the function $f(x) = m^x \bmod N$, and the other steps are done with classical algorithms. The quantum circuit in Fig. 8.4 implements the quantum part of the algorithm where U_f and $\mathcal{F}$ stand for the map

$$U_f |x\rangle |0\rangle = |x\rangle |m^x \bmod N\rangle$$

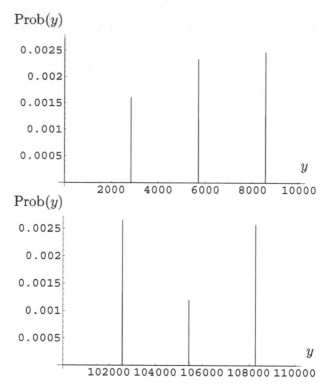

FIGURE 8.2

(a) Probability distribution $\text{Prob}(y)$ for $0 \le y \le 10,000$. (b) Same graph for the range $100,000 \le y \le 110,000$.

and the QFT, respectively.

It is instructive to recollect step 2 with our example of $799 = 17 \cdot 47$, for which $n = 20$. We take $m = 7$ as before.

STEP 2.0: The initial state is

$$|\psi_0\rangle = |0\rangle|0\rangle. \tag{8.18}$$

STEP 2.1: The QFT on the first register results in

$$|\psi_1\rangle = \frac{1}{\sqrt{Q}} \sum_{x=0}^{Q-1} |x\rangle|0\rangle, \tag{8.19}$$

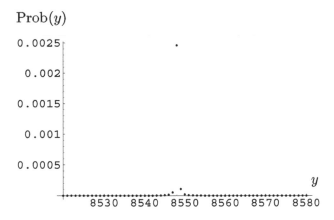

FIGURE 8.3
Probability distribution $\text{Prob}(y)$ for $8,520 \leq y \leq 8,580$.

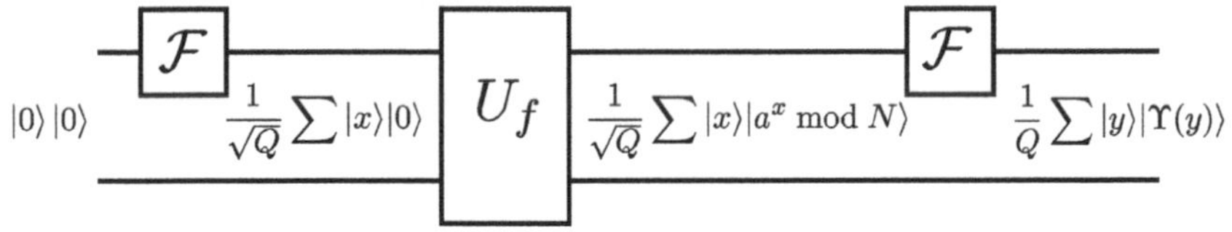

FIGURE 8.4
Quantum circuit to find the order of $f(x) = m^x \bmod N$.

where $Q = 2^{20} = 1048576$.

STEP 2.2: Application of U_f on $|\psi_1\rangle$ produces

$$|\psi_2\rangle = \frac{1}{\sqrt{Q}} \sum_{x=0}^{Q-1} |x\rangle |7^x \bmod 799\rangle$$

$$= \frac{1}{\sqrt{Q}} \Big[|0\rangle |1\rangle + |1\rangle |7\rangle + |2\rangle |49\rangle + |3\rangle |343\rangle + |4\rangle |4\rangle + |5\rangle |28\rangle$$

$$+ \ldots + |368\rangle |1\rangle + |369\rangle |7\rangle + |370\rangle |49\rangle + \ldots$$

$$+ |Q-2\rangle |756\rangle + |Q-1\rangle |498\rangle \Big]. \tag{8.20}$$

Note that there are only $P = 368$ different states in the second register.

STEP 2.3: The QFT with $\omega = e^{2\pi i/Q}$, $Q = 2^n$, is applied to the first register. This results in

$$|\psi_3\rangle = \frac{1}{\sqrt{Q}} \sum_{x=0}^{Q-1} \frac{1}{\sqrt{Q}} \sum_{y=0}^{Q-1} \omega^{-xy} |y\rangle |7^x \bmod 799\rangle$$

$$\equiv \frac{1}{Q} \sum_{y=0}^{Q-1} |y\rangle |\Upsilon(y)\rangle, \tag{8.21}$$

where

$$|\Upsilon(y)\rangle = \sum_{x=0}^{Q-1} \omega^{-xy} |7^x \bmod 799\rangle$$

$$= \sum_{x=0}^{Q-1} e^{-2\pi i x y / Q} |7^x \bmod 799\rangle$$

$$= |1\rangle + \omega^{-y}|7\rangle + \omega^{-2y}|49\rangle + \omega^{-3y}|343\rangle + \ldots$$
$$+ \omega^{-368y}|1\rangle + \omega^{-369y}|7\rangle + \omega^{-370y}|49\rangle + \omega^{-371y}|343\rangle + \ldots$$
$$+ \ldots +$$
$$+ \omega^{-736y}|1\rangle + \omega^{-737y}|7\rangle + \omega^{-738y}|49\rangle + \omega^{-739y}|343\rangle + \ldots$$
$$+ \ldots +$$
$$+ \omega^{-1048432y}|1\rangle + \omega^{-1048433y}|7\rangle + \omega^{-1048434y}|49\rangle + \omega^{-1048435y}|343\rangle$$
$$\ldots + \omega^{-1048575y}|498\rangle$$
$$= (1 + \omega^{-368y} + \omega^{-736y} + \ldots + \omega^{-1048432y})|1\rangle$$
$$+ (\omega^{-y} + \omega^{-369y} + \omega^{-737y} + \ldots + \omega^{-1048433y})|7\rangle$$
$$+ (\omega^{-2y} + \omega^{-370y} + \omega^{-738y} + \ldots + \omega^{-1048434y})|49\rangle$$
$$+ (\omega^{-3y} + \omega^{-371y} + \omega^{-739y} + \ldots + \omega^{-1048435y})|343\rangle$$
$$+ \ldots$$
$$+ (\omega^{-87y} + \omega^{-455y} + \omega^{-823y} + \ldots)|794\rangle. \tag{8.22}$$

There are 368 ket vectors in the above expansion. The coefficient of each vector becomes sizeable when and only when y is approximately a multiple of 2849. For example,

$$\sum_{k=0}^{2849} \omega^{-368ky} = 0.608696 + 0.000262611i, \quad \left| \sum_{k=0}^{2849} \omega^{-368ky} \right| = 0.608696$$

for $y = 1$ while

$$\sum_{k=0}^{2849} \omega^{-368ky} = 2315.79 + 1408.03i, \quad \left| \sum_{k=0}^{2849} \omega^{-368ky} \right| = 2710.25$$

for $y = 8548$. The previous result is recovered as

$$\text{Prob}(8548) = 368 \left(\frac{2710.25}{Q} \right)^2 = 0.00245848. \tag{8.23}$$

The order P may be inferred by repeating measurements. However, the number of measurements required to guess P grows rapidly as N becomes larger and larger. We certainly need a technique with which we may find P with a single measurement, which is the subject of the next section.

8.5 Continued Fractions and Order Finding

We introduce a few symbols. For $\forall x \in \mathbb{R}$, we define the **ceiling** $\lceil x \rceil = \inf\{n \in \mathbb{Z} | x \leq n\}$. In other words, $\lceil x \rceil = n$, where $n - 1 < x \leq n$. For example, $\lceil 2 \rceil = 2$, $\lceil 2.6 \rceil = 3$, $\lceil -4.5 \rceil = -4$ and $\lceil -5 \rceil = -5$. Similarly the **floor** is defined as $\lfloor x \rfloor = \sup\{n \in \mathbb{Z} | n \leq x\}$ for $\forall x \in \mathbb{R}$. The floor function $\lfloor x \rfloor$ is also called the integer part of x. For example, $\lfloor 4.5 \rfloor = 4$, $\lfloor 2 \rfloor = 2$ and $\lfloor -4.7 \rfloor = -5$. The floor function has been already introduced in §6.4.

Let us consider **continued fraction expansion** of a rational number. Continued fraction expansion exists also for an irrational number, but it does not terminate. The continued fraction expansion of $x \in \mathbb{Q}$ is

$$x = a_0 + \cfrac{1}{a_1 + \cfrac{1}{a_2 + \cfrac{1}{\cdots + \cfrac{1}{a_q}}}}, \tag{8.24}$$

where $a_i \in \mathbb{N}$ for $j \geq 1$. This number is also written as

$$x = [a_0, a_1, \ldots, a_q]. \tag{8.25}$$

Let us consider $x = 17/47$, for example. x is written in a continued fraction form as

$$\frac{17}{47} = 0 + \cfrac{1}{\cfrac{47}{17}} = 0 + \cfrac{1}{2 + \cfrac{13}{17}}$$

$$= 0 + \cfrac{1}{2 + \cfrac{1}{1 + \cfrac{4}{13}}} = 0 + \cfrac{1}{2 + \cfrac{1}{1 + \cfrac{1}{3 + \cfrac{1}{4}}}} = [0, 2, 1, 3, 4].$$

Let us summarize what we have done to obtain this expansion. We first find the integer part of x as $a_0 = \lfloor 17/47 \rfloor = 0$ and the fractional part as $r_0 = x - a_0 = 17/47$. We invert $17/47$ and find that the integral part of $1/r_0$ is $a_1 = \lfloor 1/r_0 \rfloor = \lfloor 47/17 \rfloor = 2$, and the fractional part is $r_1 = 1/r_0 - \lfloor 1/r_0 \rfloor = 13/17$. The interger part of $1/r_1$ is $a_2 = \lfloor 1/r_1 \rfloor = 1$, and the fractional part is $r_2 = 1/r_1 - \lfloor 1/r_1 \rfloor = 4/13$. The integer part of $1/r_2$ is $\lfloor 1/r_2 \rfloor = 3$, and the fractional part is $r_3 = 1/r_2 - \lfloor 1/r_2 \rfloor = 1/4$. The expansion terminates when r_j has the numerator 1 ($j = 3$ in the present example).

EXERCISE 8.3 Find the continued fraction expansions of $x = 61/45$ and $x = 121/13$.

The algorithm to obtain a continued fraction expansion is summarized as

1. Let x be a rational number to be expanded. Let $a_0 = \lfloor x \rfloor$ and $r_0 = 1/x - a_0$.

2. Let $m = 1$ and set $a_m = \left\lfloor \dfrac{1}{r_{m-1}} \right\rfloor$ and $r_m = \dfrac{1}{r_{m-1}} - a_m$. Repeat this step until $r_M = 0$ is reached. M is always finite when x is a rational number.

3. The continued fraction expansion of x is

$$x = a_0 + \cfrac{1}{a_1 + \cfrac{1}{a_2 + \cfrac{1}{\ddots \cfrac{1}{a_{M-1} + \cfrac{1}{a_M}}}}} = [a_0, a_1, \ldots, a_{M-1}, a_M].$$

Given $x = [a_0, a_1, \ldots, a_M]$, the continued fraction $[a_0, a_1, \ldots, a_j]$ $(j \le M)$ is called the jth **convergent** of x. The Mth convergent is x itself. Note that $[a_0, a_1, \ldots, a_M] = [a_0, a_1, \ldots, a_M - 1, 1]$. Thus the number M may be made either even or odd.

With the above preliminaries, we come back to Shor's algorithm. Suppose y is obtained upon the measurement of the first register in Shor's algorithm. Then y/Q is a rational number close to n/P with some $n \in \mathbb{N}$. Now we show that the following algorithm finds the order P of m^x mod N.

1. Find the continued fraction expansion $[a_0, a_1, \ldots, a_M]$ of y/Q. We always have $a_0 = 0$ since $y/Q < 1$.

2. Let $p_0 = a_0$ and $q_0 = 1$.

3. Let $p_1 = a_1 p_0 + 1$ and $q_1 = a_1 q_0$.

4. Let $p_i = a_i p_{i-1} + p_{i-2}$ and $q_i = a_i q_{i-1} + q_{i-2}$ for $2 \le i \le M$. We obtain the sequence $(p_0, q_0), (p_1, q_1), \ldots, (p_M, q_M)$. It is shown below that p_j/q_j is the jth convergent of y/Q.

5. Find the smallest k $(0 \le k \le M)$ such that $|p_k/q_k - y/Q| \le 1/(2Q)$. Such k is unique.

6. The order is found as $P = q_k$.

EXAMPLE 8.2 Let us consider our favorite example. Let $N = 799$, $Q = 2^{20} = 1048576$ and $m = 7$. The error bound is $1/(2Q) = 4.76837 \times 10^{-7}$. Suppose we obtain $y = 8548$ as a measurement outcome of the first register. We expect that y/Q is an approximation of n/P for some $n \in \mathbb{N}$.

1. The continued fraction expansion of $8548/1048576$ is $[0, 122, 1, 2, 44, 5, 3]$ and $M = 6$.

2. Let $p_0 = a_0 = 0$ and $q_0 = 1$.

3. We obtain $p_1 = a_1 p_0 + 1 = 122 \times 0 + 1 = 1$ and $q_1 = a_1 q_0 = 122 \times 1 = 122$. We find $|p_1/q_1 - y/Q| = |1/122 - 8548/1048576| = 4.47133 \times 10^{-5} > 1/(2Q)$.

4. $p_2 = a_2 p_1 + p_0 = 1$, $q_2 = a_2 q_1 + q_0 = 123$ and $|p_2/q_2 - y/Q| = |1/123 - 8548/1048576| = 2.19 \times 10^{-5} > 1/(2Q)$.

5. $p_3 = a_3 p_2 + p_1 = 3$, $q_3 = a_3 q_2 + q_1 = 368$ and $|p_3/q_3 - y/Q| = |3/368 - 8548/1048576| = 1.65856 \times 10^{-7} \le 1/(2Q)$. We have obtained $k = 3$.

6. The order is found to be $P = q_3 = 368$.

EXERCISE 8.4 Suppose $y = 37042$ is the measurement outcome in the above example. Find the order P by repeating the above algorithm. Suppose $y = 65536$ has been obtained in the next measurement. Apply the above algorithm. What is the "order" you find?

We must know when we obtain the correct order and when not. Here is the sufficient condition.

The correct order P is obtained in the above algorithm when the measurement outcome y belongs to the set

$$\mathcal{C} = \left\{ y \left| \exists d \in \{1, 2, \ldots, P-1\}, \left| \frac{d}{P} - \frac{y}{Q} \right| \le \frac{1}{2Q}, \gcd(P, d) = 1 \right. \right\}. \quad (8.26)$$

The set $\mathcal{C}$ is not an empty set. In fact, for any $P < Q$ there always exists $y \in \{0, 1, 2, \ldots, Q-1\}$ such that $-P/2 < Q - yP \le P/2$, from which we obtain

$$\left| \frac{1}{P} - \frac{y}{Q} \right| \le \frac{1}{2Q}.$$

Therefore, the set $\mathcal{C}$ contains at least one element y, for which $d/P = 1/P$. It is important to note that it is impossible to tell whether a particular outcome y is in $\mathcal{C}$ or not since P is not known in advance.

Let us look at the second case $y = 65536$ in the exercise above. Although we verify that $|23/368 - 65536/Q| = 0$, we also have $\gcd(368, 23) = 23$. The next smallest $|d/P - y/Q|$ is attained when $d = 22$ and $d = 24$, for which case $|d/368 - 65536/Q| = 0.00271 > 1/(2Q)$. Therefore an integer $d \in \mathcal{C}$ does not exist for this case.

We outline the proof of the factor-finding algorithm based on the continued fraction expansion. We have $y/Q = [a_0, a_1, \ldots, a_M]$ in our mind in the following lemmas.

LEMMA 8.1 The sequences $a_0, a_1, \ldots, a_M$ and $(p_0, q_0), (p_1, q_1), \ldots, (p_M, q_M)$ obtained in the algorithm introduced in this section satisfy

$$[a_0] = \frac{p_0}{q_0}, \quad [a_0, a_1] = \frac{p_1}{q_1}, \ldots, [a_0, a_1, \ldots, a_M] = \frac{p_M}{q_M}. \tag{8.27}$$

Proof. We prove this by induction. It is easily verified that $[a_0] = p_0/q_0$ and $[a_0, a_1] = p_1/q_1$. Suppose $[a_0, a_1, \ldots, a_k] = p_k/q_k$. Then

$$
\begin{aligned}
[a_0, a_1, \ldots, a_k, a_{k+1}] &= [a_0, a_1, \ldots, a_k + 1/a_{k+1}] \\
&= \frac{(a_k + 1/a_{k+1})p_{k-1} + p_{k-2}}{(a_k + 1/a_{k+1})q_{k-1} + q_{k-2}} \\
&= \frac{a_{k+1}(a_k p_{k-1} + p_{k-2}) + p_{k-1}}{a_{k+1}(a_k q_{k-1} + q_{k-2}) + q_{k-1}} \\
&= \frac{a_{k+1}p_k + p_{k-1}}{a_{k+1}q_k + q_{k-1}} = \frac{p_{k+1}}{q_{k+1}}.
\end{aligned}
$$

∎

LEMMA 8.2 All the fractions $p_1/q_1, p_2/q_2, \ldots, p_M/q_M$ are irreducible.

Proof. We first prove

$$p_{k-1}q_k - q_{k-1}p_k = (-1)^k \tag{8.28}$$

for $1 \leq k \leq M$ by induction. This is obviously satisfied for $k = 1$ since $p_0 q_1 - q_0 p_1 = 0 - 1$ and for $k = 2$ since $p_1 q_2 - q_1 p_2 = (a_1 a_2 + 1) - a_1 a_2 = 1$. Suppose $p_{k-2}q_{k-1} - q_{k-2}p_{k-1} = (-1)^{k-1}$ is satisfied. Then

$$
\begin{aligned}
p_{k-1}q_k - q_{k-1}p_k &= p_{k-1}(a_k q_{k-1} + q_{k-2}) - q_{k-1}(a_k p_{k-1} + p_{k-2}) \\
&= p_{k-1}q_{k-2} - q_{k-1}p_{k-2} = (-1)^k.
\end{aligned}
$$

Now suppose p_k and q_k are not coprime and let $\gcd(p_k, q_k) = d_k \geq 2$. Put $p_k = d_k p'_k$ and $q_k = d_k q'_k$, where $\gcd(p'_k, q'_k) = 1$. Then $p_{k-1}q_k - q_{k-1}p_k = d_k(p_{k-1}q'_k - q_{k-1}p'_k) = (-1)^k$. The second equality is a contradiction since $d_k \geq 2$. ∎

LEMMA 8.3 Let p and q be positive integers such that $\gcd(p, q) = 1$ and x be a positive rational number. The rational number p/q is a convergent of x if they satisfy the inequality

$$\left| \frac{p}{q} - x \right| \leq \frac{1}{2q^2}. \tag{8.29}$$

Proof. Let $p/q = [a_0, a_1, \ldots, a_m]$. We assume, without loss of generality, that m is even. Note that $p = p_m$ and $q = q_m$.

Let $\delta = 2q^2(x - p/q)$. We find $|\delta| \leq 2q^2/(2q^2) = 1$ by assumption. The statement of the lemma is trivial when $\delta = 0$. Suppose $0 < \delta \leq 1$. (Redefine δ by $-\delta$ if $\delta < 0$.) There is a rational number α such that

$$x = \frac{\alpha p_m + q_{m-1}}{\alpha q_m + q_{m-1}} = [a_0, a_1, \dots, a_m, \alpha].$$

The condition $\alpha \geq 1$ must be satisfied for $[a_0, a_1, \dots, a_m, \alpha]$ to be a continued fraction expansion of x. By inverting the above relation, we obtain

$$\alpha = \frac{p_{m-1} - xq_{m-1}}{xq_m - p_m} = \frac{2q_m(p_{m-1} - xq_{m-1})}{\delta}$$
$$= \frac{2(p_{m-1}q_m - q_{m-1}p_m)}{\delta} - \frac{q_{m-1}}{q_m}.$$

It has been shown in the proof of Lemma 8.2 that $p_{m-1}q_m - q_{m-1}p_m = 1$ for an even m, from which we obtain $\alpha = 2/\delta - q_{m-1}/q_m > 1$, where we noted $0 < \delta \leq 1$ and $q_m > q_{m-1}$. Therefore α has a continued fraction expansion $[b_0, \dots, b_n]$ with $b_0 \geq 1$ and we obtain $x = [a_0, \dots, a_m, b_0, \dots, b_n]$, from which $p/q = [a_0, \dots, a_m]$ is shown to be a convergent of x. ∎

We note that the above lemma does not claim the uniqueness of the convergent $p/q = p_m/q_m$. There may be many convergents p_k/q_k of x, which satisify

$$\left| \frac{p_k}{q_k} - x \right| \leq \frac{1}{2q_k^2}.$$

It simply claims that a rational number p/q satisfying $|p/q - x| \leq 1/2q^2$ is one of the convergents of x. It should be also noted that there may be convergents of x, which do not satisfy the above inequality.

LEMMA 8.4 For a given N, a measurement outcome y and Q, such that $N^2 \leq Q < 2N^2$, there exists a unique rational number d/P such that $0 < P < N$ and

$$\left| \frac{d}{P} - \frac{y}{Q} \right| \leq \frac{1}{2Q}. \tag{8.30}$$

Proof. Suppose there are two sets of (d, P) satisfying this condition, which we call (d_1, P_1) and (d_2, P_2). Then we find

$$\left| \frac{d_1}{P_1} - \frac{d_2}{P_2} \right| = \left| \frac{d_1}{P_1} - \frac{y}{Q} + \frac{y}{Q} - \frac{d_2}{P_2} \right|$$
$$\leq \left| \frac{d_1}{P_1} - \frac{y}{Q} \right| + \left| \frac{y}{Q} - \frac{d_2}{P_2} \right| \leq \frac{1}{2Q} + \frac{1}{2Q} = \frac{1}{Q} \leq \frac{1}{N^2}.$$

It follows from $P_i < N$ that

$$\frac{1}{N^2} \geq \left| \frac{d_1}{P_1} - \frac{d_2}{P_2} \right| = \frac{|d_1 P_2 - d_2 P_1|}{P_1 P_2} > \frac{|d_1 P_2 - d_2 P_1|}{N^2},$$

from which we obtain $|d_1 P_2 - d_2 P_1| < 1$, namely $d_1 P_2 = d_2 P_1 = 0$. ∎

Now we are ready to justify the order-finding algorithm.

PROPOSITION 8.2 Suppose $y \in \mathcal{C}$. Then the number P obtained by the algorithm in this section is the correct order of the modular exponential function $m^x \bmod N$.

Proof. Let $[a_0, a_1, \ldots, a_M]$ be the continued fraction expansion of y/Q. Since $y \in \mathcal{C}$, there exists $d \in \{1, 2, \ldots, P-1\}$ such that

$$\gcd(P, d) = 1, \quad \left| \frac{d}{P} - \frac{y}{Q} \right| \le \frac{1}{2Q}.$$

Such d must be unique due to Lemma 8.4. By noticing that $P < N$ and $P^2 < N^2 \le Q$, we also obtain

$$\left| \frac{d}{P} - \frac{y}{Q} \right| \le \frac{1}{2Q} \le \frac{1}{2N^2} < \frac{1}{2P^2}.$$

It then follows from Lemma 8.3 that d/P is one of the convergents of $y/Q = [a_0, a_1, \ldots, a_M]$. Let us call this convergent p_m/q_m.

Now we show that this m is the smallest k which satisfies

$$\left| \frac{p_k}{q_k} - \frac{y}{Q} \right| \le \frac{1}{2Q}.$$

It is obvious that this inequality is satisfied for any p_k/q_k such that $m \le k \le M$. However, Lemma 8.4 tells us that there is only one p_k/q_k which satisfies $q_k < N$ and the above inequality. Therefore $q_k > N$ for $k > m$ and we have shown $d/P = p_m/q_m$. Since $\gcd(d, P) = \gcd(p_m, q_m) = 1$ due to Lemma 8.2, we must have $q_m = P$. ∎

In summary, a single measurement of the first register provides the correct order if the measurement outcome y belongs to the set $\mathcal{C}$. We have seen in Example 8.2 that $y = 8548$ satisfies $|3/P - y/Q| \le 1/(2Q)$ and hence $y \in \mathcal{C}$. It should be kept in mind that P, and hence $\mathcal{C}$, is not known in advance.

8.6 Modular Exponential Function

The block diagram of the quantum circuit to find the order of $f(x) = m^x \bmod N$ is depicted in Fig. 8.4. It has been shown in §6.4 that an n-qubit QFT circuit may be implemented with $\Theta(n^2)$ elementary gates. We now work out the implementation of the other component in Shor's algorithm, the modular exponential function,

$$U_f |x\rangle |0\rangle = |x\rangle |m^x \bmod N\rangle.$$

There are several proposals for the implementation of this function; some save computational steps while the other saves the number of qubits required. Here we follow the standard implementation given in [7] and [4].

Implementation of a modular exponential function is divided into several steps. We need to implement

1. Adder, which outputs $a + b$ given non-negative integers a and b.

2. Modular adder, which outputs $a + b \bmod N$.

3. Modular multiplexer, which outputs $ab \bmod N$.

4. Modular exponential function, which outputs $m^x \bmod N$.

8.6.1   Adder

Let $a = a_{n-1}2^{n-1} + a_{n-2}2^{n-2} + \ldots + a_1 2 + a_0$ and $b = b_{n-1}2^{n-1} + b_{n-2}2^{n-2} + \ldots + b_1 2 + b_0$ be non-negative integers and $s = s_n 2^n + s_{n-1}2^{n-1} + \ldots + s_1 2 + s_0$ be their sum. Let c_k denote the carry bit. The binary numbers s_k satisfy the following recursion relations,

$$s_0 = a_0 \oplus b_0, \quad c_0 = a_0 b_0, \quad s_k = a_k \oplus b_k \oplus c_{k-1},$$
$$c_k = a_k b_k \oplus a_k c_{k-1} \oplus b_k c_{k-1}, \quad (1 \le k \le n), \tag{8.31}$$

where $\oplus$ is addition mod 2, $a_k b_k = a_k \wedge b_k$ and we formally put $a_n = b_n = 0$. Note that $s_n = c_{n-1}$. Carry bit c_k is 1 only when at least two of a_k, b_k and c_{k-1} are 1. This condition is expressed in the final relation in Eq. (8.31).

It is instructive to examine how two-digit numbers are added quantum mechanically. Let $a = a_1 2 + a_0$ and $b = b_1 2 + b_0$, for which $s = s_2 2^2 + s_1 2 + s_0$. It follows from Eq. (8.31) that

$$s_0 = a_0 \oplus b_0, \quad c_0 = a_0 b_0, \quad s_1 = a_1 \oplus b_1 \oplus a_0 b_0,$$
$$s_2 = c_1 = a_1 b_1 \oplus a_0 b_0 (a_1 \oplus b_1). \tag{8.32}$$

Now we want to implement a quantum circuit, which we call ADD(2), carrying out the above algebra. Our implementation is generalized to an n-qubit adder ADD(n) subsequently. We have to align qubits in such a way that the gate acts on them in a nice way. It turns out that the following order is the most convenient one in our implementation;

$$\text{ADD}(2)|0, a_0, b_0, 0, a_1, b_1, 0\rangle = |0, a_0, s_0, 0, a_1, s_1, s_2\rangle, \tag{8.33}$$

where the first and the second 0 are scratch qubits to deal with carry bits, while the third 0 is ultimately replaced by the sum s_2. We choose a_i to remain their input values while b_i $(i = 0, 1)$ are updated to s_i. We further decompose ADD(2) into a one-bit adder SUM and carry bit gate CARRY.

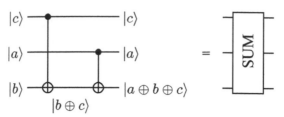

FIGURE 8.5
Quantum circuit to sum two binary bits and a carry bit. It will be also denoted
as a black box called SUM.

The action of SUM is defined as

$$\text{SUM}|c_{k-1}, a_k, b_k\rangle = |c_{k-1}, a_k, a_k \oplus b_k \oplus c_{k-1}\rangle, \qquad (8.34)$$

where c_{k-1} is the carry bit from the last bit. We drop all the subscripts
to simplify our notation hereafter. By recalling that the CNOT gate, whose
action is $U_{\text{CNOT}}|a, b\rangle = |a, a \oplus b\rangle$, works as a mod 2 adder, we immediately
find the circuit given in Fig. 8.5 implements the SUM gate. Let us make sure
it works OK. We abuse the notation so that CNOT_{ij} denotes the CNOT gate
with the control bit i and the target bit j whose action on a qubit $k(\neq i, j)$
is trivial. For example CNOT_{12} really means $\text{CNOT}_{12} \otimes I$. According to this
notation, we obtain

$$\text{SUM}|c, a, b\rangle = \text{CNOT}_{23}\text{CNOT}_{13}|c, a, b\rangle = \text{CNOT}_{23}|c, a, b \oplus c\rangle$$
$$= |c, a, a \oplus b \oplus c\rangle.$$

Next we consider the CARRY gate whose action is defined as

$$\text{CARRY}|c_{k-1}, a_k, b_k, 0\rangle = |c_{k-1}, a_k, a_k \oplus b_k, c_k = a_k b_k \oplus a_k c_{k-1} \oplus b_k c_{k-1}\rangle.$$
$$(8.35)$$

By considering that the carry bit c_k is 1 if and only if two or more of a_k, b_k
and c_{k-1} are 1, we find that the quantum circuit given in Fig. 8.6 implements
the CARRY gate. In fact, we verify

$$\text{CCNOT}_{13;4}\text{CNOT}_{23}\text{CCNOT}_{23;4}|c, a, b, 0\rangle$$
$$= \text{CCNOT}_{13;4}\text{CNOT}_{23}|c, a, b, ab\rangle = \text{CCNOT}_{13;4}|c, a, a \oplus b, ab\rangle$$
$$= |c, a, a \oplus b, ab \oplus c(a \oplus b)\rangle = |c, a, a \oplus b, ab \oplus ac \oplus bc\rangle$$
$$= \text{CARRY}|c, a, b, 0\rangle,$$

where we have dropped the subscripts for a, b and c for simplicity. Here
$\text{CCNOT}_{ij;k}$ stands for the CCNOT gate with the control bits i and j and
the target bit k. We have explicitly written nontrivial gates only, and all the
other qubits are acted by the unit matrix I as before.

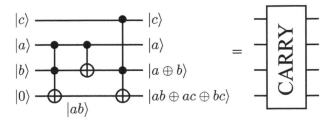

FIGURE 8.6
Quantum circuit which implements the CARRY gate.

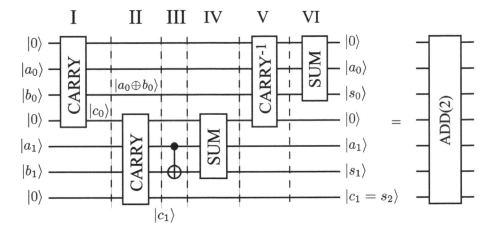

FIGURE 8.7
Quantum circuit to implement the ADD(2) gate, which adds two 2-bit numbers.

EXERCISE 8.5 Suppose the fourth input qubit is not 0 but a binary number c'. Show that the output state of the CARRY gate shown in Fig. 8.6 on this state is

$$\text{CARRY}|c, a, b, c'\rangle = |c, a, a \oplus b, ab \oplus ac \oplus bc \oplus c'\rangle. \tag{8.36}$$

Now we are ready to implement ADD(2) with these gates, which is explicitly shown in Fig. 8.7. Note that carry bits are required to compute s_i. Therefore we evaluate c_0 and c_1 first to compute s_1 and s_2 The bit b_0 is updated to $a_0 \oplus b_0$ during this process. We need to apply the inverse gate CARRY^{-1} on $|0, a_0, a_0 \oplus b_0, c_0\rangle$ to put it back to $|0, a_0, b_0, 0\rangle$ so that the state produces $|0, a_0, s_0, 0\rangle$ after applying the SUM gate of the layer VI. Thus the prescription is

Layer I Compute $c_0 = a_0 b_0$ from a_0 and b_0 with the first CARRY.

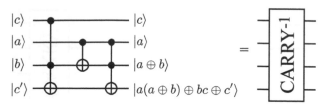

FIGURE 8.8
Inverse of the CARRY gate.

Layer II Compute $c_1 = s_2$ from a_1, b_1 and c_0 with the second CARRY.

Layer III If follows from Eq. (8.35) that the input b_1 is updated to $a_1 \oplus b_1$ in the layer II and we need to put this bit back to b_1 to evaluate s_1. This is done by the CNOT gate as $\mathrm{CNOT}|a_1, a_1 \oplus b_1\rangle = |a_1, a_1 \oplus b_1 \oplus a_1\rangle = |a_1, b_1\rangle$. We need b_1 for the next step.

Layer IV Apply the SUM gate on $|c_0, a_1, b_1\rangle$ to obtain $s_1 = a_1 \oplus b_1 \oplus c_0$.

Layer V We need to compute $s_0 = a_0 \oplus b_0$, for which we have to retrieve b_0. (Note that b_0 is mapped to $a_0 \oplus b_0$ in the second layer.) This is done by applying the CARRY^{-1} gate. This gate also puts the carry bit c_0 back to the initial state $|0\rangle$ for further use.

Layer VI Finally the SUM gate is applied on the input bits $|0, a_0, b_0\rangle$ to produce $|0, a_0, s_0\rangle$.

Before we verify the circuit in Fig. 8.7 indeed implements the ADD(2) gate, we look at the newly introduced CARRY^{-1} gate in some details. We obtain, from the implementation of the CARRY gate in Fig. 8.6, that

$$\mathrm{CARRY}^{-1} = (\mathrm{CCNOT}_{13;4}\mathrm{CNOT}_{23}\mathrm{CCNOT}_{23;4})^{\dagger}$$
$$= \mathrm{CCNOT}^{\dagger}_{23;4}\mathrm{CNOT}^{\dagger}_{23}\mathrm{CCNOT}^{\dagger}_{13;4}$$
$$= \mathrm{CCNOT}_{23;4}\mathrm{CNOT}_{23}\mathrm{CCNOT}_{13;4}, \qquad (8.37)$$

where we noted that $\mathrm{CCNOT}^{\dagger} = \mathrm{CCNOT}$ and $\mathrm{CNOT}^{\dagger} = \mathrm{CNOT}$. Therefore CARRY^{-1} is obtained by reversing the order of the constituent controlled gates in the CARRY gate as depicted in Fig. 8.8.

EXERCISE 8.6 Show explicitly that

$$\mathrm{CARRY}^{-1}|c, a, b, c'\rangle = |c, a, a \oplus b, a(a \oplus b) \oplus (bc) \oplus c'\rangle. \qquad (8.38)$$

Now we are ready to verify the implementation of the ADD(2) in Fig. 8.7. We denote the unitary matrix correspoinding to the kth layer by U_k. We have

$$\mathrm{ADD}(2)|0, a_0, b_0, 0, a_1, b_1, 0\rangle$$

$$= U_{\mathrm{VI}}U_{\mathrm{V}}U_{\mathrm{IV}}U_{\mathrm{III}}U_{\mathrm{II}}U_{\mathrm{I}}|0, a_0, b_0, 0, a_1, b_1, 0\rangle$$
$$= U_{\mathrm{VI}}U_{\mathrm{V}}U_{\mathrm{IV}}U_{\mathrm{III}}U_{\mathrm{II}}|0, a_0, a_0 \oplus b_0, a_0 b_0, a_1, b_1, 0\rangle$$
$$= U_{\mathrm{VI}}U_{\mathrm{V}}U_{\mathrm{IV}}U_{\mathrm{III}}|0, a_0, a_0 \oplus b_0, a_0 b_0, a_1, a_1 \oplus b_1, a_1 b_1 \oplus a_1 a_0 b_0 \oplus b_1 a_0 b_0\rangle$$
$$= U_{\mathrm{VI}}U_{\mathrm{V}}U_{\mathrm{IV}}|0, a_0, a_0 \oplus b_0, a_0 b_0, a_1, b_1, a_1 b_1 \oplus a_1 a_0 b_0 \oplus b_1 a_0 b_0\rangle$$
$$= U_{\mathrm{VI}}U_{\mathrm{V}}|0, a_0, a_0 \oplus b_0, a_0 b_0, a_1, a_1 \oplus b_1 \oplus a_0 b_0, a_1 b_1 \oplus a_1 a_0 b_0 \oplus b_1 a_0 b_0\rangle$$
$$= U_{\mathrm{VI}}|0, a_0, b_0, 0, a_1, a_1 \oplus b_1 \oplus a_0 b_0, a_1 b_1 \oplus a_1 a_0 b_0 \oplus b_1 a_0 b_0\rangle$$
$$= |0, a_0, a_0 \oplus b_0, 0, a_1, a_1 \oplus b_1 \oplus a_0 b_0, a_1 b_1 \oplus a_1 a_0 b_0 \oplus b_1 a_0 b_0\rangle.$$

The last line of the above equation is identified with $|0, a_0, s_0, 0, a_1, s_1, s_2\rangle$.

The adder for n-bit numbers, which we call ADD(n), is obtained immediately by generalizing the above implementation. Figure 8.9 shows the ADD(n) gate. The most significant sum bit s_n of $a + b$ is evaluated first. We need to calculate all the carry bits c_i for this purpose, and the left layers of the gates before s_n is obtained are devoted for calculating the carry bits. Then we reverse all the operations, except for the most significant bit, to restore $\{a_i\}$ and $\{b_i\}$ with which $\{s_i\}$ are evaluated as $b_i \mapsto s_i = a_i \oplus b_i \oplus c_{i-1}$. The readers should verify the circuit indeed implements n-bit addition.

8.6.2   Modular Adder

Let us consider the **modular adder**, denoted MODADD(n), which evaluates $a + b \bmod N$ for given inputs a and b, where $a, b < N$. We need to introduce an n-qubit subtraction circuit to this end. It is shown that ADD(n)$^{-1}$, the inverse of ADD(n) does the job. Instead of giving a general proof, we will be satisfed with demonstration of this statement for the simplest case, in which both a and b are 2-bit numbers, $a = a_1 2 + a_0, b = b_1 2 + b_0$.

Subtraction is carried out by introducing a **two's complement** in a similar way as in a classical computer. Let $a = a_{n-1} 2^{n-1} + \ldots a_1 2 + a_0$ be a positive n-bit number. A negative number $-a$ is stored in a computer memory as its two's complement, which is defined as $2^{n+1} - a$. By noting that $2^{n+1} = (2^n + 2^{n-1} + \ldots + 2 + 1) + 1$, we obtain

$$2^{n+1} - a = 2^n + (1 - a_{n-1})2^{n-1} + \ldots (1 - a_1)2 + (1 - a_0) + 1. \tag{8.39}$$

Note that $2^{n+1} - a$ is an $(n+1)$-bit number for a postive number a. In summary, the two's complement of a number $a = a_{n-1}a_{n-2}\ldots a_1 a_0$ is obtained by flipping each bit as $a_k \to 1 - a_k$, adding 2^n and finally adding 1. Let $a = 2$, in decimal notation, for example. Then its binary expression is 10 and the two's complement of -2 is $101 + 1 = 110$.

EXERCISE 8.7 Let $n = 3$. Find the two's complements of negative numbers from -7 to -1.

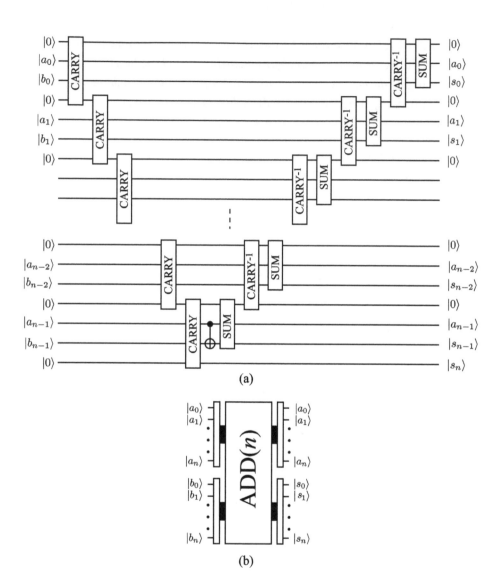

(a)

(b)

FIGURE 8.9

(a) Quantum circuit to implement the ADD(n) gate, which adds two n-bit numbers a and b. The result is encoded in the qubits $\{|s_k\rangle\}$. (b) The black box representation of ADD(n). Note the order of the input bits and the output bits. We have explicitly added $|a_n\rangle = |b_n\rangle = |0\rangle$.

Let us concentrate on the case in which $n = 2$. Then the two's complement of $-a$ is $2^3 - a = 2^2 + (1 - a_1)2 + (1 - a_0) + 1$. It is also written as

$$2^3 - a = 2^2 + (a_0 \oplus a_1)2 + a_0$$
$$= (a_0 \oplus a_1 \oplus a_0 a_1)2^2 + (a_0 \oplus a_1)2 + a_0, \qquad (8.40)$$

where we noted that $a > 0$ and $a_0 \oplus a_1 \oplus (a_0 a_1) = 1$. Let us evaluate $b - a$, where both a and b are 2-bit numbers. We carry out this subtraction as $b + (2^3 - a)$ by making use of the two's complement of $-a$. We find

$$b + (2^3 - a) = (b_1 2 + b_0) + \left[(a_0 \oplus a_1 \oplus a_0 a_1)2^2 + (a_0 \oplus a_1)2 + a_0 \right]$$
$$= \left[(a_0 \oplus a_1 \oplus a_0 a_1) \oplus (a_0 \oplus a_1)b_1 \oplus a_0 b_0 (a_0 \oplus a_1 \oplus b_1) \right]2^2$$
$$+ (a_0 \oplus a_1 \oplus b_1 \oplus a_0 b_0)2 + (a_0 \oplus b_0)$$
$$= s_2 2^2 + s_1 2 + s_0, \qquad (8.41)$$

where

$$s_2 = (a_0 \oplus a_1 \oplus a_0 a_1) \oplus (a_0 \oplus a_1)b_1 \oplus a_0 b_0 (a_0 \oplus a_1 \oplus b_1)$$
$$= a_0 \oplus a_1 \oplus a_0 a_1 \oplus a_0 b_1 \oplus a_1 b_1 \oplus a_0 b_0 \oplus a_0 a_1 b_1 \oplus a_0 b_0 b_1,$$
$$s_1 = a_0 \oplus a_1 \oplus b_1 \oplus a_0 b_0, \qquad (8.42)$$
$$s_0 = a_0 \oplus b_0.$$

Now we show that

$$\text{ADD}(2)^{-1} |0, a_0, b_0, 0, a_1, b_1, 0\rangle = |0, a_0, s_0, 0, a_1, s_1, s_2\rangle. \qquad (8.43)$$

In fact, we verify

$$\text{ADD}(2)^{-1} |0, a_0, b_0, 0, a_1, b_1, 0\rangle$$
$$= U_{\text{I}} U_{\text{II}} U_{\text{III}} U_{\text{IV}} U_{\text{V}} U_{\text{VI}} |0, a_0, b_0, 0, a_1, b_1, 0\rangle$$
$$= U_{\text{I}} U_{\text{II}} U_{\text{III}} U_{\text{IV}} U_{\text{V}} |0, a_0, a_0 \oplus b_0, 0, a_1, b_1, 0\rangle$$
$$= U_{\text{I}} U_{\text{II}} U_{\text{III}} U_{\text{IV}} |0, a_0, b_0, a_0 \oplus a_0 b_0, a_1, b_1, 0\rangle$$
$$= U_{\text{I}} U_{\text{II}} U_{\text{III}} |0, a_0, b_0, a_0 \oplus a_0 b_0, a_1, a_1 \oplus b_1 \oplus a_0 \oplus a_0 b_0, 0\rangle$$
$$= U_{\text{I}} U_{\text{II}} |0, a_0, b_0, a_0 \oplus a_0 b_0, a_1, b_1 \oplus a_0 \oplus a_0 b_0, 0\rangle$$
$$= U_{\text{I}} |0, a_0, b_0, a_0 \oplus a_0 b_0, a_1, a_1 \oplus b_1 \oplus a_0 b_0,$$
$$\quad a_0 \oplus a_1 \oplus a_0 a_1 \oplus a_0 b_0 \oplus a_0 b_1 \oplus a_1 b_1 \oplus a_0 b_0 a_1 \oplus a_0 b_0 b_1, 0\rangle$$
$$= |0, a_0, a_0 \oplus b_0, 0, a_1, a_0 \oplus a_1 \oplus b_1 \oplus a_0 b_0,$$
$$\quad a_0 \oplus a_1 \oplus a_0 a_1 \oplus a_0 b_1 \oplus a_1 b_1 \oplus a_0 b_0 \oplus a_0 a_1 b_0 \oplus a_0 b_0 b_1\rangle$$
$$= |0, a_0, s_0, 0, a_1, s_1, s_2\rangle$$

where we noticed that $U_k^{\dagger} = U_k$ in all the layers in the circuit. This shows that the action of $\text{ADD}(2)^{-1}$ yields $b + (2^{n+1} - a)$ as promised.

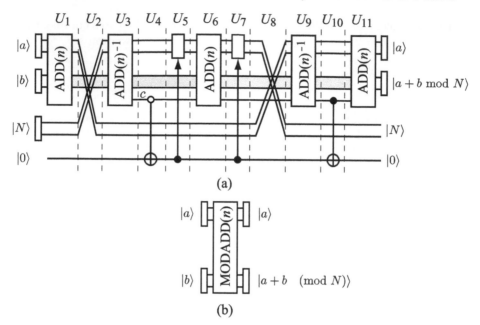

(a)

(b)

FIGURE 8.10

(a) Modular adder which computes $a + b \bmod N$. The gray line shows the information flow of the second register, corresponding to the input b and the output $a + b \bmod N$. U_k denotes the unitary operation of the kth layer. The white circle in U_4 denotes a negated control node, which flips the bottom qubit when the control bit is 0, while does nothing when it is 1. (b) Modular adder is symbolically denoted as MODADD(n). The third register $|N\rangle$ and the carry qubit $|0\rangle$ are omitted.

The n-qubit subtraction is similarly obtained by inverting ADD(n) as

$$\text{ADD}(n)^{-1}|0, a_0, b_0, 0, a_1, b_1, 0, \ldots, a_{n-1}, b_{n-1}, 0\rangle$$
$$= |0, a_0, s_0, 0, a_1, s_1, 0, \ldots, a_{n-1}, s_{n-1}, s_n\rangle. \tag{8.44}$$

The output bits $s_0 \sim s_{n-1}$ represent $b - a$, while the last digit s_n is 1 if $b - a < 0$ and 0 if $b - a > 0$.

EXERCISE 8.8 Let a and b be 2-bit numbers. Verify that s_2 in Eq. (8.42) is 1 when $b < a$, while 0 when $b > a$.

Now we are ready to implement the modular adder. We find the quantum circuit depicted in Fig. 8.10 indeed performs modular addition $a + b \bmod N$, where a, b and N are n-bit numbers satisfying $0 < a, b < N$. Therefore, we have either $0 < a + b < N$ or $N \leq a + b < 2N$. We write the input state as $|a, b, N, 0\rangle$, where the last qubit $|0\rangle$ will be used as a scratch space. Let us verify its operations.

- The gate U_1 is an ordinary adder;

$$U_1|a, b, N, 0\rangle = |a, a + b, N, 0\rangle. \tag{8.45}$$

- The gate U_2 swaps the first and the third registers;

$$U_2|a, a + b, N, 0\rangle = |N, a + b, a, 0\rangle. \tag{8.46}$$

- The gate U_3 subtracts N from $a + b$;

$$U_3|N, a + b, a, 0\rangle = |N, a + b - N, c, a, 0\rangle. \tag{8.47}$$

The $(n + 1)$st output bit c of the second register is 1 when $a + b < N$ and 0 when $a + b \geq N$.

- The gate U_4 changes the scratch bit $|0\rangle$ to $|1\rangle$ if $c = 0$, while it is left unchanged when $c = 1$;

$$U_4|N, a + b - N, c, a, 0\rangle = |N, a + b - N, c, a, 1 - c\rangle. \tag{8.48}$$

- The gate U_5 resets the first register $|N\rangle$ to $|0\rangle$ when the temporary qubit is $|1\rangle$ while it remains in the state $|N\rangle$ when it is $|0\rangle$;

$$U_5|N, a + b - N, c, a, 1 - c\rangle = \begin{cases} |N, a + b - N, c, a, 0\rangle & (c = 1) \\ |0, a + b - N, c, a, 1\rangle & (c = 0) \end{cases}$$
$$= |cN, a + b - N, c, a, 1 - c\rangle. \tag{8.49}$$

Unitarity of U_5 requires that $U_5|0, a+b-N, c, a, 1\rangle = |N, a+b-N, c, a, 1\rangle$.

- The gate U_6 is a simple adder;

$$U_6|cN, a+b-N, c, a, 1-c\rangle = |cN, a+b-(1-c)N, 1-c, a, 1-c\rangle. \tag{8.50}$$

Note that the carry bit c has been flipped.

- The gate U_7 is the same as the layer 5. The first register is mapped to $|N\rangle$ when $c = 0$, and it remains in $|N\rangle$ when $c = 1$;

$$U_7|cN, a+b-(1-c)N, 1-c, a, 1-c\rangle = |N, a+b-(1-c)N, 1-c, a, 1-c\rangle. \tag{8.51}$$

- The gate U_8 swaps $|a\rangle$ and $|N\rangle$ so that

$$U_8|N, a+b-(1-c)N, 1-c, a, 1-c\rangle = |a, a+b-(1-c)N, 1-c, N, 1-c\rangle. \tag{8.52}$$

- The gate U_9 subtracts a from $a + b - (1 - c)N$;

$$U_9|a, a + b - (1 - c)N, 1 - c, N, 1 - c\rangle = |a, b - (1 - c)N, 1 - c, N, 1 - c\rangle.$$
(8.53)

- The gate U_{10} transforms the temporary qubit to $(1 - c) \oplus (1 - c) = 0$;

$$U_{10}|a, b - (1 - c)N, 1 - c, N, 1 - c\rangle = |a, b - (1 - c)N, 1 - c, N, 0\rangle. \quad (8.54)$$

- The final gate U_{11} adds the first and the second registers;

$$U_{11}|a, b - (1 - c)N, 1 - c, N, 0\rangle = |a, a + b - (1 - c)N, 1 - c, N, 0\rangle. \quad (8.55)$$

In case $a + b \geq N$, for which $c = 0$, the second register in the output state is $|a + b - N\rangle$. If, in contrast, $a + b < N$, for which $c = 1$, the second register is $|a + b\rangle$. These results are conveniently written as $|a + b \bmod N\rangle$.

We call this quantum circuit an n-qubit modular adder, denoted MODADD(n); see Fig. 8.10 (b).

8.6.3   Modular Multiplexer

It turnes out that a **controlled-modular multiplexer**

$$\text{CMODMULTI}(n)|c, x, 0, 0\rangle = \begin{cases} |c, x, 0, ax \bmod N\rangle & (c = 1) \\ |c, x, 0, x\rangle & (c = 0) \end{cases} \quad (8.56)$$

is requried to construct the modular exponential gate, which computes $a^x \bmod N$, rather than the ordinary modular multiplexer circuit.

Fig. 8.11 depicts the **controlled-modular multiplexer circuit**. The control bit is denoted as $|c\rangle$, while two registers are initially set to $|x\rangle$ and $|0\rangle$. There is also a temporary register which is also set initially to $|0\rangle$. We need to evaluate $ax \bmod N$ for various x with the numbers a and N fixed. Therefore a and N are hardwired as parts of the circuit, while x is one of the input parameters. Let us verify it works as expected.

Suppose $|c\rangle = |1\rangle$ first. Let $x = x_{n-1}2^{n-1} + \ldots x_1 2 + x_0$ and $a = a_{n-1}2^{n-1} + \ldots a_1 2 + a_0$ be binary expressions of positive integers x and a, respectively. The product of these numbers is

$$ax = ax_{n-1}2^{n-1} + \ldots + ax_1 2 + ax_0 = \sum_{x_k = 1} a2^k.$$

The RHS tells us that the $ax \bmod N$ is obtained by adding $a2^k \bmod N$ with respect to those k for which $x_k = 1$. The modular adder MODADD(n) in Fig. 8.11 adds $a2^i \bmod N$ for such i as $x_i = 1$ to the second register whose initial state is $|0\rangle$. Each MODADD(n) is accompanied with a pair of controlled

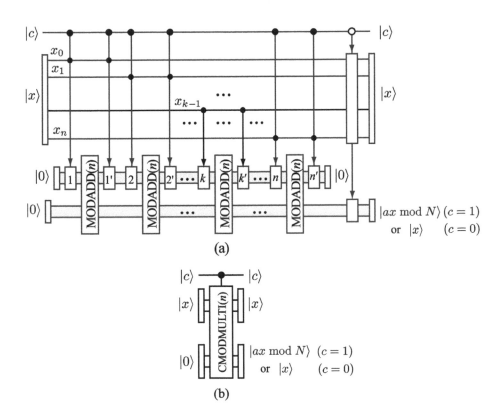

(a)

(b)

FIGURE 8.11
(a) Quantum circuit of an n-qubit controlled-modular multiplexer CMODMULTI(n). There are n layers of MODADD(n), where the kth layer adds $a2^{k-1} \bmod N$ to the second register when $c = 1$ and $x_{k-1} = 1$. The numbers a and N are fixed and are hardwired. The output of the second register is $ax \bmod N$. This circuit is denoted as (b), where the temporary register has no external input and output ports and is not shown explicitly.

gates. Let us concentrate on the kth MODADD(n) gate. The first controlled gate, denoted as k in Fig. 8.11 associated with the kth MODADD(n) gate stores $a2^{k-1}$ in the temporary register if $x_{k-1} = 1$, while it stores 0 if $x_{k-1} = 0$. MODADD(n) adds $x_{k-1}2^{k-1}$ to $\sum_{i=0}^{k-2} ax_i 2^i$ and updates the second register to $\sum_{i=0}^{k-1} ax_i 2^i$. The second controlled gate immediately after the kth MODADD(n), denoted as k', undoes the operation of the first controlled gate when $x_{k-1} = 1$, so that the first register is now put back to 0 for recycled use for the $(k+1)$st MODADD(n). Both controlled gates leave the temporary gate to $|0\rangle$ when $x_{k-1} = 0$. In this way, we finally obtain $ax \bmod N$ in the second register.

Let $c = 0$ next. Then the temporary register remains in $|0\rangle$ and each MODADD(n) just adds 0 to the second register. We will obtain $|0, x, 0\rangle$ after the nth MODADD(n) has been applied. We need to act a set of controlled gates to copy the first register $|x\rangle$ to the second register so that the output is $|0, x, x\rangle$. This is done by n CNOT gates, which act on each pair of qubits of the first and the second registers as $U_{\text{CNOT}}|0, x_i, 0_i\rangle = |0, x_i, 0_i \oplus x_i\rangle = |0, x_i, x_i\rangle$.

In summary, the CMODMULTI gate works as

$$\text{CMODMULTI}|c, x, 0\rangle = \begin{cases} |1, x, ax \bmod N\rangle & (c = 1) \\ |0, x, x\rangle & (c = 0). \end{cases} \tag{8.57}$$

8.6.4 Modular Exponential Function

Let $a = a_{n-1}2^{n-1} + \ldots + a_1 2 + a_0$ and $x = x_{n-1}2^{n-1} + \ldots + x_1 2 + x_0$ be two natural numbers. We need to implement a quantum circuit which outputs $a^x \bmod N$.

Figure 8.12 shows the quantum circuit which implements the modular exponential function. Let us verify how it works to produce the correct result.

Let $x = x_{n-1}2^{n-1} + x_{n-2}2^{n-2} + \ldots + x_1 2 + x_0$. Then the power a^x is expressed as

$$a^x = a^{x_{n-1}2^{n-1}} \times a^{x_{n-2}2^{n-2}} \times \ldots a^{x_1 2} \times a^{x_0} = \prod_{\substack{k=0 \\ x_k=1}}^{n-1} a^{2^k}. \tag{8.58}$$

Namely, terms of the form a^{2^k} are multiplied with respect to k for which $x_k = 1$ to evaluate a^x. The modular exponential $a^x \bmod N$ is obtained by multiplying a^{2^k} from $k = 0$ to $k = n - 1 \bmod N$ for those k satisfying $x_k = 1$.

- Let us look at the first layer composed of gates I and II in Fig. 8.12. The output of the gate I is

$$\text{CMODMULTI}|x, 1, 0\rangle = |x, 1, a^{x_0} \bmod N\rangle.$$

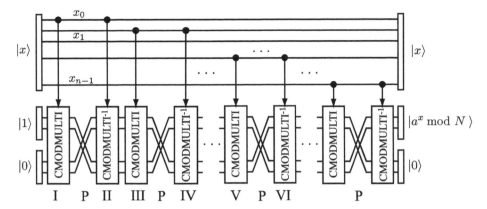

FIGURE 8.12
Quantum cirucuit which implements the modular exponential function $a^x \bmod N$. The numbers a and N are hardwired. The number x is initially input to the first register, while the second register is set to $|1\rangle$. The result of the modular exponential function is stored in the second register. There is an extra temporary register which is set initially to $|0\rangle$.

The output states $|1\rangle$ and $|a^{x_0} \bmod N\rangle$ are exchanged by the permutation gate P before they are fed to the gate II so that

$$P|x, 1, a^{x_0} \bmod N\rangle = |x, a^{x_0} \bmod N, 1\rangle.$$

Now the gate II acts on this state is

$$\text{CMODMULTI}^{-1}|x, a^{x_0} \bmod N, 1\rangle = |x, a^{x_0} \bmod N, 0\rangle.$$

- The next layer is comprised of two gates: III and IV. The output of the gate III is

$$\text{CMODMULTI}|x, a^{x_0} \bmod N, 0\rangle = |x, 1, a^{x_1 2} a^{x_0} \bmod N\rangle,$$

which is followed by P and then CMODMULTI^{-1} to produce

$$\text{CMODMULTI}^{-1}P|x, 1, a^{x_1 2} a^{x_0} \bmod N\rangle = |x, a^{x_1 2} a^{x_0} \bmod N, 0\rangle.$$

- The factor $a^{x_k 2^k}$ is multiplied to $\prod_{j=0}^{k-1} a^{x_j 2^j} \bmod N$ each time a new layer comprised of a pair of gates CMODMULTI and CMODMULTI^{-1} is applied. Eventually we obtain the output

$$\text{MODEXP}|x, 1, 0\rangle = |x, a^x \bmod N, 0\rangle.$$

8.6.5 Computational Complexity of Modular Exponential Circuit

We evaluate the complexity of the modular exponential circuit before we close this chapter. The evaluation is divided into several steps.

- Let us first look at ADD(n) given in Fig. 8.9. It follows from this figure that ADD(n) requires n SUM, $n - 1$ CARRY^{-1}, n CARRY and one CNOT. Therefore the complexity $T_{\text{ADD}(n)}$ of ADD(n) is

$$T_{\text{ADD}(n)} = nT_S + (2n - 1)T_C + 1, \qquad (8.59)$$

where T_S and T_C are numbers of elementary gates in SUM and CARRY, which are clearly independent of n.

- Figure 8.10 shows that the modular adder MODADD(n) is composed of three ADD(n), two ADD(n)$^{-1}$, two permutation gates, two controlled gates and four CNOT gates. Therefore the complexity is

$$T_{\text{MODADD}(n)} = 5T_{\text{ADD}(n)} + 2T_P + 2T_U + 4, \qquad (8.60)$$

where T_P and T_U are numbers of elementary gates in P and the controlled-U gate, respectively. It can be shown that these gates can be implemented with the Toffoli gates and each Toffoli gate is implemented with a polynomial number of elementary gates. The power of polynomial depends on the actual implementation. Implementation with a minimal number of qubits requires $O(n^2)$ elementary gates, but it may be reduced to $O(n)$ if some extra qubits are added.

- The controlled modular multiplexer depicted in Fig. 8.11 requires n MODADD(n), n controlled gates of type k (see Fig. 8.11) and n controlled gates of type k', two CNOT gates and one "copy" gate which works as $|x0\rangle \rightarrow |xx\rangle$ when the control bit is $c = 0$. The copy gate, controlled gates of types k and k' are all implemented with Toffoli gates. The number of elementary gates required to implement CMODMULTI(n) is therefore

$$T_{\text{CMODMULT}(n)} = nT_{\text{MODADD}(n)} + nT_k + nT_{k'} + T_{\text{copy}} + 2, \qquad (8.61)$$

where T_k and $T_{k'}$ are the numbers of elementary gates in the controlled gates of types k and k', respectively, and T_{copy} is the number of elementary gates in the copy gate.

- The modular exponential gate MODEXP(n) depicted in Fig. 8.12 requires n CMODMULTI(n), n CMODMULTI(n)$^{-1}$ and n exchange gates P. The complexity is

$$T_{\text{MODEXP}(n)} = 2nT_{\text{CMODMULTI}(n)} + nT_P. \qquad (8.62)$$

In summary, the number of elementary gates in the modular exponential circuit MODEXP(n) is of polynomial order in n. The maximum power of the polynomial depends on actual circuit implementation and algorithms employed to design the circuit. It may happen that adding extra qubits makes the maximum power smaller. Alternatively, the number of qubits required to implement a given algorithm may be reduced by sacrificing the number of elementary gates.

It was also shown in Chapter 6 that the quantum Fourier transform circuit is implemented with $O(n^2)$ number of elementary gates. Therefore prime number factorization for a number $N \sim 2^n$ is carried out with a polynomial number of steps in n in contrast with a classical algorithm which requires an exponentially large number of steps.

References

[1] P. W. Shor, *Proc. of 35th Annual Symposium on the Foundations of Computer Science*, 124, IEEE Computer Society Press, Los Alamitos (1994).

[2] P. W. Shor, SIAM J. Comput., **26**, 1484 (1997).

[3] S. J. Lomonaco, Jr. in *Quantum Computation: A Grand Mathematical Challenge for the Twenty-First Century and the Millennium*, ed. S. J. Lomonaco, Jr., AMS (2002). e-print quant-ph/0010034.

[4] Y. Uesaka, *Mathematical Principle of Quantum Computation*, Corona Publishing, Tokyo, in Japanese (2000).

[5] See F. Bornemann, Notices of the AMS, **50**, 545 (2003), for example.

[6] R. L. Rivest, A. Shamir and L. M. Adleman, Comm. ACM, **21**, 120 (1978).

[7] V. Vedral, A. Barenco and A. Ekert, Phys. Rev. A **54**, 147 (1996).

9

Decoherence

A quantum system is always in interaction with its environment. This interaction inevitably alters the state of the quantum system, which causes loss of information encoded in the system. The system under consideration is not a *closed* system any more when interaction with the outside world is in action. We formulate the theory of an *open* quantum system in this chapter by regarding the combined system of the quantum system and its environment as a closed system and subsequently trace out the environmental degrees of freedom. Let ρ_S and ρ_E be the density matrices of the system and the environment, respectively. Even when the initial state is an uncorrelated state $\rho_S \otimes \rho_E$, the system-environment interaction entangles the total system so that the total state develops to an inseparable entangled state in general. Decoherence is a process in which environment causes various changes in the quantum system, which manifests itself as undesirable noise.

We study quantum error correcting codes (QECC), which is one of the strategies to fight against decoherence, in the next chapter.

9.1   Open Quantum System

Let us start our exposition with some mathematical background materials. We will closely follow [1] and [2].

We deal with general quantum states described by density matrices. We are interested in a general evolution of a quantum system, which is described by a powerful tool called a **quantum operation**. One of the simplest quantum operations is a unitary time evolution of a closed system. Let ρ_S be a density matrix of a closed system at $t = 0$ and let $U(t)$ be the time evolution operator. Then the corresponding quantum map $\mathcal{E}$ is defined as

$$\mathcal{E}(\rho_S) = U(t)\rho_S U(t)^\dagger. \tag{9.1}$$

One of our primary aims in this section is to generalize this map to cases of open quantum systems.

9.1.1   Quantum Operations and Kraus Operators

Suppose a system of interest is coupled with its **environment**. We must specify the details of the environment and the coupling between the system and
the environment to study the effect of the environment on the behavior of the
system. Let H_S, H_E and H_{SE} be the system Hamiltonian, the environment
Hamiltonian and their interaction Hamiltonian, respectively. We assume the
system-environment interaction is weak enough so that this separation into
the system and its environment makes sense. To avoid confusion, we often
call the system of interest a **principal system**. The total Hamiltonian H_T
is then

$$H_T = H_S + H_E + H_{SE}. \tag{9.2}$$

Correspondingly, we denote the system Hilbert space and the environment
Hilbert space as $\mathcal{H}_S$ and $\mathcal{H}_E$, respectively, and the total Hilbert space as
$\mathcal{H}_T = \mathcal{H}_S \otimes \mathcal{H}_E$. The condition of weak system-environment interaction
may be lifted in some cases. Let us consider a qubit propagating through
a noisy quantum channel, for example. "Propagating" does not necessarily
mean propagating in space. The qubit may be spatially fixed and subject
to time-dependent noise. When the noise is localized in space and time, the
input and the output qubit states belong to a well-defined Hilbert space $\mathcal{H}_S$,
and the above separation of the Hamiltonian is perfectly acceptable even for
strongly interacting cases. We consider, in the following, how the principal
system state ρ_S at $t = 0$ evolves in time in the presence of its environment.
A map which describes a general change of the state from ρ_S to $\mathcal{E}(\rho_S)$ is
called a **quantum operation**. We have already noted that the unitary time
evolution is an example of a quantum operation. Other quantum operations
include state change associated with measurement and state change due to
noise. The latter quantum map is our primary interest in this chapter.

The state of the total system is described by a density matrix ρ. Suppose
ρ is uncorrelated initally at time $t = 0$,

$$\rho(0) = \rho_S \otimes \rho_E, \tag{9.3}$$

where ρ_S (ρ_E) is the initial density matrix of the principal system (environment). The total system is assumed to be closed and to evolve with a unitary
matrix $U(t)$ as

$$\rho(t) = U(t)(\rho_S \otimes \rho_E)U(t)^\dagger. \tag{9.4}$$

Note that the resulting state is not a tensor product state in general. We are
interested in extracting information on the state of the system at some later
time $t > 0$.

Even under these circumstances, however, we may still define the system
density matrix $\rho_S(t)$ by taking partial trace of $\rho(t)$ over the environment
Hilbert space as

$$\rho_S(t) = \text{tr}_E[U(t)(\rho_S \otimes \rho_E)U(t)^\dagger]. \tag{9.5}$$

We may forget about the environment by taking a trace over $\mathcal{H}_E$. This is an example of a quantum operation, $\mathcal{E}(\rho_S) = \rho_S(t)$. Details of the environment dynamics are made irrelevant at this stage. Let $\{|e_j\rangle\}$ be a basis of the system Hilbert space and $\{|\varepsilon_a\rangle\}$ be that of the environment Hilbert space. We may take the basis of $\mathcal{H}_T$ to be $\{|e_j\rangle \otimes |\varepsilon_a\rangle\}$. The initial density matrices may be written as

$$\rho_S = \sum_j p_j |e_j\rangle\langle e_j|, \ \rho_E = \sum_a r_a |\varepsilon_a\rangle\langle \varepsilon_a|.$$

Action of the time evolution operator on a basis vector of $\mathcal{H}_T$ is explicitly written as

$$U(t)|e_j, \varepsilon_a\rangle = \sum_{k,b} U_{kb;ja} |e_k, \varepsilon_b\rangle, \tag{9.6}$$

where $|e_j, \varepsilon_a\rangle = |e_j\rangle \otimes |\varepsilon_a\rangle$, for example. Using this expression, the density matrix $\rho(t)$ is written as

$$U(t)(\rho_S \otimes \rho_E)U(t)^\dagger = \sum_{j,a} p_j r_a U(t)|e_j, \varepsilon_a\rangle\langle e_j, \varepsilon_a|U(t)^\dagger$$

$$= \sum_{j,a,k,b,l,c} p_j r_a U_{kb;ja} |e_k, \varepsilon_b\rangle\langle e_l, \varepsilon_c|U^*_{lc;ja}. \tag{9.7}$$

The partial trace over $\mathcal{H}_E$ is carried out to yield

$$\rho_S(t) = \mathrm{tr}_E[U(t)(\rho_S \otimes \rho_E)U(t)^\dagger]$$

$$= \sum_{j,a,k,b,l} p_j r_a U_{kb;ja} |e_k\rangle\langle e_l|U^*_{lb;ja}$$

$$= \sum_{j,a,b} p_j \left(\sum_k \sqrt{r_a} U_{kb;ja}|e_k\rangle\right) \left(\sum_l \sqrt{r_a}\langle e_l|U^*_{lb;ja}\right). \tag{9.8}$$

To write down the quantum operation in a closed form, we assume the initial environment state is a pure state, which we take, without loss of generality, $\rho_E = |\varepsilon_0\rangle\langle\varepsilon_0|$. Even when ρ_E is a mixed state, we may always complement $\mathcal{H}_E$ with a fictitious Hilbert space to "purify" ρ_E (see §2.4.2). With this assumption, $\rho_S(t)$ is written as

$$\rho_S(t) = \mathrm{tr}_E[U(t)(\rho_S \otimes |\varepsilon_0\rangle\langle\varepsilon_0|)U(t)^\dagger]$$

$$= \sum_a (I \otimes \langle\varepsilon_a|)U(t)(\rho_S \otimes |\varepsilon_0\rangle\langle\varepsilon_0|)|U(t)^\dagger(I \otimes |\varepsilon_a\rangle)$$

$$= \sum_a (I \otimes \langle\varepsilon_a|)U(t)(I \otimes |\varepsilon_0\rangle)\rho_S(I \otimes \langle\varepsilon_0|)U(t)^\dagger(I \otimes |\varepsilon_a\rangle).$$

We will drop $I\otimes$ from $I \otimes \langle\varepsilon_a|$ hereafter, whenever it does not cause confusion. Let us define the **Kraus operator** $E_a(t) : \mathcal{H}_S \to \mathcal{H}_S$ by

$$E_a(t) = \langle\varepsilon_a|U(t)|\varepsilon_0\rangle. \tag{9.9}$$

Then we may write

$$\mathcal{E}(\rho_S) = \rho_S(t) = \sum_a E_a(t)\rho_S E_a(t)^\dagger. \qquad (9.10)$$

This is called the **operator-sum representation (OSR)** of a quantum operation $\mathcal{E}$. Note that $\{E_a\}$ satisfies the **completeness relation**

$$\left[\sum_a E_a(t)^\dagger E_a(t)\right]_{kl} = \left[\sum_a \langle\varepsilon_0|U(t)^\dagger|\varepsilon_a\rangle\langle\varepsilon_a|U(t)|\varepsilon_0\rangle\right]_{kl}$$
$$= \langle\varepsilon_0|U(t)^\dagger U(t)|\varepsilon_0\rangle_{lk} = I_{kl} = \delta_{kl}, \qquad (9.11)$$

where I is the unit matrix in $\mathcal{H}_S$. This is equivalent with the trace-preserving property of $\mathcal{E}$ as

$$1 = \text{tr}_S\rho_S(t) = \text{tr}_S(\mathcal{E}(\rho_S)) = \text{tr}_S\left(\sum_a E_a^\dagger E_a\rho_S\right)$$

for any $\rho_S \in \mathcal{S}(\mathcal{H}_S)$. The completeness relation and trace-preserving property are satisfied since our total system is a closed system. A general quantum map does not necessarily satisfy these properties [3].

At this stage, it turns out to be useful to relax the condition that $U(t)$ be a time evolution operator. Instead, we assume U to be any operator including an arbitrary unitary gate. Let us consider a two-qubit system on which the CNOT gate acts. Suppose the principal system is the control qubit while the environment is the target qubit. Then we find

$$E_0 = (I \otimes \langle 0|)U_{\text{CNOT}}(I \otimes |0\rangle) = P_0, \quad E_1 = (I \otimes \langle 1|)U_{\text{CNOT}}(I \otimes |0\rangle) = P_1,$$

where $P_i = |i\rangle\langle i|$, and consequently

$$\mathcal{E}(\rho_S) = P_0\rho_S P_0 + P_1\rho_S P_1 = \rho_{00}P_0 + \rho_{11}P_1 = \begin{pmatrix} \rho_{00} & 0 \\ 0 & \rho_{11} \end{pmatrix}, \qquad (9.12)$$

where

$$\rho_S = \begin{pmatrix} \rho_{00} & \rho_{01} \\ \rho_{10} & \rho_{11} \end{pmatrix}.$$

We interchangeably regard U as a time evolution operator or as a black box unitary operator. The unitarity condition may be relaxed when measurements are included as quantum operations, for example.

Another generalization of a quantum operation is a map $\rho_E \to \rho_S(t)$, for example. Let us take the initial state $|e_0\rangle\langle e_0| \otimes \rho_E$ at $t = 0$. Then after application of $U(t)$ on the total system, we obtain

$$\rho_S(t) = \text{tr}_E\left[U(t)(|e_0\rangle\langle e_0| \otimes \rho_E)U(t)^\dagger\right]. \qquad (9.13)$$

Therefore quantum operations do not necessarily map density matrices to density matrices belonging to the same state space.

Tracing out the extra degrees of freedom makes it impossible to invert a quantum operation. Given an initial principal system state ρ_S, there are infinitely many U that yield the same $\mathcal{E}(\rho_S)$. Therefore even though it is possible to compose two quantum operations, the set of quantum operations is not a group but merely a **semigroup**.[*]

9.1.2 Operator-Sum Representation and Noisy Quantum Channel

Operator-sum representation (OSR) introduced in the previous subsection seems to be rather abstract. Here we give an interpretation of OSR as a noisy quantum channel. Suppose we have a set of unitary matrices $\{U_a\}$ and a set of non-negative real numbers $\{p_a\}$ such that $\sum_a p_a = 1$. By choosing U_a randomly with probability p_a and applying it to ρ_S, we introduce the expectation value of the resulting density matrix as

$$\mathcal{M}(\rho_S) = \sum_a p_a U_a \rho_S U_a^\dagger, \tag{9.14}$$

which we call a **mixing process** [4]. This takes place, for example, when a flying qubit is sent through a noisy quantum channel which transforms the qubit density matrix by U_a with probability p_a. Note that no environment has been introduced in the above definition, and hence no partial trace is involved.

Now the correspondence between $\mathcal{E}(\rho_S)$ and $\mathcal{M}(\rho_S)$ should be clear. Let us define

$$E_a \equiv \sqrt{p_a} U_a. \tag{9.15}$$

Then Eq. (9.14) is rewritten as

$$\mathcal{M}(\rho_S) = \sum_a E_a \rho_S E_a^\dagger, \tag{9.16}$$

and the equivalence has been shown. Operators E_a are identified with the Kraus operators. The system transforms, under the action of U_a, as

$$\rho_S \to \frac{E_a \rho_S E_a^\dagger}{\mathrm{tr}\left(E_a \rho_S E_a^\dagger\right)}. \tag{9.17}$$

Conversely, given a noisy quantum channel $\{U_a, p_a\}$ we may introduce an "environment" with the Hilbert space $\mathcal{H}_E$ as follows. Let $\mathcal{H}_E = \mathrm{Span}(|\varepsilon_a\rangle)$

---

[*]A set S is called a semigroup if S is closed under a product satisfying associativity $(ab)c = a(bc)$. If S has a unit element e, such that $ea = ae = a, \forall a \in S$, it is called a monoid.

be a Hilbert space with the dimension equal to the number of the unitary matrices $\{U_a\}$, where $\{|\varepsilon_a\rangle\}$ is an orthonormal basis. Define formally the environment density matrix $\rho_E = \sum_a p_a |\varepsilon_a\rangle\langle\varepsilon_a|$ and

$$U \equiv \sum_a U_a \otimes |\varepsilon_a\rangle\langle\varepsilon_a| \tag{9.18}$$

which acts on $\mathcal{H}_S \otimes \mathcal{H}_E$. It is easily verified from the orthonormality of $\{|\varepsilon_a\rangle\}$ that U is indeed a unitary matrix. Now a partial trace over the fictitious $\mathcal{H}_E$ yields

$$\mathcal{E}(\rho_S) = \mathrm{tr}_E[U(\rho_S \otimes \rho_E)U^\dagger]$$

$$= \sum_a (I \otimes \langle\varepsilon_a|) \left(\sum_b U_b \otimes |\varepsilon_b\rangle\langle\varepsilon_b| \right) \left(\rho_S \otimes \sum_c p_c |\varepsilon_c\rangle\langle\varepsilon_c| \right)$$

$$\times \left(\sum_d U_d \otimes |\varepsilon_d\rangle\langle\varepsilon_d| \right) (I \otimes |\varepsilon_a\rangle)$$

$$= \sum_a p_a U_a \rho_S U_a^\dagger = \mathcal{M}(\rho_S), \tag{9.19}$$

showing that the mixing process is also decribed by a quantum operation with a fictitious environment.

EXERCISE 9.1 Show that U defined as Eq. (9.18) is unitary.

9.1.3   Completely Positive Maps

All linear operators we have encountered so far map vectors to vectors. A quantum operation maps a density matrix to another density matrix linearly.[†] A linear operator of this kind is called a **superoperator**. Let Λ be a superoperator acting on the system density matrices, $\Lambda : \mathcal{S}(\mathcal{H}_S) \to \mathcal{S}(\mathcal{H}_S)$. The operator Λ is easily extended to an operator acting on $\mathcal{H}_T$ by $\Lambda_T = \Lambda \otimes I_E$, which acts on $\mathcal{S}(\mathcal{H}_S \otimes \mathcal{H}_E)$. Note, however, that Λ_T is not necessarily a map $\mathcal{S}(\mathcal{H}_T) \to \mathcal{S}(\mathcal{H}_T)$. It may happen that $\Lambda_T(\rho)$ is not a density matrix any more. We have already encountered this situation when we introduced partial transpose operation in §2.4.1. Let $\mathcal{H}_T = \mathcal{H}_1 \otimes \mathcal{H}_2$ be a two-qubit Hilbert space, where $\mathcal{H}_k$ is the kth qubit Hilbert space. It is clear that the transpose operation $\Lambda_t : \rho_1 \to \rho_1^t$ on a single-qubit state ρ_1 preserves the density matrix properties, i.e., non-negativity, Hermiticity and unit trace property. For a two-qubit density matrix ρ_{12}, however, this is not always the case. In fact,

---

[†]Of course, the space of density matrices is not a linear vector space. What is meant here is a linear operator, acting on the vector space of Hermitian matrices, also acts on the space of density matrices and maps a density matrix to another density matrix.

we have seen that $\Lambda_t \otimes I : \rho_{12} \to \rho_{12}^{\text{pt}}$ defined by Eq. (2.42) maps a density matrix to a matrix which is not a density matrix when ρ_{12} is inseparable.

A map Λ which maps a positive operator acting on $\mathcal{H}_S$ to another positive operator acting on $\mathcal{H}_S$ is said to be positive. Moreover, it is called a **completely positive map (CP map)**, if its extension $\Lambda_T = \Lambda \otimes I_n$ remains a positive operator for an arbitrary $n \in \mathbb{N}$.

THEOREM 9.1 A linear map Λ is CP if and only if there exists a set of operators $\{E_a\}$ such that $\Lambda(\rho)$ can be written as

$$\Lambda(\rho_S) = \sum_a E_a \rho_S E_a^\dagger. \tag{9.20}$$

We require not only that Λ be CP but also that $\Lambda(\rho)$ be a density matrix, i.e.,

$$\text{tr}\,\Lambda(\rho_S) = \text{tr}\left(\sum_a E_a \rho E_a^\dagger\right) = \text{tr}\left(\sum_a E_a^\dagger E_a \rho\right) = 1. \tag{9.21}$$

This condition is satisfied for any ρ if and only if

$$\sum_a E_a^\dagger E_a = I_S. \tag{9.22}$$

Therefore, any quantum operation obtained by tracing out the environment degrees of freedom is CP and preserves trace.

9.2 Measurements as Quantum Operations

We have already seen that a unitary evoluation $\rho_S \to U\rho_S U^\dagger$ is a quantum operation and that the mixing process $\rho_S \to \sum_i p_i U_i \rho_S U_i^\dagger$ is a quantum operation. We will see further examples of quantum operations in this section and the next. This section deals with measurements as quantum operations.

9.2.1 Projective Measurements

Suppose we measure an observable $A = \sum_i \lambda_i P_i$, where $P_i = |\lambda_i\rangle\langle\lambda_i|$ is the projection operator corresponding to the eigenvector $|\lambda_i\rangle$. We have seen in Chapter 2 that the probability of observing λ_i upon a measurement of A in a state ρ is

$$p(i) = \langle\lambda_i|\rho|\lambda_i\rangle = \text{tr}\,(P_i\rho), \tag{9.23}$$

and the state changes as $\rho \to P_i\rho P_i/p(i)$. This process happens with a probability $p(i)$. Thus we may regard the measurement process as a quantum operation

$$\rho_S \rightarrow \sum_i p(i) \frac{P_i \rho_S P_i}{p(i)} = \sum_i P_i \rho_S P_i, \tag{9.24}$$

where the set $\{P_i\}$ satisifes the completeness relation $\sum_i P_i P_i^\dagger = I$.

The projective measurement is a special case of a quantum operation in which the Kraus operators are $E_i = P_i$.

9.2.2 POVM

We have been concerned with projective measurements so far. However, it should be noted that they are not a unique type of measurements. Here we will be concerned with the most general framework of measurement and show that it is a quantum operation.

Suppose a system and an environment, prepared in a product state $|\psi\rangle|e_0\rangle$, are acted by a unitary operator U, which applies an operator M_i on the system and, at the same time, put the environment to $|e_i\rangle$ for various i. It is written explicitly as

$$|\Psi\rangle = U|\psi\rangle|e_0\rangle = \sum_i M_i|\psi\rangle|e_i\rangle. \tag{9.25}$$

The system and its environment are correlated in this way. This state must satisfy the normalization condition since U is unitary;

$$\langle\psi|\langle e_0|U^\dagger U|\psi\rangle|e_0\rangle = \sum_{i,j}\langle\psi|\langle e_i|M_i^\dagger M_j \otimes I|\psi\rangle|e_j\rangle$$

$$= \langle\psi|\sum_i M_i^\dagger M_i|\psi\rangle = 1. \tag{9.26}$$

Since $|\psi\rangle$ is arbitrary, we must have

$$\sum_i M_i^\dagger M_i = I_S, \tag{9.27}$$

where I_S is the unit matrix acting on the system Hilbert space $\mathcal{H}_S$. Operators $\{M_i^\dagger M_i\}$ are said to form a **POVM (positive operator-valued measure)**. Suppose we measure the environment with a measurement operator

$$O = I_S \otimes \sum_i \lambda_i|e_i\rangle\langle e_i| = \sum_i \lambda_i \left(I_S \otimes |e_i\rangle\langle e_i|\right).$$

We obtain a measurement outcome λ_k with a probability

$$p(k) = \langle\Psi|(I_S \otimes |e_k\rangle\langle e_k|)|\Psi\rangle$$

$$= \sum_{i,j}\langle\psi|\langle e_i|M_i^\dagger (I_S \otimes |e_k\rangle\langle e_k|)M_j|\psi\rangle|e_j\rangle$$

$$= \langle\psi|M_k^\dagger M_k|\psi\rangle, \tag{9.28}$$

where $|\Psi\rangle = U|\psi\rangle|e_0\rangle$. The combined system immediately after the measurement is

$$\frac{1}{\sqrt{p(k)}}(I_S \otimes |e_k\rangle\langle e_k|)U|\psi\rangle|e_0\rangle = \frac{1}{\sqrt{p(k)}}(I_S \otimes |e_k\rangle\langle e_k|)\sum_i M_i|\psi\rangle|e_i\rangle$$

$$= \frac{1}{\sqrt{p(k)}}M_k|\psi\rangle|e_k\rangle. \qquad (9.29)$$

Let $\rho_S = \sum_i p_i|\psi_i\rangle\langle\psi_i|$ be an arbitrary density matrix of the principal system. It follows from the above observation for a pure state $|\psi\rangle\langle\psi|$ that the reduced density matrix immediately after the measurement is

$$\sum_k p(k)\frac{M_k\rho_S M_k^\dagger}{p(k)} = \sum_k M_k\rho_S M_k^\dagger. \qquad (9.30)$$

This shows that POVM measurement is a quantum operation in which the Kraus operators are given by the generalized measurement operators $\{M_i\}$. The projective measurement is a special class of POVM, in which $\{M_i\}$ are the projection operators.

9.3 Examples

Now we examine several important examples which have relevance in quantum information theory. Decoherence appears as an error in quantum information processing. The next chapter is devoted to strategies to fight against some errors introduced in this section.

9.3.1 Bit-Flip Channel

A bit-flip channel is defined by a quantum operation

$$\mathcal{E}(\rho_S) = (1-p)\rho_S + p\sigma_x\rho_S\sigma_x, \ 0 \le p \le 1. \qquad (9.31)$$

In other words, the input ρ_S is bit-flipped ($|0\rangle \mapsto |1\rangle$ and $|1\rangle \mapsto |0\rangle$) with a probability p while it remains in its input state with a probability $1-p$. The Kraus operators are easily read off as

$$E_0 = \sqrt{1-p}I, \ E_1 = \sqrt{p}\sigma_x. \qquad (9.32)$$

Let us next consider a quantum circuit which models this channel. Consider a closed two-qubit system with a Hilbert space $\mathbb{C}^2 \otimes \mathbb{C}^2$. We call the first qubit the "(principal) system" and the second qubit the "environment." We show that the circuit depicted in Fig. 9.1 indeed models the bit-flip channel provided

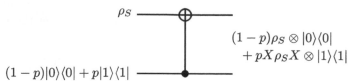

FIGURE 9.1
Quantum circuit modelling a bit-flip channel. The gate is the inverted CNOT gate $I \otimes |0\rangle\langle 0| + \sigma_x \otimes |1\rangle\langle 1|$.

that the second qubit is in a mixed state $(1 - p)|0\rangle\langle 0| + p|1\rangle\langle 1|$. The circuit is nothing but the inverted CNOT gate

$$V = I \otimes |0\rangle\langle 0| + \sigma_x \otimes |1\rangle\langle 1|.$$

The output of this circuit is

$$
\begin{aligned}
V \left(\rho_S \otimes [(1 - p)|0\rangle\langle 0| + p|1\rangle\langle 1|] \right) V^\dagger \\
= (1 - p)\rho_S \otimes |0\rangle\langle 0| + p\sigma_x \rho_S \sigma_x |1\rangle\langle 1|,
\end{aligned}
\tag{9.33}
$$

from which we obtain

$$\mathcal{E}(\rho_S) = (1 - p)\rho_S + p\sigma_x \rho_S \sigma_x \tag{9.34}$$

after tracing over the environment Hilbert space.

The choice of the second qubit input state is far from unique and so is the choice of the circuit. Suppose the initial state of the environment is a pure state

$$|\psi_E\rangle = \sqrt{1 - p}|0\rangle + \sqrt{p}|1\rangle, \tag{9.35}$$

for example. Then the output of the circuit in Fig. 9.1 is

$$\mathcal{E}(\rho_S) = \mathrm{tr}_E[V\rho_S \otimes |\psi_E\rangle\langle\psi_E|V^\dagger] = (1 - p)\rho_S + p\sigma_x \rho_S \sigma_x, \tag{9.36}$$

which is the same result as before.

Let us see what transformation this quantum operation brings about in ρ_S. We parametrize ρ_S using the Bloch vector as

$$\rho_S = \frac{1}{2}\left(I + \sum_{k=x,y,z} c_k \sigma_k \right), \quad (c_k \in \mathbb{R}), \tag{9.37}$$

where $\sum_k c_k^2 \leq 1$. We obtain

$$
\begin{aligned}
\mathcal{E}(\rho_S) &= (1 - p)\rho_S + p\sigma_x \rho_S \sigma_x \\
&= \frac{1 - p}{2}(I + c_x\sigma_x + c_y\sigma_y + c_z\sigma_z) + \frac{p}{2}(I + c_x\sigma_x - c_y\sigma_y - c_z\sigma_z) \\
&= \frac{1}{2}\begin{pmatrix} 1 + (1 - 2p)c_z & c_x - i(1 - 2p)c_y \\ c_x + i(1 - 2p)c_y & 1 - (1 - 2p)c_z \end{pmatrix}.
\end{aligned}
\tag{9.38}
$$

Observe that the radius of the Bloch sphere is reduced along the y- and the z-axes so that the radius in these directions is $|1 - 2p|$. Equation (9.38) shows that the quantum operation has produced a mixture of the Bloch vector states (c_x, c_y, c_z) and $(c_x, -c_y, -c_z)$ with weights $1 - p$ and p, respectively. Figure 9.2 (a) shows the Bloch sphere which represents the input qubit states. The Bloch sphere shrinks along the y- and z-axes, which results in the ellipsoid shown in Fig. 9.2 (b).

9.3.2 Phase-Flip Channel

The phase-flip channel is defined by a quantum operation

$$\mathcal{E}(\rho_S) = (1 - p)\rho_S + p\sigma_z\rho_S\sigma_z, \ 0 \leq p \leq 1. \tag{9.39}$$

In other words, the input ρ_S is phase-flipped ($|0\rangle \mapsto |0\rangle$ and $|1\rangle \mapsto -|1\rangle$) with a probability p while it remains in its input state with a probability $1 - p$. The corresponding Kraus operators are

$$E_0 = \sqrt{1 - p}I, \ E_1 = \sqrt{p}\sigma_z. \tag{9.40}$$

Consider again a closed two-qubit system, whose first qubit is called the "(principal) system" and the second qubit its "environment." A quantum circuit which models the phase-flip channel is shown in Fig. 9.3. Let ρ_S be the first qubit input state and $(1-p)|0\rangle\langle0|+p|1\rangle\langle1|$ be the second qubit input state. The circuit is the inverted controlled-σ_z gate

$$V = I \otimes |0\rangle\langle0| + \sigma_z \otimes |1\rangle\langle1|.$$

The output of this circuit is

$$V\left(\rho_S \otimes [(1 - p)|0\rangle\langle0| + p|1\rangle\langle1|]\right)V^\dagger$$
$$= (1 - p)\rho_S \otimes |0\rangle\langle0| + p\sigma_z\rho_S\sigma_z \otimes |1\rangle\langle1|, \tag{9.41}$$

from which we obtain

$$\mathcal{E}(\rho_S) = (1 - p)\rho_S + p\sigma_z\rho_S\sigma_z. \tag{9.42}$$

The second qubit input state may be a pure state

$$|\psi_E\rangle = \sqrt{1 - p}|0\rangle + \sqrt{p}|1\rangle, \tag{9.43}$$

for example. Then we find

$$\mathcal{E}(\rho_S) = \mathrm{tr}_E[V\rho_S \otimes |\psi_E\rangle\langle\psi_E|V^\dagger]$$
$$= E_0\rho_S E_0^\dagger + E_1\rho_S E_1^\dagger, \tag{9.44}$$

where the Kraus operators are

$$E_0 = \langle0|V|\psi_E\rangle = \sqrt{1 - p}I, \ E_1 = \langle1|V|\psi_E\rangle = \sqrt{p}\sigma_z. \tag{9.45}$$

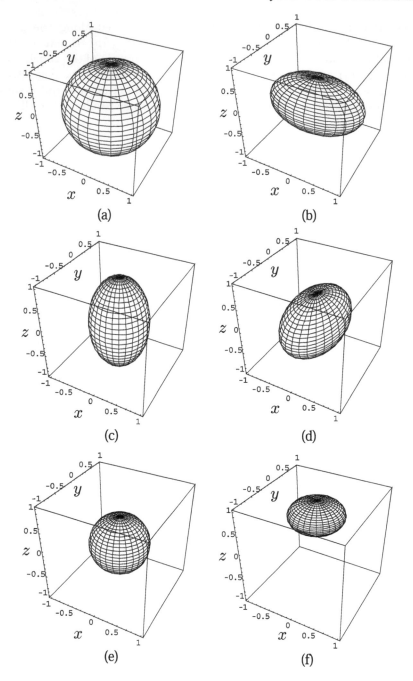

FIGURE 9.2
Bloch sphere of the input state ρ_S (a) and output states (b) $\sim$ (f) of various noisy channels. (b) Bit-flip channel, (c) phase-flip channel, (d) bit-phase flip channel, (e) depolarizing channel and (f) amplitude damping channel. The probability $p = 0.2$ is common to all the noisy channels except in (f), for which $p = 0.7$.

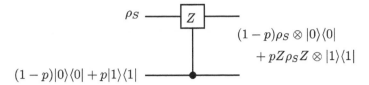

FIGURE 9.3
Quantum circuit modelling a phase-flip channel. The gate is the inverted controlled-σ_z gate.

Let us work out the transformation this quantum operation brings about to ρ_S. We parametrize ρ_S using the Bloch sphere as before. We obtain

$$\mathcal{E}(\rho_S) = (1-p)\rho_S + p\sigma_z\rho_S\sigma_z$$

$$= \frac{1-p}{2}(I + c_x\sigma_x + c_y\sigma_y + c_z\sigma_z) + \frac{p}{2}(I - c_x\sigma_x - c_y\sigma_y + c_z\sigma_z)$$

$$= \frac{1}{2}\begin{pmatrix} 1+c_z & (1-2p)(-c_x - ic_y) \\ (1-2p)(c_x + ic_y) & 1-c_z \end{pmatrix}. \tag{9.46}$$

Observe that the off-diagonal components decay while the diagonal components remain the same. Equation (9.46) shows that the quantum operation has produced a mixture of the Bloch vector states (c_x, c_y, c_z) and $(-c_x, -c_y, c_z)$ with weights $1 - p$ and p, respectively. The initial state has a definite phase $\phi = \tan^{-1}(c_y/c_x)$ in the off-diagonal components. The phase after the quantum operation is applied is a mixture of states with ϕ and $\phi + \pi$. This process is called the **phase relaxation process**, or the T_2 process in the context of NMR. The radius of the Bloch sphere is reduced along the x- and the y-axes to $|1 - 2p|$. Figure 9.2 (c) shows the effect of the phase-flip channel on the Bloch sphere for $p = 0.2$.

EXERCISE 9.2 Let us define the bit-phase flip channel by a quantum operation

$$\mathcal{E}(\rho_S) = (1-p)\rho_S + pY\rho_S Y, \tag{9.47}$$

where $Y = -i\sigma_y$. It follows from the observation $XZ = Y$ that bit-phase flip may be considered as a combination of bit-flip and phase-flip.

Find a quantum circuit which models the bit-phase flip channel. Show that the effect of the channel on the Bloch sphere is given by Fig. 9.2 (d).

9.3.3 Depolarizing Channel

A depolarizing channel maps the input state ρ to a maximally mixed state with a probability p and leaves as it is with a probability $1 - p$;

$$\mathcal{E}(\rho_S) = (1-p)\rho_S + p\frac{I}{2}. \tag{9.48}$$

We introduce the decomposition $\rho_S = (I + c_x\sigma_x + c_y\sigma_y + c_z\sigma_z)/2$ to write the OSR form of the channel. We observe first that

$$\sigma_x\rho_S\sigma_x = \frac{1}{2}(I + c_x\sigma_x - c_y\sigma_y - c_z\sigma_z)$$

$$\sigma_y\rho_S\sigma_y = \frac{1}{2}(I - c_x\sigma_x + c_y\sigma_y - c_z\sigma_z)$$

$$\sigma_z\rho_S\sigma_z = \frac{1}{2}(I - c_x\sigma_x - c_y\sigma_y + c_z\sigma_z),$$

from which, we obtain

$$2I = \rho_S + \sum_{k=x,y,z} \sigma_k\rho_S\sigma_k. \tag{9.49}$$

Substituting this into Eq. (9.48), we obtain the OSR of the depolarizing channel

$$\mathcal{E}(\rho_S) = \left(1 - \frac{3}{4}p\right)\rho_S + \frac{p}{4}\sum_k \sigma_k\rho_S\sigma_k. \tag{9.50}$$

The Kraus operators are read off as

$$E_0 = \sqrt{1 - \frac{3}{4}p}I, E_k = \sqrt{\frac{p}{4}}\sigma_k \ (k = x, y, z). \tag{9.51}$$

It is convenient to rescale the probability as $p' = 3p/4$ $(0 \le p' \le 3/4)$ to obtain

$$\mathcal{E}(\rho_S) = (1 - p')\rho_S + \frac{p'}{3}\sum_k \sigma_k\rho_S\sigma_k. \tag{9.52}$$

The channel transports ρ_S without any change with probability $1 - p'$ while X, Y and Z acts each with probability $p'/3$.

The fact that there are four Kraus operators suggests that the environment Hilbert space is at least four-dimensional. In fact, we can construct a quantum circuit model shown in Fig. 9.4. It is a Fredkin gate with the bottom control bit. The input state is

$$\rho_S \otimes \frac{I}{2} \otimes [(1 - p)|0\rangle\langle 0| + p|1\rangle\langle 1|].$$

The action of the Fredkin gate on the input state yields the output

$$\rho = (I_4 \otimes |0\rangle\langle 0| + U_{\text{SWAP}} \otimes |1\rangle\langle 1|)\left(\rho_S \otimes \frac{I}{2} \otimes [(1 - p)|0\rangle\langle 0| + p|1\rangle\langle 1|]\right)$$

$$(I_4 \otimes |0\rangle\langle 0| + U_{\text{SWAP}} \otimes |1\rangle\langle 1|)$$

$$= (1 - p)\rho_S \otimes \frac{I}{2} \otimes |0\rangle\langle 0| + p\frac{I}{2} \otimes \rho_S \otimes |1\rangle\langle 1|, \tag{9.53}$$

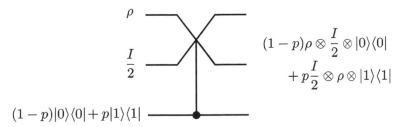

$$(1-p)\rho \otimes \frac{I}{2} \otimes |0\rangle\langle 0|$$
$$+ p\frac{I}{2} \otimes \rho \otimes |1\rangle\langle 1|$$

FIGURE 9.4
Quantum circuit which models a depolarizing channel. The gate is an inverted Fredkin gate, in which the control bit is the third qubit.

where use has been made of the relation

$$U_{\mathrm{SWAP}}(\rho_1 \otimes \rho_2)U_{\mathrm{SWAP}} = \rho_2 \otimes \rho_1. \tag{9.54}$$

Tracing out the two-qubit environment produces

$$\mathrm{tr}_E \rho = (1-p)\rho_S + p\frac{I}{2} \tag{9.55}$$

as promised.

The effect of this channel on the Bloch sphere of the qubit is obtained from the expression

$$\mathcal{E}(\rho_S) = p\frac{I}{2} + \frac{1-p}{2}(I + \sum_k c_k \sigma_k) = \frac{I}{2} + \frac{1-p}{2} \sum_k c_k \sigma_k. \tag{9.56}$$

It is found that the radius is uniformly reduced from 1 to $1-p$. Figure 9.2 (e) shows the Bloch sphere of the output state (9.56) with $p = 0.2$.

9.3.4 Amplitude-Damping Channel

Our final example is the amplitude-damping channel, which is described by a one-way decay only. A qubit decays from $|1\rangle$ to $|0\rangle$ with a probability p but not in the other way around. This process describes the T_1 process in NMR, for example.

Let

$$\mathcal{E}(\rho_S) = E_0 \rho_S E_0^\dagger + E_1 \rho_S E_1^\dagger \tag{9.57}$$

be the OSR of the quantum operation. This downward decay process is naturally described by a Kraus operator

$$E_1 = \sqrt{p} \begin{pmatrix} 0 & 1 \\ 0 & 0 \end{pmatrix}. \tag{9.58}$$

The other Kraus operator E_0 is fixed by the requirement $\sum_k E_k^\dagger E_k = I$ as

$$E_0 = \begin{pmatrix} 1 & 0 \\ 0 & \sqrt{1-p} \end{pmatrix}. \tag{9.59}$$

Let us consider the effect of this channel on the Bloch sphere. By substituting the Bloch sphere representation of ρ_S, we obtain

$$\begin{aligned}
\mathcal{E}(\rho_S) &= p \begin{pmatrix} 0 & 1 \\ 0 & 0 \end{pmatrix} \rho_S \begin{pmatrix} 0 & 0 \\ 1 & 0 \end{pmatrix} + \begin{pmatrix} 1 & 0 \\ 0 & \sqrt{1-p} \end{pmatrix} \rho_S \begin{pmatrix} 1 & 0 \\ 0 & \sqrt{1-p} \end{pmatrix} \\
&= \begin{pmatrix} \rho_{00} + p\rho_{11} & \sqrt{1-p}\,\rho_{01} \\ \sqrt{1-p}\,\rho_{10} & \rho_{11} - p\rho_{11} \end{pmatrix} \\
&= \frac{1}{2} \begin{pmatrix} 1 + [p + (1-p)c_z] & \sqrt{1-p}(c_x - ic_y) \\ \sqrt{1-p}(c_x + ic_y) & 1 - [p + (1-p)c_z] \end{pmatrix}.
\end{aligned} \tag{9.60}$$

Note that the center of the Bloch sphere is shifted toward the north pole ($|0\rangle$) by p under the influence of the channel and, at the same time, the radius in the x- and y-directions are reduced by $\sqrt{1-p}$ and the radius in the z-direction by $1 - p$. Figure 9.2 (f) shows this effect graphically.

9.4 Lindblad Equation

Let $\mathcal{H}_S$ be a closed system with the Hamiltonian H_S. The system density matrix ρ_S evolves according to the Liouville-von Neumann equation

$$\frac{\partial \rho_S}{\partial t} = \frac{1}{i}[H, \rho_S]. \tag{9.61}$$

What is the corresponding equation for an open quantum system? We have found in §9.1 that the principal system density matrix $\rho_S(t)$ at later time $t > 0$ is obtained by tracing out the environment degrees of freedom as

$$\rho_S(t) = \operatorname{tr}_E[U(t)(\rho_S \otimes \rho_E)U(t)^\dagger], \tag{9.62}$$

where $U(t)$ is the time evolution operator of the total system and the initial state is assumed to be a tensor product state. Then the dynamics of $\rho_S(t)$ is

$$\frac{\partial \rho_S(t)}{\partial t} = \frac{\partial}{\partial t}\operatorname{tr}_E[U(t)(\rho_S \otimes \rho_E)U(t)^\dagger] = \frac{1}{i}\operatorname{tr}_E[H, U(t)(\rho_S \otimes \rho_E)U(t)^\dagger], \tag{9.63}$$

where $H = H_S + H_E + H_{SE}$ is the total Hamiltonian and we noted that time derivative commutes with tr_E. This equation, albeit exact, is not closed and is of little use in actual applications. Now we must introduce an approximation in the above exact equation, with which we may extract useful information on the system dynamics.

One of the most popular assumptions is the **Markovian approximation**. The state $\rho_S(t)$ depends on the previous history of the system in general. For some problems, however, this memory effect may be negligible, and the dynamics of $\rho_S(t)$ is determined by $\rho_S(t)$ itself. For example we may assume that the behavior of the system is Markovian, if we are interested in a time scale much longer than the environment correlation times. If this is the case, the master equation should look like

$$\frac{\partial}{\partial t}\rho_S(t) = \mathcal{L}\rho_S(t) \tag{9.64}$$

with a superoperator $\mathcal{L}$.

9.4.1 Quantum Dynamical Semigroup

Let

$$\Phi_t : \rho_S \rightarrow \rho_S(t), \quad \Phi_0 = I \tag{9.65}$$

be the quantum operation associated with a time evolution operator $U(t)$. We require that Φ_t be trace-preserving and completely positive. The map Φ_t has the OSR

$$\Phi_t(\rho_S) = \sum_a E_a(t)\rho_S E_a^\dagger(t), \quad \sum_a E_a^\dagger(t)E_a(t) = I. \tag{9.66}$$

Suppose the reduced Liouville-von Neumann equation satisfies the Markovian property. Then it is formally solved to yield

$$\Phi_t(\rho_S) = e^{\mathcal{L}t}\rho_S, t \in [0, \infty). \tag{9.67}$$

This map satisfies the semigroup property

$$\Phi_{t_2}(\Phi_{t_1}(\rho_S)) = \Phi_{t_1+t_2}(\rho_S). \tag{9.68}$$

The set of maps $\{\Phi_t\}$ is called the **quantum dynamical semigroup**.

9.4.2 Lindblad Equation

Let $\dim\mathcal{H}_S = d$. A superoperator $\mathcal{L}$ has components $\mathcal{L}_{ij,kl}$ with $1 \leq i, j, k, l \leq d$ showing that it acts on a d^2-dimensional vector space. Let us introduce the Hilbert-Schmidt scalar product in the space of matrices as

$$(A, B)_{\mathrm{HS}} = \mathrm{tr}\,(A^\dagger B), \tag{9.69}$$

where A and B are $d \times d$ matrices. Verify that this definition satisfies all the axioms of a scalar product.

The basis of the vector space of matrices may be $\{e_{ij} = |i\rangle\langle j|\}$ where

$$(e_{ij})_{pq} = \delta_{ip}\delta_{jq}.$$

Instead of the bi-index notation $\{e_{ij}\}$, we properly assign a single index to the basis vectors to write $\{e_J\}$ ($1 \leq J \leq d^2$). It turns out to be convenient to introduce a special basis vector

$$e_{d^2} = \frac{1}{\sqrt{d}}I_d \qquad (9.70)$$

and take the other $d^2 - 1$ basis vectors traceless. The basis vectors satisfy

$$(e_J, e_K)_{\text{HS}} = \delta_{JK}, \quad \text{tr}\, e_J = \begin{cases} 0, & 1 \leq J \leq d^2 - 1 \\ \sqrt{d}, & J = d^2. \end{cases} \qquad (9.71)$$

Let us write the OSR of Φ_t as

$$\Phi_t(\rho_S) = \sum_a E_a(t)\rho_S E_a^\dagger(t), \qquad (9.72)$$

where we have written the time dependence of the Kraus operator E_a explicitly. Now we expand E_a in terms of $\{e_J\}$ as

$$E_a(t) = \sum_J e_J(e_J, E_a(t))_{\text{HS}} \qquad (9.73)$$

and substitute this into Eq. (9.72) to obtain

$$\Phi_t(\rho_S) = \sum_{J,K} c_{JK}(t)e_J\rho_S e_K^\dagger, \qquad (9.74)$$

where

$$c_{JK}(t) = \sum_a (e_J, E_a(t))_{\text{HS}}(e_K, E_a(t))_{\text{HS}}^*. \qquad (9.75)$$

The coefficient matrix (c_{JK}) is positive and Hermitian.

Now we are ready to derive the Lindblad equation by employing these tools. By separating the summation over J and K in Eq. (9.74) into $1 \leq J, K \leq d^2-1$ and $J, K = d^2$, we obtain

$$\mathcal{L}\rho = \lim_{\tau \to 0} \frac{\Phi_\tau(\rho_S) - \rho_S}{\tau}$$

$$= \lim_{\tau \to 0} \frac{c_{d^2 d^2}(\tau)/d - 1}{\tau}\rho_S + \left(\lim_{\tau \to 0} \sum_{J=1}^{d^2-1} \frac{c_{Jd^2}(\tau)}{\sqrt{d}\tau}e_J\right)\rho_S$$

$$+\rho_S\left(\lim_{\tau \to 0} \sum_{K=1}^{d^2-1} \frac{c_{d^2K}(\tau)}{\sqrt{d}\tau}e_K^\dagger\right) + \lim_{\tau \to 0} \sum_{J,K=1}^{d^2-1} \frac{c_{JK}(\tau)}{\tau}e_J\rho_S e_K^\dagger. \qquad (9.76)$$

Now we introduce notations

$$c_0 = \lim_{\tau \to 0} \frac{c_{d^2d^2}(\tau)/d - 1}{\tau}, \quad B = \lim_{\tau \to 0} \sum_{J=1}^{d^2-1} \frac{c_{Jd^2}(\tau)}{\sqrt{d\tau}} e_J,$$

$$B^\dagger = \lim_{\tau \to 0} \sum_{K=1}^{d^2-1} \frac{c_{d^2K}(\tau)}{\sqrt{d\tau}} e_K^\dagger, \quad \alpha_{JK} = \lim_{\tau \to 0} \frac{c_{JK}(\tau)}{\tau} \tag{9.77}$$

to rewrite Eq. (9.76) as

$$\mathcal{L}\rho_S = c_0\rho_S + B\rho_S + \rho_S B^\dagger + \sum_{J,K=1}^{d^2-1} \alpha_{JK} e_J \rho_S e_K^\dagger$$

$$= \frac{1}{i}[\tilde{H}, \rho_S] + (G\rho_S + \rho_S G) + \sum_{J,K} \alpha_{JK} e_J \rho_S e_K^\dagger, \tag{9.78}$$

where

$$\tilde{H} = \frac{1}{2i}(B - B^\dagger), \quad G = \frac{1}{2}(B + B^\dagger + c_0). \tag{9.79}$$

Although we are tempted to identify $\tilde{H}$ with the Hamiltonian in the von-Neumann equation, it is not necessarily justifiable in the present case. It is possible to eliminate G in favor of α_{st} if we notice the trace-preserving property $\operatorname{tr} \rho_S = \operatorname{tr}(\Phi_t \rho_S) - 1$. In fact, it follows from $\operatorname{tr}(\mathcal{L}\rho_S) = \operatorname{tr} \partial_t \Phi_t(\rho_S) = \partial_t \operatorname{tr} \Phi_t(\rho_S) = 0$ that

$$0 = \operatorname{tr}(\mathcal{L}\rho_S) = \operatorname{tr}\left[\left(2G + \sum_{J,K=1}^{d^2-1} \alpha_{JK} e_K^\dagger e_J\right)\rho_S\right] \tag{9.80}$$

for an arbitrary ρ_S. Now G is solved as

$$G = -\frac{1}{2}\sum_{J,K} \alpha_{JK} e_K^\dagger e_J. \tag{9.81}$$

Substituting this into Eq. (9.78), we obtain

$$\mathcal{L}\rho_S = \frac{1}{i}[\tilde{H}, \rho_S] + \sum_{J,K=1}^{d^2-1} \alpha_{JK}\left(e_J \rho_S e_K^\dagger - \frac{1}{2}e_K^\dagger e_J \rho_S - \frac{1}{2}\rho_S e_K^\dagger e_J\right). \tag{9.82}$$

The first term in the right hand side generates the unitary time evolution, while the second term brings about incoherent time evolution in the system dynamics.

The matrix α is positive and Hermitian and can be diagonalized with a properly chosen unitary matrix W. Let $W\alpha W^\dagger = D = \operatorname{diag}(\gamma_1, \ldots \gamma_{d^2})$, where γ_i is a positive real number. Let us define operators

$$L_K = \sum_{J=1}^{d^2-1} e_J W_{JK}^\dagger \tag{9.83}$$

so that $e_J = \sum_K L_K W_{KJ}$. Then Eq. (9.82) is put in the form

$$\mathcal{L}\rho_S = \frac{1}{i}[\tilde{H}, \rho_S] + \sum_{K=1}^{N} \gamma_K \left(L_K \rho_S L_K^\dagger - \frac{1}{2} L_K^\dagger L_K \rho_S - \frac{1}{2} \rho_S L_K^\dagger L_K \right), \quad (9.84)$$

where $N \le d^2 - 1$. Note that if some elements γ_K vanish, then the number of L_K involved reduces from $d^2 - 1$ by this amount. The operators L_K are called the **Lindblad operators**, while Eq. (9.84) is called the **Lindblad equation** [5, 6]. Let us compare the term $\gamma_K L_K \rho_S L_K^\dagger$ in Eq. (9.84) with Eq. (9.14). The operator L_K may be compared with U_a and γ_K with p_a.

Note that the Lindblad equation is invariant under the transformation

$$L_K \to L_K + c_K, \quad \tilde{H} \to \tilde{H} + \frac{1}{2i} \sum_K \left(c_K^* L_K - c_K L_K^\dagger \right) + c_0. \quad (9.85)$$

By making use of this freedom, it is always possible to make L_K traceless. Moreover, if we absorb γ_K in the definition of L_K as $\sqrt{\gamma_K} L_K$, the diagonal matrix $W \alpha W^\dagger$ reduces to the unit matrix, and the remaining freedom

$$\sqrt{\gamma_K} L_K \to \sum_J U_{KJ} \sqrt{\gamma_J} L_J \quad (9.86)$$

becomes manifest.

9.4.3 Examples

We consider two examples taken from [7].

The first example is **spontaneous emission**. Let us consider a two-level atom whose Hamiltonian is

$$H = -\frac{\omega_0}{2} \sigma_z, \quad (9.87)$$

where ω_0 is the energy difference between the ground state $|0\rangle = (1,0)^t$ and the excited state $|1\rangle = (0,1)^t$. Suppose there is a single Lindblad operator

$$L = \sqrt{\Gamma} \begin{pmatrix} 0 & 1 \\ 0 & 0 \end{pmatrix}, \quad (9.88)$$

which corresponds to relaxation process $|1\rangle \to |0\rangle$. It is reasonable to assume $\tilde{H}$ in Eq. (9.84) reduces to H when the system-environment coupling is small, for which case the Lindblad equation should reduce to the Liouville-von Neumann equation. The Lindblad equation is then

$$\frac{\partial}{\partial t} \begin{pmatrix} \rho_{00} & \rho_{01} \\ \rho_{10} & \rho_{11} \end{pmatrix} = -i[H, \rho] + L\rho L^\dagger - \frac{1}{2}\left(L^\dagger L \rho + \rho L^\dagger L \right)$$

$$= i\omega_0 \begin{pmatrix} 0 & \rho_{01} \\ -\rho_{10} & 0 \end{pmatrix} + \Gamma \begin{pmatrix} \rho_{11} & -\frac{1}{2}\rho_{01} \\ -\frac{1}{2}\rho_{10} & -\rho_{11} \end{pmatrix}, \quad (9.89)$$

where we have dropped the subscript S to simplify the notation. The above equation is solved with the initial condition $\rho(0)$ as

$$\rho_{00}(t) = \rho_{00}(0) + \rho_{11}(0)(1 - e^{-\Gamma t}), \quad \rho_{01}(t) = \rho_{01}(0)e^{(i\omega_0 - \Gamma/2)t},$$

$$\rho_{10}(t) = \rho_{10}(0)e^{(-i\omega_0 - \Gamma/2)t}, \quad \rho_{11}(t) = \rho_{11}(0)e^{-\Gamma t}. \tag{9.90}$$

The population ρ_{11} in the excited state $|1\rangle$ decays with the characteristic time $T_1 = 1/\Gamma$, while the off-diagonal components (coherence) ρ_{01} and ρ_{10} decay with the time constant $T_2 = 2/\Gamma$. This shows that the coherence decay process is slower than the amplitude damping process.

The second example is the **Bloch equation** in NMR. Let us take the Hamiltonian (9.87) again. We introduce three Lindblad operators

$$L_+ = \sqrt{\Gamma_+} \begin{pmatrix} 0 & 1 \\ 0 & 0 \end{pmatrix}, \quad L_- = \sqrt{\Gamma_-} \begin{pmatrix} 0 & 0 \\ 1 & 0 \end{pmatrix}, \quad L_z = \sqrt{\Gamma_z} \begin{pmatrix} 1 & 0 \\ 0 & -1 \end{pmatrix}, \tag{9.91}$$

where L_+ describes relaxation process $|1\rangle \to |0\rangle$, L_- to excitation process $|0\rangle \to |1\rangle$ and L_z to dephasing process without energy transfer between the spin and the environment.

The Lindblad equation is

$$\frac{\partial}{\partial t} \begin{pmatrix} \rho_{00} & \rho_{01} \\ \rho_{10} & \rho_{11} \end{pmatrix} = i\omega_0 \begin{pmatrix} 0 & \rho_{01} \\ -\rho_{10} & 0 \end{pmatrix} + \Gamma_+ \begin{pmatrix} \rho_{11} & -\rho_{01}/2 \\ -\rho_{10}/2 & -\rho_{11} \end{pmatrix}$$

$$+ \Gamma_- \begin{pmatrix} -\rho_{00} & -\rho_{01}/2 \\ -\rho_{10}/2 & \rho_{00} \end{pmatrix} + \Gamma_z \begin{pmatrix} 0 & -2\rho_{01} \\ -2\rho_{10} & 0 \end{pmatrix}. \tag{9.92}$$

Let us find the equilibrium values ρ^{eq} of $\rho(t)$. It should be clear that the off-diagonal components (the coherence) disappear as $t \to \infty$. As for the diagonal components, the condition $\partial \rho(t)/\partial t = 0$ leads to $\Gamma_+ \rho_{11}^{\mathrm{eq}} - \Gamma_- \rho_{00}^{\mathrm{eq}} = 0$, namely,

$$\frac{\rho_{11}^{\mathrm{eq}}}{\rho_{00}^{\mathrm{eq}}} = \frac{\Gamma_-}{\Gamma_+} = e^{-\omega_0/k_B T}, \tag{9.93}$$

where we noted that the diagonal components of $\rho(t)$ represent the state populations, and the equilibrium population ratio follows the Boltzmann distribution. Straightforward but tedious calculations yield the solution

$$\rho_{00}(t) = \rho_{00}^{\mathrm{eq}} + \left(\rho_{00} - \rho_{00}^{\mathrm{eq}}\right) e^{-t/T_1},$$

$$\rho_{01}(t) = \rho_{01} e^{(i\omega_0 - 1/T_2)t}, \quad \rho_{10}(t) = \rho_{10} e^{(-i\omega_0 - 1/T_2)t}, \tag{9.94}$$

$$\rho_{11}(t) = \rho_{11}^{\mathrm{eq}} + \left(\rho_{11} - \rho_{11}^{\mathrm{eq}}\right) e^{-t/T_1},$$

where

$$T_1^{-1} = \Gamma_+ + \Gamma_-, \quad T_2^{-1} = \frac{\Gamma_+ + \Gamma_-}{2} + 2\Gamma_z. \tag{9.95}$$

It is exepcted that the L_+ process involves the emission of rf photons (see Chapter 12) and the L_- process the absorption of photons. Therefore we assume

$$\Gamma_+ = \gamma[1 + n(\omega_0)], \quad \Gamma_- = \gamma n(\omega_0). \tag{9.96}$$

The function $n(\omega)$ is found by making use of Eq. (9.93) as

$$n(\omega_0) = \frac{1}{1 - e^{-\omega_0/k_B T}}. \tag{9.97}$$

Then T_1 and T_2 are given as

$$T_1^{-1} = \gamma e^{\omega_0/k_B T} \coth\left(\frac{\omega_0}{2k_B T}\right), \quad T_2^{-1} = \frac{T_1^{-1}}{2} + 2\Gamma_z. \tag{9.98}$$

EXERCISE 9.3 Let $\rho(t) = (I + \sum_{k=x,y,z} c_k(t)\sigma_k)/2$ be the density matrix of a qubit, where $\{c_k(t)\}$ are the components of the Bloch vector.

Write down the Lindblad equation for $\{c_k(t)\}$ with the Hamiltonian (9.87) and the Lindblad operators (9.91). Find the solution $\{c_k(t)\}$ of the Lindblad equation.

References

[1] M. A. Neilsen and I. L. Chuang, *Quantum Computation and Quantum Information*, Cambridge University Press (2000).

[2] K. Hornberger, e-print quant-ph/0612118 (2006).

[3] H. Barnum, M. A. Nielsen and B. Schumacher, Phys. Rev. A **57**, 4153 (1998).

[4] Y. Kondo, *et al.*, J. Phys. Soc. Jpn. **76** (2007) 074002.

[5] G. Lindblad, Commun. Math. Phys. **48**, 119 (1976).

[6] V. Gorini, A. Kossakowski and E. C. G. Sudarshan, J. Math. Phys., **17**, 821 (1976).

[7] A. J. Fisher, Lecture note available at
http://www.cmmp.ucl.ac.uk/~ajf/course_notes.pdf

10

Quantum Error Correcting Codes

10.1 Introduction

It has been shown in the previous chapter that interactions between a quantum system with environment cause undesirable changes in the state of the quantum system. In the case of qubits, they appear as bit-flip and phase-flip errors, for example. To reduce such errors, we must build in some sort of error correcting mechanism in the algorithm.

Before we introduce quantum error correcting codes, we have a brief look at the simplest version of error correcting code in classical bits. Suppose we transmit a series of 0's and 1's through a noisy classical channel. Each bit is assumed to flip independently with a probability p. Thus a bit 0 sent through the channel will be received as 0 with probability $1 - p$ and as 1 with probability p. To reduce channel errors, we may invoke the majority vote. Namely, we encode logical 0 by 000 and 1 by 111, for example. When 000 is sent through this channel, it will be received as 000 with probability $(1 - p)^3$, as 100, 010 or 001 with probability $3p(1 - p)^2$, as 011, 101 or 110 with probability $3p^2(1 - p)$ and finally as 111 with probability p^3. Note that the summation of all the probabilities is 1 as it should be. By taking the majority vote, we correctly reproduce the desired result 0 with probability $p_0 = (1 - p)^3 + 3p(1 - p)^2 = (1 - p)^2(1 + 2p)$ and fail with probability $p_1 = 3p^2(1 - p) + p^3 = (3 - 2p)p^2$. We obtain $p_0 \gg p_1$ for sufficiently small $p \geq 0$. In fact, we find $p_0 = 0.972$ and $p_1 = 0.028$ for $p = 0.1$. The success probability p_0 increases as p approaches 0, or alternatively, if we use more bits to encode 0 or 1.

EXERCISE 10.1 Suppose five bits are employed to encode a bit. Find the success probability when the bit is sent through a noisy channel, one by one, in which a bit is flipped with probability p. Assume the received bits are decoded according to majority vote.

This method cannot be applicable to qubits, however, due to the no-cloning theorem. We have to somehow think out the way to overcome this theorem.

General references for this chapter are [1, 2, 3] and [4].

10.2   Three-Qubit Bit-Flip Code and Phase-Flip Code

It is instructive to introduce a simple example of **quantum error correcting codes (QECC)** before we consider more general theory. We closely follow Steane [2] here.

10.2.1   Bit-Flip QECC

Suppose Alice wants to send a qubit or a serise of qubits to Bob through a noisy quantum channel. Let $|\psi\rangle = a|0\rangle + b|1\rangle$ be the state she wants to send. If she is to transmit a serise of qubits, she sends them one by one and the following argument applies to each of the qubits. Let p be the probability with which a qubit is flipped ($|0\rangle \leftrightarrow |1\rangle$), and we assume there are no other types of errors in the channel. In other words, the operator X is applied to the qubit with probability p, and consequently the state is mapped to

$$|\psi\rangle \rightarrow |\psi'\rangle = X|\psi\rangle = a|1\rangle + b|0\rangle. \tag{10.1}$$

We have already seen in the previous section that this channel is described by a quantum operation (9.31).

10.2.1.1   Encoding

To reduce the error probability, we want to mimic somehow the classical counterpart without using a clone machine. Let us recall that the action of a CNOT gate is

$$\text{CNOT} : |j0\rangle \rightarrow |jj\rangle, \quad j \in \{0,1\} \tag{10.2}$$

and therefore it duplicates the control bit $j \in \{0,1\}$ when the initial target bit is set to $|0\rangle$. We use this fact to *triplicate* the basis vectors as

$$|\psi\rangle|00\rangle = (a|0\rangle + b|1\rangle)|00\rangle \rightarrow |\psi\rangle_L = a|000\rangle + b|111\rangle, \tag{10.3}$$

where $|\psi\rangle_L$ denotes the encoded state. The state $|\psi\rangle_L$ is called the **logical qubit**, while each constituent qubit is called the **physical qubit**. We borrow terminologies from classical error correcting code (ECC) and call the set

$$C = \{a|000\rangle + b|111\rangle \,|\, a, b \in \mathbb{C}, |a|^2 + |b|^2 = 1\} \tag{10.4}$$

the **code** and call each member of C a **codeword**. It is important to note that the state $|\psi\rangle$ is not triplicated but only the basis vectors are triplicated. This redundancy makes it possible to detect errors in $|\psi\rangle_L$ and correct them as we see below.

A quantum circuit which implements the encoding (10.3) is easily found from our experience in the CNOT gate. Let us consider the circuit shown in

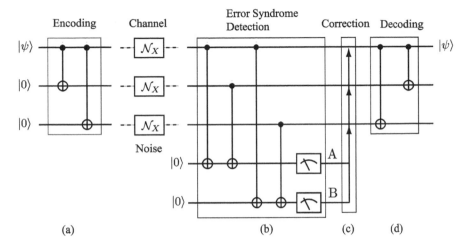

FIGURE 10.1

Quantum circuits to (a) encode, (b) detect bit-flip error syndrome, (c) make correction to a relevant qubit and (d) decode. The gate $\mathcal{N}_X$ stands for the bit-flip noise. The circuit (a) belongs to Alice, while the circuits (b), (c) and (d) belong to Bob.

Fig. 10.1 (a) whose inputs are the state $|\psi\rangle$ and two ancillary qubits in the state $|00\rangle$. When the target bit is $|\psi\rangle = |0\rangle$, the ancillary qubits are left unchanged and the output is $|000\rangle$, while when the input is $|\psi\rangle = |1\rangle$, the output is $|111\rangle$. By superposition principle, the input $|\psi\rangle|00\rangle = (a|0\rangle + b|1\rangle)|00\rangle$ results in the output $|\psi\rangle_L = a|000\rangle + b|111\rangle$ as promised.

10.2.1.2 Transmission

Now the state $|\psi\rangle_L$ is sent through a quantum channel which introduces bit-flip error with a rate p for each qubit independently. We assume p is sufficiently small so that not many errors occur during the qubit transmission. The received state depends on in which physical qubit(s) the bit-flip occurred. Table 10.1 lists possible received states and the probabilities with which these states are received.

10.2.1.3 Error Syndrome Detection and Correction

Now Bob has to extract from the received state which error occurred during the qubit transmission. For this purpose, Bob prepares two ancillary qubits in the state $|00\rangle$ as depicted in Fig. 10.1 (b) and applies four CNOT operations whose control bits are the encoded qubits while the target qubits are Bob's two ancillary qubits. Let $|x_1 x_2 x_3\rangle$ be a basis vector Bob has received and let A (B) be the output state of the first (second) ancilla qubit. It is seen from

TABLE 10.1

Bob receives the following
states with given probabilities.

Received state	Probability
$a\lvert000\rangle + b\lvert111\rangle$	$(1-p)^3$
$a\lvert100\rangle + b\lvert011\rangle$	$p(1-p)^2$
$a\lvert010\rangle + b\lvert101\rangle$	$p(1-p)^2$
$a\lvert001\rangle + b\lvert110\rangle$	$p(1-p)^2$
$a\lvert110\rangle + b\lvert001\rangle$	$p^2(1-p)$
$a\lvert101\rangle + b\lvert010\rangle$	$p^2(1-p)$
$a\lvert011\rangle + b\lvert100\rangle$	$p^2(1-p)$
$a\lvert111\rangle + b\lvert000\rangle$	p^3

TABLE 10.2

States after error extraction is made and the
probabilities with which these states are produced.

State after error syndrome extraction	Probability
$(a\lvert000\rangle + b\lvert111\rangle)\lvert00\rangle$	$(1-p)^3$
$(a\lvert100\rangle + b\lvert011\rangle)\lvert11\rangle$	$p(1-p)^2$
$(a\lvert010\rangle + b\lvert101\rangle)\lvert10\rangle$	$p(1-p)^2$
$(a\lvert001\rangle + b\lvert110\rangle)\lvert01\rangle$	$p(1-p)^2$
$(a\lvert110\rangle + b\lvert001\rangle)\lvert01\rangle$	$p^2(1-p)$
$(a\lvert101\rangle + b\lvert010\rangle)\lvert10\rangle$	$p^2(1-p)$
$(a\lvert011\rangle + b\lvert100\rangle)\lvert11\rangle$	$p^2(1-p)$
$(a\lvert111\rangle + b\lvert000\rangle)\lvert00\rangle$	p^3

Fig. 10.1 (b) that $A = x_1 \oplus x_2$ and $B = x_1 \oplus x_3$. Let $a\lvert100\rangle + b\lvert011\rangle$ be the received logical qubit, for example. Note that the first qubit state in each of the basis vectors is different from the second and the third qubit states. These differences are detected by the pairs of CNOT gates in Fig. 10.1 (b). The error extracting sequence transforms the ancillary qubits as

$$(a\lvert100\rangle + b\lvert011\rangle)\lvert00\rangle \to a\lvert10011\rangle + b\lvert01111\rangle = (a\lvert100\rangle + b\lvert011\rangle)\lvert11\rangle.$$

Both of the ancillary qubits are flipped since $x_1 \oplus x_2 = x_1 \oplus x_3 = 1$ for both $\lvert100\rangle$ and $\lvert011\rangle$. The set of two bits is called the **(error) syndrome**, and it tells Bob in which physical qubit the error occurred during transmission. It is important to note that (i) the syndrome is independent of a and b and (ii) the received state $a\lvert100\rangle + b\lvert011\rangle$ is left unchanged; we have detected an error without measuring the received state! These features are common to all QECC.

We list the results of other cases in Table 10.2. Note that among eight possible states, there are exactly two states with the same ancilla state. Does it mean this error extraction scheme does not work? Now let us compare the probabilities associated with the same ancillary state. When the ancillary

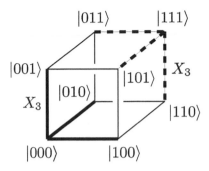

FIGURE 10.2
Encoded basis vectors $|000\rangle$ and $|111\rangle$. The thick solid line shows the action
of $X = \sigma_x$ on the vector $|000\rangle$, while the thick broken line shows the action of
X on $|111\rangle$. The action of $X_3 = I \otimes I \otimes X$ takes $|000\rangle$ to $|001\rangle$ and $|111\rangle$ to
$|110\rangle$ as depicted. Observe that the intersection between thick solid lines and
thick broken lines is an empty set. The Hamming distance, to be defined in
§10.4.1, between 000 and 111 is 3, which guarantees any single bit-flip error
may be corrected.

state is $|10\rangle$, for example, there are two possible received states $a|010\rangle + b|101\rangle$
and $a|101\rangle + b|010\rangle$. The former is received with probability $p(1-p)^2$ and the
latter with $p^2(1-p)$. Therefore the latter probability is negligible compared
to the former for sufficiently small p.

It is instructive to visualize what errors do to the encoded basis vectors as
depicted in Fig. 10.2. The encoded basis vectors $|000\rangle$ and $|111\rangle$ are assigned
to the opposite vertices. An action of X_i, the operator $X = \sigma_x$ acting on the
ith qubit, takes these basis vectors to the nearest neighbor vetices, which differ
from the correct basis vectors in the ith position. Note that the intersection
of the sets of vectors obtained by a single action of X_i on $|000\rangle$ and $|111\rangle$ is an
empty set. Therefore an action of a single error operator X can be corrected
with no ambiguity. It should be clear from this observation that two-qubit
encoding cannot correct single qubit error.

Now Bob measures his ancillary qubits and obtains two bits of classical
information (syndrome). Bob applies correcting procedure to the received
state according to the error syndrome he has obtained. Ignoring multiple
error states with small probabilities, we immediately find that the following
action must be taken:

Error syndrome	Correction to be made	
(00)	identity operation (nothing is required)	
(01)	apply X_3	(10.5)
(10)	apply X_2	
(11)	apply X_1	

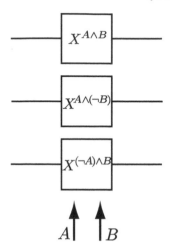

FIGURE 10.3
Explicit form of the bit-flip error correction circuit. It is understood that $X^0 = I$ and $X^1 = \sigma_x$. $\neg A$ stands for the negation of the bit A.

Suppose the syndrome is 01, for example. The state Bob received is likely to be $a|001\rangle + b|110\rangle$. Bob recovers the initial state Alice has sent by applying $X_3 = I \otimes I \otimes X$ on the received state:

$$X_3(a|001\rangle + b|110\rangle) = a|000\rangle + b|111\rangle.$$

Figure 10.3 depicts the explicit form of the error correction circuit.
If Bob receives the state $a|110\rangle + b|001\rangle$, unfortunately, he will obtain

$$X_3(a|110\rangle + b|001\rangle) = a|111\rangle + b|000\rangle.$$

In fact, for any error syndrome, Bob obtains either $a|000\rangle + b|111\rangle$ or $a|111\rangle + b|000\rangle$. The latter case occurs when and only when more than one qubit is flipped, and hence it is less likely to happen for small error rate p.

10.2.1.4 Decoding

Now that Bob has corrected an error, what is left for him is to decode the encoded state. This is nothing but the inverse transformation of the encoding (10.3). It can be seen from Fig. 10.1 (d) that

$$\text{CNOT}_{12}\text{CNOT}_{13}(a|000\rangle + b|111\rangle) = a|000\rangle + b|100\rangle = (a|0\rangle + b|1\rangle)|00\rangle. \quad (10.6)$$

Suppose Bob has received a state with more than one qubit flipped; we will obtain

$$\text{CNOT}_{12}\text{CNOT}_{13}(a|111\rangle + b|000\rangle) = a|100\rangle + b|000\rangle = (a|1\rangle + b|0\rangle)|00\rangle. \quad (10.7)$$

The probability with which this error happens is found from Table 10.1 as

$$P(\text{error}) = 3p^2(1-p) + p^3 = 3p^2 - 2p^3. \tag{10.8}$$

This error rate is less than p, the error probability of a single channel, provided that $p < 1/2$. In contrast, success probability has been enhanced from $1-p$ to $1 - P(\text{error}) = 1 - 3p^2 + 2p^3$. Let $p = 0.1$, for example. Then the error rate is lowered to $P(\text{error}) = 0.028$, while the success probability is enhanced from 0.9 to 0.972.

10.2.1.5 Miracle of Entanglement

This example, albeit simple, contains almost all fundamental ingredients of QECC. We prepare some redundant qubits which somehow "triplicate" the original qubit state to be sent without violating the no-cloning theorem. Then the encoded qubits are sent through a noisy channel, which causes a bit-flip in at most one of the qubits. The received state, which may be subject to an error, is then entangled with ancillary qubits which detect what kind of an error occurred during the state transmission. This results in an entangled state

$$\sum_k |\text{A bit-flip error in the } k\text{th qubit}\rangle \otimes |\text{corresponding error syndrome}\rangle,$$

$$\tag{10.9}$$

set aside the state without errors. The wave function, upon the measurement of the ancillary qubits, collapses to a state with a bit-flip error corresponding to the observed error syndrome. In a sense, syndrome measurement singles out a particular error state which produces the observed syndrome. This principle "measure the syndrome and single out an error" will be used repeatedly in the rest of this chapter.

Once the syndrome is found, it is an easy task to transform the received state back to the original state. Note that everything is done without knowing what the origial state is.

10.2.1.6 Continuous Rotations

We have considered noise X so far. Suppose noise in the channel is characterized by a continuous paramter α as

$$U_\alpha = e^{i\alpha X} = \cos\alpha I + i\sin\alpha X, \tag{10.10}$$

which maps a state $|\psi\rangle$ to

$$U_\alpha|\psi\rangle = \cos\alpha|\psi\rangle + i\sin\alpha X|\psi\rangle. \tag{10.11}$$

It is assumed that U_α acts on each qubit of the logical qubit independently with probability p. The probability with which the logical qubit is received without affected by U_α is $(1-p)^3$ as before.

Suppose U_α acts on the first qubit, for example. Bob then receives

$$(U_\alpha \otimes I \otimes I)(a|000\rangle + b|111\rangle)$$
$$= \cos\alpha(a|000\rangle + b|111\rangle) + i\sin\alpha(a|100\rangle + b|011\rangle).$$

The output of the error syndrome detection circuit, before the syndrome measurement is made, is an entangled state

$$\cos\alpha(a|000\rangle + b|111\rangle)|00\rangle + i\sin\alpha(a|100\rangle + b|111\rangle)|11\rangle \qquad (10.12)$$

(see Table 10.2). Measurement of the error syndrome yields either (00) or (11). In the former case the state collapses to $|\psi\rangle = a|000\rangle + b|111\rangle$, and this happens with a probability $p\cos^2\alpha$. In the latter case, on the other hand, the received state collapses to $X|\psi\rangle = a|100\rangle + b|011\rangle$, and this happens with a probability $p\sin^2\alpha$. Bob applies I (X) to the first qubit to correct the error when the syndrome readout is 00 (11). Now the error probability is given by $P(\text{error}) = p\sin^2\alpha$.

It is clear that error U_α may act on the second or the third qubit. In this way, continuous rotation U_α for any α may be corrected. In general, linearity of a quantum circuit guarantees that any QECC, which corrects the bit-flip error X, corrects continuous error U_α too.

10.2.2   Phase-Flip QECC

Let us consider a phase-flip channel. Phase flip $Z : |x\rangle \mapsto (-1)^x|x\rangle$, ($x \in \{0,1\}$) occurs with probability p for each qubit independently when it is sent through a channel. We consider how to correct phase-flip error mimicking the prescription given for a bit-flip error.

EXERCISE 10.2 (1) Show that $U_H Z U_H = X$, where U_H is the matrix representation of the Hadamard gate.
(2) Show that $Z = \sigma_z$ maps $|+\rangle$ to $|-\rangle$ and $|-\rangle$ to $|+\rangle$, where

$$|\pm\rangle = \frac{1}{\sqrt{2}}(|0\rangle \pm |1\rangle).$$

(3) Show that $X|+\rangle = |+\rangle$ and $X|-\rangle = -|-\rangle$.

What we have learned from the above exercise is that the phase-flip error in the basis $\{|0\rangle, |1\rangle\}$ is nothing but the bit-flip error in the basis $\{|+\rangle, |-\rangle\}$. This observation suggests that we encode $|\psi\rangle = a|0\rangle + b|1\rangle$ into $|\Psi_1\rangle = a|+{+}{+}\rangle + b|{-}{-}{-}\rangle$ by applying $U_H^{\otimes 3}$ on the state $|\Psi_0\rangle = a|000\rangle + b|111\rangle$. Now $|\Psi_1\rangle$ is sent through a noisy channel with possible phase-flip errors. We assume at most a single qubit is subject to the error.

Suppose the phase-flip error acts on the first physical qubit, for example, when $|\Psi_1\rangle$ is sent through the channel. Bob then receives a state $|\Psi_1'\rangle =

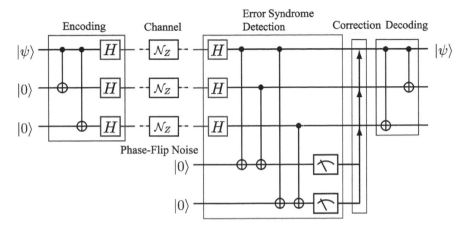

FIGURE 10.4
Three-qubit phase-flip error QECC.

$a|-++\rangle + b|+--\rangle$. We found in the above exercise that a phase-flip error acts as a bit-flip error in the basis $|\pm\rangle$; $U_H Z U_H = X$. Therefore, we need to put the basis back from $|\pm\rangle$ to $\{|0\rangle, |1\rangle\}$ by applying the second Walsh-Hadamard transform $U_H^{\otimes 3}$ so that a phase-flip error is recognized as a bit-flip error and we can employ the same error syndrome detection circuit as well as the error correction circuit as those used for the bit-flip error QECC.

Collecting these results, the quantum circuit for the phase-flip QECC is constructed as shown in Fig. 10.4.

EXERCISE 10.3 Show that the circuit depicted in Fig. 10.4 is able to correct also a continuous error $U_\beta = e^{i\beta Z}$ acting on one of the qubits.

10.3 Shor's Nine-Qubit Code

Let us consider a more general noisy channel in which all possible single-qubit errors occur. Namely, the following errors are now active in the channel;

$$\text{Bit-Flip Error} \quad X : \begin{pmatrix} a \\ b \end{pmatrix} \mapsto \begin{pmatrix} b \\ a \end{pmatrix}$$

$$\text{Phase-Flip Error} \quad Z : \begin{pmatrix} a \\ b \end{pmatrix} \mapsto \begin{pmatrix} a \\ -b \end{pmatrix} \tag{10.13}$$

$$\text{Phase- and Bit-Flip Error} \quad Y : \begin{pmatrix} a \\ b \end{pmatrix} \mapsto \begin{pmatrix} -b \\ a \end{pmatrix}.$$

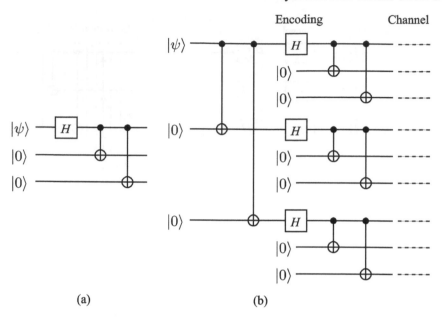

FIGURE 10.5

(a) Quantum circuit to transform $|0\rangle \rightarrow |+\rangle$ and $|1\rangle \rightarrow |-\rangle$. (b) Encoding circuit for Shor's nine-qubit QECC, which maps $|\psi\rangle = a|0\rangle + b|1\rangle$ to $a|0\rangle_L + b|1\rangle_L$.

Shor's nine-qubit QECC which corrects the above three types of noise is constructed by concatenating the phase-flip QECC into the bit-flip QECC, which were introduced in the previous section [5].

10.3.1   Encoding

We encode $|0\rangle$ and $|1\rangle$ as

$$|0\rangle \rightarrow |0\rangle_L \equiv |+++\rangle, \quad |1\rangle \rightarrow |1\rangle_L \equiv |---\rangle, \tag{10.14}$$

where

$$|+\rangle = \frac{1}{\sqrt{2}}(|000\rangle + |111\rangle), \quad |-\rangle = \frac{1}{\sqrt{2}}(|000\rangle - |111\rangle). \tag{10.15}$$

The encoding circuit for a codeword $a|0\rangle_L + b|1\rangle_L$ is shown in Fig. 10.5.

EXERCISE 10.4 Show that the circuit in Fig. 10.5 (a) maps

$$|0\rangle|00\rangle \mapsto |+\rangle, \quad |1\rangle|00\rangle \mapsto |-\rangle. \tag{10.16}$$

It follows from Exercise 10.4 that the quantum circuit given in Fig. 10.5 (b) maps $(a|0\rangle + b|1\rangle)|00000000\rangle$ to $a|+++\rangle + b|---\rangle$. In fact

$$(a|0\rangle + b|1\rangle)|00\rangle \rightarrow a|000\rangle + b|111\rangle$$

$$\rightarrow \frac{a}{\sqrt{8}}[(|0\rangle + |1\rangle)|00\rangle]^{\otimes 3} + \frac{b}{\sqrt{8}}[(|0\rangle - |1\rangle)|00\rangle]^{\otimes 3}$$

$$\rightarrow \frac{a}{\sqrt{8}}(|000\rangle + |111\rangle)^{\otimes 3} + \frac{b}{\sqrt{8}}(|000\rangle - |111\rangle)^{\otimes 3}$$

$$= a|+++\rangle + b|---\rangle. \tag{10.17}$$

Now encoded logical qubits are sent through the noisy channel which causes errors (10.13).

10.3.2 Transmission

Noise depicted by Eq. (10.13) is in action while the encoded qubit is sent through the quantum channel. Let p be the probability with which each physical qubit is flipped by one of the error operators in (10.13).

The logical qubit is sent intact with probability $(1-p)^9$. The probability with which one of the physical qubits is affected is $9p(1-p)^8$, which may be corrected by the following process. Then the probability that more than one qubit are flipped is

$$1 - (1-p)^9 - 9p(1-p)^8 = 1 - (1+8p)(1-p)^8 \simeq 36p^2, \tag{10.18}$$

where $p \ll 1$ is assumed.

10.3.3 Error Syndrome Detection and Correction

Thanks to the superposition principle, we may consider errors in the basis vectors $|0\rangle_L = |+++\rangle$ and $|1\rangle_L = |---\rangle$ separately. We concentrate on $|0\rangle_L$ first and consider the following example. Suppose a bit-flip error occurs in the fifth qubit during transmission. Then the received state will be

$$|+\rangle \frac{1}{\sqrt{2}}(|010\rangle + |101\rangle)|+\rangle$$

instead of the expected $|+++\rangle$. To correct this error we have to identify at which qubit the error occurred during transmission. To this end, we introduce two ancilla qubits to each group of three qubits as shown in Fig. 10.6. These ancillary qubits are all set to the initial state $|0\rangle$ in the beginning. Suppose the input state to the ith group of three qubits is $|x_1 x_2 x_3\rangle$. Then the ancilla output states $|A_i\rangle$ and $|B_i\rangle$ are given as $A_i = x_1 \oplus x_2$ and $B_i = x_1 \oplus x_3$. Let's come back to our example. Error did not occur in the first group of three qubits, and hence we measure $A_1 = B_1 = 0$ for both input states $|000\rangle$ and $|111\rangle$. Therefore we obtain $A_1 = B_1 = 0$. Similarly we find $A_3 = B_3 = 0$ for the third group. For the second group of three qubits, on the other hand, we find $A_2 = 1$ and $B_2 = 0$, from which we find a bit-flip error occurred in the fifth qubit (the second qubit in the second group of three qubits). We simply need to apply the X gate to the fifth qubit to correct this error.

EXERCISE 10.5 Suppose bit-flip error occurred in (1) the first, (2) the second or (3) the third qubit of the first group. What is the syndrome (A_1, B_1) in each case?

Next, let us look at the phase-flip errors. The following exercise suggests that the syndrome $(A_i, B_i)_{1 \leq i \leq 3}$ is not able to detect phase-flip errors.

EXERCISE 10.6 Suppose phase-flip error occurred in the second qubit:

$$|0\rangle_L \to \frac{1}{\sqrt{2}}(|000\rangle - |111\rangle)|+\rangle|+\rangle,$$

$$|1\rangle_L \to \frac{1}{\sqrt{2}}(|000\rangle + |111\rangle)|-\rangle|-\rangle.$$

Show that the output bits A_i and B_i are both 0 for all $1 \leq i \leq 3$.

Let us recall the phase-flip error correcting code introduced in the previous section, in which Z acts as a bit-flip operator in the basis $\{|\pm\rangle\}$. Let us check the syndrome (A_4, B_4). Suppose phase-flip error acted on one of the qubits in the second group of three qubits. Now $|0\rangle_L$ is received as $(|+\rangle(|000\rangle - |111\rangle)|+\rangle)/\sqrt{2} = |+-+\rangle$ while $|1\rangle_L$ as $(|+\rangle(|000\rangle + |111\rangle)|+\rangle)/\sqrt{2} = |-+-\rangle$. Each group of three qubits, after it goes through the Hadamard gates in the middle layer of Fig. 10.6, transforms as

$$|+\rangle \mapsto \tfrac{1}{2}(|000\rangle + |011\rangle + |101\rangle + |110\rangle)$$

$$|-\rangle \mapsto \tfrac{1}{2}(|001\rangle + |010\rangle + |100\rangle + |111\rangle).$$

It is important to note that $|+\rangle$ is mapped to a superposition of states with even number of 1 and $|-\rangle$ to one with odd number of 1. Let $|x_1 x_2 x_3\rangle$ be a particular component of a state after the first Hadamard gates operate on the input state. Then $x_1 \oplus x_2 \oplus x_3 = 0$ when the input state is $|+\rangle$, while $x_1 \oplus x_2 \oplus x_3 = 1$ when the input state is $|-\rangle$. Therefore the syndrome (A_4, B_4) for the example in our hands take values $A_4 = 1$ and $B_4 = 0$, from which we find a phase-flip occurred in the second group of three qubits.

Readers should verify the phase-flip error syndrome is summarized as shown in Table 10.3. By measuring the syndrome (A_4, B_4), Bob knows for sure on which group of qubits the phase-flip operator acted, assuming there exists only a single error.

Now Bob transforms the above state to $|\pm\rangle$ by applying the rightmost layer of the Hadamard gates in Fig. 10.6. Error syndrome has been extracted between two successive operations of the Hadamard gates. For our example, Bob has syndrome (10), from which he knows phase-flip error occurred in the second group. Now he operates Z on one of the three qubits in the group to recover the input state $a|+++\rangle + b|---\rangle$.

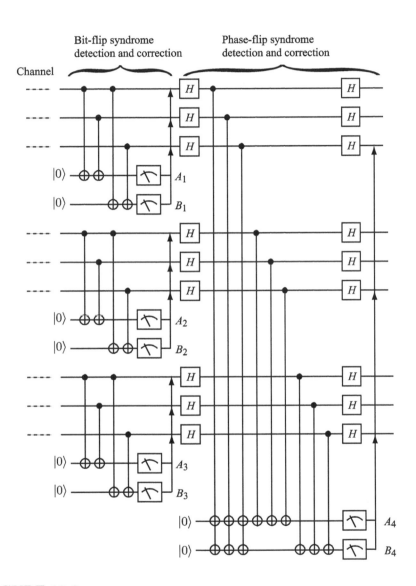

FIGURE 10.6

Quantum circuit to detect bit-flip and phase-flip error syndrome and correct the error.

TABLE 10.3
States after error extraction is made.

Input to the detecting circuit	Ancillas state $|A_4, B_4\rangle$
$a|+++\rangle + b|---\rangle$	$|00\rangle$
$a|-++\rangle + b|+--\rangle$	$|11\rangle$
$a|+-+\rangle + b|-+-\rangle$	$|10\rangle$
$a|++-\rangle + b|--+\rangle$	$|01\rangle$
$a|--+\rangle + b|++-\rangle$	$|01\rangle$
$a|-+-\rangle + b|+-+\rangle$	$|10\rangle$
$a|+--\rangle + b|-++\rangle$	$|11\rangle$
$a|---\rangle + b|+++\rangle$	$|00\rangle$

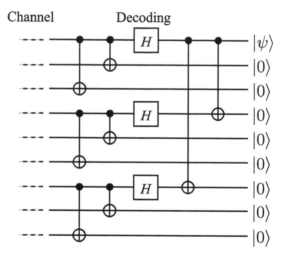

FIGURE 10.7
Decoding circuit for Shor's nine-qubit QECC, which maps $a|+++\rangle + b|---\rangle$
to $(a|0\rangle + b|1\rangle)|00000000\rangle$.

10.3.4   Decoding

The decoding circuit given in Fig. 10.7 is the inverse of the encoding circuit in
Fig. 10.5 (a). Readers should verify that it indeed maps $a|+++\rangle + b|---\rangle$
to $(a|0\rangle + b|1\rangle)|00000000\rangle$.

Although we have not proved that Shor's nine-qubit QECC also corrects Y-
errors, its validity should be clear from the identity $Y = XZ$ (see the following
exercise).

EXERCISE 10.7 Suppose the Y-error occurs in the first qubit of the third
group of three qubits. The state is fed into Bob's circuit in Fig. 10.6.
(1) What is the reading of the syndrome $(A_i, B_i)_{1 \le i \le 3}$?
(2) Find the state after the first layer of the Hadamard gates is applied to the

received state. What is the syndrome (A_4, B_4)?

(3) Repeat (1) and (2) when the error is a continuous rotation $e^{-\gamma Y}$.

The results of this section show that Shor's nine-bit QECC corrects any one-qubit errors, so far as it acts on a single qubit only. In general a QECC which corrects I, X, Y and Z corrects every single-qubit error.

10.4 Seven-Qubit QECC

Shor's nine-qubit QECC is by no means the most efficient QECC. Logical qubits may be implemented with fewer physical qubits as we show in the following sections. We need to summarize classical error correcting codes before we introduce general theory of QECC. See [6] for more extensive accounts of the classical error correcting codes. The set of bit states has been denoted simply as $\{0, 1\}$ so far. Here we want to look at the same set with its field operations, multiplication and addition mod 2. For this reason, we will often use the symbol GF(2) to denote the same set equipped with the field operations.

Our presentation closely follows [3, 4, 1] in this and the next sections.

10.4.1 Classical Theory of Error Correcting Codes

Let us recall the simplest classical error correcting scheme, in which the bit 0 is encoded as 000 and 1 as 111. Three bits ($n = 3$) are used to send a single logical bit ($k = 1$), in which case we say the redundancy is $n - k = 2$ bits.

We assume the noisy transmission channel is symmetric, i.e., $i \in$ GF(2) is received as i with probability $1 - p$, while i is received as $1 \oplus i$ with probability p.

Supppose Alice encodes k bits of information into a code c which is made of n ($> k$) bits and sends it to a receiver Bob through a noisy channel. Bob checks if there were errors in the transmission process by using a **parity check matrix** H, which is designed in such a way that

$$Hc^t = 0 \tag{10.19}$$

when there are no errors during transmission. In contrast, at least one of the bits is flipped during transmission if $Hc^t \neq 0$. The output Hc^t is called the **syndrome**. Bit-flip occurred if a syndrome does not vanish.

Let us consider sending $k = 4$ bits of information by using a code of the length $n = 7$. It is assumed that at most a single bit error occurs during transmission of a codeword. We need to identify which bit has erroneously transmitted, for which three bits are required. We assign 000 for error-free transmission, 001 for error in x_1, 010 for error in x_2 and so forth. Let us write the list of three bits, except for the error-free case 000, as

$$001$$
$$010$$
$$\cdots$$
$$111$$

and take the transpose of the resulting matix to obtain the parity check matrix

$$H = \begin{pmatrix} 0\,0\,0\,1\,1\,1\,1 \\ 0\,1\,1\,0\,0\,1\,1 \\ 1\,0\,1\,0\,1\,0\,1 \end{pmatrix}. \tag{10.20}$$

Let $c_1 = (0, 1, 1, 0, 0, 1, 1)$ be a code Bob receives. He applies H on c^t to obtain

$$Hc_1^t = \begin{pmatrix} 0 \\ 0 \\ 0 \end{pmatrix}.$$

If $c_2 = (0, 1, 1, 1, 0, 1, 1)$ is received, on the other hand, the output of the parity check matrix is

$$Hc_2^t = \begin{pmatrix} 1 \\ 0 \\ 0 \end{pmatrix}$$

and Bob knows there is an error in the code he received. Moreover, he knows from the syndrome that the bit flip error acted on the fourth bit x_4, as we show now. Let $c = (x_1, x_2, \ldots, x_7)$ be the bit string Bob has received. Application of H produces the syndrome

$$Hc^t = \begin{pmatrix} x_4 \oplus x_5 \oplus x_6 \oplus x_7 \\ x_2 \oplus x_3 \oplus x_6 \oplus x_7 \\ x_1 \oplus x_3 \oplus x_5 \oplus x_7 \end{pmatrix}. \tag{10.21}$$

All the components in the RHS should be even numbers if all the bits are received without error. It is found from the syndrome Bob obtained for c_2 that x_1, x_2, x_3, x_5, x_6 and x_7 are received without error because at most a single error is assumed. Error must have occurred in a bit x_4, which is in the first component of the RHS of Eq. (10.21) but not in the second nor the third components. Bob applies a NOT gate on the fourth bit of the codeword he receives to recover the correct codeword.

EXERCISE 10.8 Suppose Bob receives a code $(0, 0, 0, 1, 1, 1, 0)$. What is the syndrome he will obtain after applying the parity check matrix H. Recover the correct code by making error correction. Repeat this for codes $(1, 1, 0, 1, 0, 0, 0)$ and $(1, 1, 0, 0, 1, 1, 1)$.

It should be noted that the syndrome depends only on which bit has suffered from error and not on the original code Alice has sent.

It should be clear that the kernel (mod 2) of H, denoted $C = \ker(H)$, is the set of correct codewords. The condition $Hc^t = 0$ introduces three relations among seven variables $\{x_i\}$, from which we obtain $\dim C = 7 - 3 = 4$, in agreement with the fact $k = 4$. The order of the set C is therefore $2^4 = 16$. The code space C is generated by the matrix

$$
M = \begin{pmatrix} 0\,0\,0\,1\,1\,1\,1 \\ 0\,1\,1\,0\,0\,1\,1 \\ 1\,0\,1\,0\,1\,0\,1 \\ 1\,1\,1\,1\,1\,1\,1 \end{pmatrix}. \tag{10.22}
$$

The matrix M is called the **generating matrix** of the error correcting code C. This particular code generated by the above M is called the **Hamming code**. Note that M is obtained from H by adding the fourth row, with all the components unity. Vectors in C are written as vM, where $v = (i_1, i_2, i_3, i_4)$, $i_k \in \mathrm{GF}(2)$. The table for v and vM is

v	vM	v	vM
(0000)	(0000000)	(1000)	(0001111)
(0001)	(1111111)	(1001)	(1110000)
(0010)	(1010101)	(1010)	(1011010)
(0011)	(0101010)	(1011)	(0100101)
(0100)	(0110011)	(1100)	(0111100)
(0101)	(1001100)	(1101)	(1000011)
(0110)	(1100110)	(1110)	(1101001)
(0111)	(0011001)	(1111)	(0010110).

(10.23)

Let us pick out an element of the code C, which contains even number of 1. They are

$$
\{(0000000), (1010101), (0110011), (1100110),
$$
$$
(0001111), (1011010), (0111100), (1101001)\}. \tag{10.24}
$$

Observe that they are (mod 2)-orthogonal to all the members of C. For example,

$$
(1010101) \cdot (0111100) = 0 \bmod 2,
$$
$$
(1010101) \cdot (0010110) = 0 \bmod 2.
$$

The set of eight elements in (10.24) is denoted as $C^\perp$. Note that they are generated by v of the form $(i_1, i_2, i_3, 0)$ as $(i_1, i_2, i_3, 0)M$, which justifies $\dim C^\perp = 8$. Note also that $C^\perp \subset C$ in the present case.

Our choice of M as a generating matrix for the code is clear by now. Let us write H and M as

$$
H = \begin{pmatrix} h_1 \\ h_2 \\ h_3 \end{pmatrix}, \quad M = \begin{pmatrix} h_1 \\ h_2 \\ h_3 \\ \mathbb{I} \end{pmatrix},
$$

where h_k is a row vector with even number of 1 and $\mathbb{I}$ is a row vector whose all entries are 1. It follows from the orthogonality of $C^{\perp}$ with C that $HM^t = 0$ and hence $Hc^t = HM^t v^t = 0$.

The elements in $C - C^{\perp}$ contain odd number of 1 and are explicitly given as

$$\{(1111111), (0101010), (1001100), (0011001),$$
$$(1110000), (0100101), (1000011), (0010110)\}. \tag{10.25}$$

Suppose $c_1 = (x_1, x_2, \ldots, x_n)$ and $c_2 = (y_1, y_2, \ldots, y_n)$ are codes in C. The **Hamming distance** $d_H(c_1, c_2)$ between c_1 and c_2 is defined as the number of indices i such that $x_i \neq y_i$. In other words, the Hamming distance measures how many bits are different in c_1 and c_2. Let $c_1 = (0101010)$ and $c_2 = (0111100)$, for example. Their Hamming distance is $d_H(c_1, c_2) = 3$. It is easy to see that the Hamming distance satisfies the axioms of distance (see the next exercise).

EXERCISE 10.9 Let d_H be the Hamming distance for a k-bit code space C. Show that

(1) $d_H(c, c) \geq 0$ for any $c \in C$.
(2) $d_H(c_1, c_2) = d_H(c_2, c_1)$ for any $c_1, c_2 \in C$.
(3) $d_H(c_1, c_3) \leq d_H(c_1, c_2) + d_H(c_2, c_3)$ for any $c_1, c_2, c_3 \in C$.

The minimal distance $d_H(C)$ of the code C is defined as the minimum Hamming distance between different elements of C:

$$d_H(C) = \min_{c, c' \in C} d_H(c, c'). \tag{10.26}$$

EXERCISE 10.10 Show that the minimal distance of the code (10.23) is 3.

A code which encodes k bits into a bit string (codeword) with length n and having the minimal distance d is denoted as (n, k, d). The Hamming code is thus characterized as $(7, 4, 3)$.

A single bit flip error generates a bit string whose Hamming distance from the original codeword is 1. Therefore we can tell for sure which codeword is closest to the bit string containing an error (see Fig. 10.8). The original codeword is recovered by applying a NOT gate on an appropriate bit in the string.

It is clear that a code with the minimal distance $d_H(C)$ is able to correct $\lfloor (d_H(C) - 1)/2 \rfloor$ bits of errors, where $\lfloor x \rfloor$ denotes the floor function [6]. It is known that the minimal distance $d_H(C)$ satisfies an inequality (**the Singleton bound**)

$$d_H(C) \leq n - k + 1, \tag{10.27}$$

where n is the codeword length and k is the number of bits to be encoded. It is also known for an error correcting code based on GF(2) that the bound is

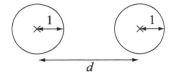

FIGURE 10.8

The crosses ($\times$) denote the closest codewords in a code C, and d is the minimal distance $d_H(C)$. A single bit error sends the codeword to a point on a circle with the radius 1, which does not belong to the code C. The minimal distance d must be at least 3 for single-error detection to make sense. Similarly, n-bit error correction requires the inequality $d - 1 \geq 2n$. Compare this figure with Fig. 10.2.

never saturated and the above equality is further limited as

$$d_H(C) \leq n - k. \tag{10.28}$$

Exercise 10.10 shows that the Hamming code saturates the inequality (10.28). We need to introduce three redundancy bits to attain the minimal distance of 3 in the code space C.

10.4.2 Seven-Qubit QECC

Steane [7] and Calderbank and Shor [8] proposed a seven-qubit QECC scheme inspired by the wisdom of classical error correcting code $(7, 4, 3)$. We index the seven qubits as $i = 0, 1, \ldots, 6$.

A possible error is the bit-flip error X and the phase-flip error Z acting independently on one of seven qubits. They act as $Y = XZ$ if they heppen to operate on a common qubit. Therefore there are $7 + 7 + 7 \times 7 + 1 = 64 = 2^6$ error types, including 1 for no errors, and at least six bits are required to classify the syndrome.

10.4.2.1 Encoding

Let us start with some algebra. Let

$$M_0 = X_4 X_3 X_2 X_1, \quad M_1 = X_5 X_3 X_2 X_0, \quad M_2 = X_6 X_3 X_1 X_0 \tag{10.29}$$

and

$$N_0 = Z_4 Z_3 Z_2 Z_1, \quad N_1 = Z_5 Z_3 Z_2 Z_0, \quad N_2 = Z_6 Z_3 Z_1 Z_0. \tag{10.30}$$

It is obvious that they satisfy $M_i^2 = N_i^2 = I$ and

$$M_i(I + M_i) = I + M_i, \quad [M_i, M_j] = [N_i, N_j] = 0. \tag{10.31}$$

A little extra work is required to show

$$[M_i, N_j] = 0 \tag{10.32}$$

by noting that $[X_i X_j, Z_i Z_j] = 0$ in spite of noncommutativity $X_i Z_i = -Z_i X_i$. We also define the following matrices,

$$\bar{X} \equiv X^{\otimes 7} = X_0 X_1 X_2 X_3 X_4 X_5 X_6, \quad \bar{Z} \equiv Z^{\otimes 7} = Z_0 Z_1 Z_2 Z_3 Z_4 Z_5 Z_6. \quad (10.33)$$

It is important to note that

$$[\bar{X}, M_i] = [\bar{X}, N_i] = 0, \quad [\bar{Z}, M_i] = [\bar{Z}, N_i] = 0. \quad (10.34)$$

We encode $|0\rangle$ and $|1\rangle$ into superpositions of seven-qubit states

$$
\begin{aligned}
|0\rangle_L &= \frac{1}{\sqrt{8}}(I + M_0)(I + M_1)(I + M_2)|0\rangle^{\otimes 7} \\
&= \frac{1}{\sqrt{8}}(|0000000\rangle + |1010101\rangle + |0110011\rangle + |1100110\rangle \\
&\quad + |0001111\rangle + |1011010\rangle + |0111100\rangle + |1101001\rangle) \quad (10.35)
\end{aligned}
$$

and

$$
\begin{aligned}
|1\rangle_L &= \frac{1}{\sqrt{8}}(I + M_0)(I + M_1)(I + M_2)|1\rangle^{\otimes 7} \\
&= \frac{1}{\sqrt{8}}(I + M_0)(I + M_1)(I + M_2)\bar{X}|0\rangle^{\otimes 7} \\
&= \frac{1}{\sqrt{8}}(|1111111\rangle + |0101010\rangle + |1001100\rangle + |0011001\rangle \\
&\quad + |1110000\rangle + |0100101\rangle + |1000011\rangle + |0010110\rangle). \quad (10.36)
\end{aligned}
$$

Observe that they are normalized and the binary basis vectors which comprise $|0\rangle_L$ have even number of 1, while those which comprise $|1\rangle_L$ have odd number of 1, and hence they are orthogonal. Therefore they satisfy $_L\langle x|y\rangle_L = \delta_{xy}$. It is important to note, using the commutation relations among $\{M_i, N_i, \bar{X}, \bar{Z}\}$, that $|0\rangle_L$ and $|1\rangle_L$ are eigenvectors of M_i and N_i with all the eigenvalues $+1$;

$$M_i|x\rangle_L = N_i|x\rangle_L = |x\rangle_L \quad (x \in \mathrm{GF}(2)). \quad (10.37)$$

It should be also noted that basis vectors comprises $|0\rangle_L$ correspond to the codewords given in Eq. (10.24) and those comprises $|1\rangle_L$ are the rest of the codewords in Eq. (10.23). Seven qubits must be required to encode a single logical qubit since there are more types of errors compared to the classical counterpart.

Let us consider the encoding circuit for the coding (10.35) and (10.36). Let us analyze the circuit shown in Fig. 10.9 to begin with. The action of the circuit on $|0\rangle|\psi\rangle$, where $|\psi\rangle$ is an arbitrary n-qubit state, is

$$\frac{1}{\sqrt{2}}\left[|0\rangle|\psi\rangle + |1\rangle(U|\psi\rangle)\right] = \frac{1}{\sqrt{2}}(I + X \otimes U)|0\rangle|\psi\rangle. \quad (10.38)$$

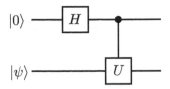

FIGURE 10.9

Hadamard gate $U_H \otimes I$ followed by the controlled-U gate acting on $|0\rangle|\psi\rangle$ produces $\frac{1}{\sqrt{2}}(I + X \otimes U)|0\rangle|\psi\rangle$, where $|\psi\rangle$ is an arbitrary n-qubit state and U is an arbitrary n-qubit gate.

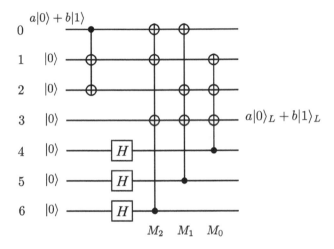

FIGURE 10.10

Encoding circuit for seven-qubit QECC [3]. H stands for the Hadamard gate.

This shows that

$$\frac{1}{\sqrt{2}}(I + M_0) = \frac{1}{\sqrt{2}}(I + X_4 X_3 X_2 X_1),$$

for example, is implemented as a controlled-$X_3 X_2 X_1$ gate where the control bit is the fourth qubit in the state $U_H|0\rangle$.

Collecting these facts, we construct the encoding circuit depicted in Fig. 10.10. The right three controlled-$X^{\otimes 3}$ gates together with the three Hadamard gates in the left implement $(I + M_0)(I + M_1)(I + M_2)$, which maps $|0\rangle^{\otimes 7}$ to $|0\rangle_L$. There is a controlled-$X_1 X_2$ gate in the very beginning of the circuit, which is inert when the input (the 0th qubit) is $|0\rangle$. Suppose the input is $|\psi\rangle = |1\rangle$. Then the input state of the first controlled-$X^{\otimes 3}$ and the Hadamard gates is $X_0 X_1 X_2|0\rangle^{\otimes 7}$. Since X_i commutes with all M_i's, we may move $X_0 X_1 X_2$ to the very end of the circuit to find the output state of this circuit is $X_0 X_1 X_2 |0\rangle_L$. However, we find $M_0 M_1 M_2 = X_3 X_4 X_5 X_6$, and hence

$$X_0 X_1 X_2 = M_0 M_1 M_2 \bar{X}.$$

Now we show

$$X_0 X_1 X_2 |0\rangle_L = M_0 M_1 M_2 \bar{X} \frac{1}{\sqrt{8}} (I + M_0)(I + M_1)(I + M_2) |0\rangle^{\otimes 7}$$

$$= \frac{1}{\sqrt{8}} (I + M_0)(I + M_1)(I + M_2) \bar{X} |0\rangle^{\otimes 7}$$

$$= |1\rangle_L, \tag{10.39}$$

which proves the circuit is indeed the seven-qubit QECC encoder.

10.4.2.2   Error Detection

Now our task is to identify what error occured during transmission by manipulating the received qubits. Note first that X_i, Y_i, Z_i commute with some of $\{M_j, N_j\}$ and anticommute with the rest. Suppose noise X_0 acts on the state $|0\rangle_L$. Then the resulting state is an eigenstate of N_1 with the eigenvalue -1,

$$N_1 X_0 |0\rangle_L = -X_0 N_1 |0\rangle_L = -X_0 |0\rangle_L.$$

It is also verified that

$$N_2 X_0 |0\rangle_L = -X_0 |0\rangle_L, \ N_0 X_0 |0\rangle_L = X_0 |0\rangle_L,$$
$$M_i X_0 |0\rangle_L = X_0 |0\rangle_L \ (i = 0, 1, 2).$$

Therefore the disrupted state $X_0 |0\rangle_L$ is characterized by the eigenvalues $(1, 1, 1; 1, -1, -1)$ of the operators $(M_0, M_1, M_2; N_0, N_1, N_2)$. This fact suggests that we may use the set of these operators as the syndrome to identify the error occured in the received qubit. To justify this conjecture, we have to show that the same eigenvalues are assigned to $X_0 |1\rangle_L$. It is easy to show from $[\bar{X}, N_i] = 0$ this is indeed the case since

$$N_1 X_0 |1\rangle_L = N_1 X_0 \bar{X} |0\rangle_L = -X_0 \bar{X} N_1 |0\rangle_L = -X_0 |1\rangle_L,$$

where we noted that $|1\rangle_L = \bar{X} |0\rangle_L$. These observations show that $X_0 |\Psi\rangle$ with $|\Psi\rangle = a |0\rangle_L + b |1\rangle_L \ (\forall a, b \in \mathbb{C})$ is an eigenvector of $(M_0, M_1, M_2; N_0, N_1, N_2)$ with the eigenvalues $(1, 1, 1; 1, -1, -1)$.

The eigenvalues associated with $X_i |\Psi\rangle$ are $+1$ for all M_i's while

$$
\begin{array}{ccccccc}
 & X_0 & X_1 & X_2 & X_3 & X_4 & X_5 & X_6 \\
N_0 & & * & * & * & * & & \\
N_1 & * & & * & * & & * & \\
N_2 & * & * & & * & & & *
\end{array}
\tag{10.40}
$$

where $*$ stands for the eigenvalue -1 while the empty entry denotes the eigenvalue $+1$. Similarly the eigenvalues associated with $Z_i|\Psi\rangle$ are $+1$ for all N_i's while

$$
\begin{array}{cccccccc}
 & Z_0 & Z_1 & Z_2 & Z_3 & Z_4 & Z_5 & Z_6 \\
M_0 & & * & * & * & * & & \\
M_1 & * & & * & * & & * & \\
M_2 & * & * & & * & & & *
\end{array}
\tag{10.41}
$$

It follows from $Y = XZ$ that the eigenvalues associated with $Y_i|\Psi\rangle$ are

$$
\begin{array}{cccccccc}
 & Y_0 & Y_1 & Y_2 & Y_3 & Y_4 & Y_5 & Y_6 \\
M_0 & & * & * & * & * & & \\
M_1 & * & & * & * & & * & \\
M_2 & * & * & & * & & & * \\
N_0 & & * & * & * & * & & \\
N_1 & * & & * & * & & * & \\
N_2 & * & * & & * & & & *
\end{array}
\tag{10.42}
$$

where use has been made of the relation $Y_i = X_i Z_i$. Note that M_i has an eigenvalue -1 whenever N_i has -1 and *vice versa*. Suppose both N_i and M_j $(i \neq j)$ have eigenvalues -1. This case corresponds to a two-qubit error $X_i Z_j$, with which we will not be concerned.

It was shown above that the set of operators $(M_0, M_1, M_2; N_0, N_1, N_2)$ reveals the syndrome. Now let us consider how they are measured. Let us first consider a simpler case. The syndromes of the bit-flip QECC are measured as shown in Fig. 10.1 (b). It is easy to see the measurement outcome A corresponds to the eigenvalue of the operator $Z_1 Z_2$, while B corresponds to that of $Z_1 Z_3$. It turns out to be convenient to switch the control qubit and the target qubit by making use of the result of Exercise 4.5. Figure 10.11 shows

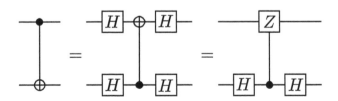

FIGURE 10.11
Switching the control qubits and the target qubit.

the equivalence of the two circuits;

$$
U_{\mathrm{CNOT}} = (I \otimes U_{\mathrm{H}})(I \otimes |0\rangle\langle 0| + Z \otimes |1\rangle\langle 1|)(I \otimes U_{\mathrm{H}}),
$$

where use has been made of the identity $U_{\mathrm{H}} X U_{\mathrm{H}} = Z$. Now the error syndrome detection circuit in Fig. 10.1 (b) takes a suggestive form given in

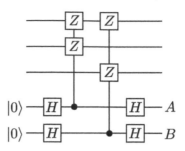

FIGURE 10.12

Error detection circuit of the bit-flip QECC with inverted controlled gates. Note that the controlled unitary gates are $Z_1 Z_2$ and $Z_1 Z_3$, whose eigenvalues fix the syndrome.

Fig. 10.12. Note that the first gate is the controlled-$Z_1 Z_2$ gate, while the second one is the controlled-$Z_1 Z_3$ gate.

The above observation leads to the seven-qubit error detection circuit shown in Fig. 10.13. The column with a product of Z's corresponds to the operator N_i as was shown for the bit-flip QECC. Let us examine the controlled-$X^{\otimes 4}$ gates next. We susupect that they might corresponds to the operators M_i. Let us examine the first controlled gate in Fig. 10.13. The relevant part of the circuit is

$$(I \otimes U_{\mathrm{H}})(I \otimes |0\rangle\langle 0| + M_0 \otimes |1\rangle\langle 1|)(I \otimes U_{\mathrm{H}})|\Psi\rangle|0\rangle$$
$$= \frac{1}{2}(I + M_0)|\Psi\rangle|0\rangle + \frac{1}{2}(I - M_0)|\Psi\rangle|1\rangle, \qquad (10.43)$$

where $|\Psi\rangle$ is an encoded seven-qubit state with a possible sigle-qubit error and the last qubit is the ancilla. Recall that $|\Psi\rangle$ is an eigenvector of M_i with the eigenvalue $+1$ or -1. In case the eigenvalue is $+1$, the output of this part is $|\Psi\rangle|0\rangle$, while it is $|\Psi\rangle|1\rangle$ if the eigenvalue is -1. Accordingly the measurement of the ancilla reveals the eigenvalue of M_0. Similarly other two gates evaluate the eigenvalues of M_1 and M_2. Now we have justified that the set of eigenvalues (i.e., the syndrome) of the received state is evaluated by the measurement of the six ancillary qubits in Fig. 10.13.

Correction of the disrupted encoded state can be done by consulting with Eqs. (10.40), (10.41) and (10.42).

It was shown above that the code space is a subspace of the seven-qubit Hilbert space, in which all the eigenvalues of the six operators $\{M_i, N_i\}$ are $+1$. In other words, encoded state $|\Psi\rangle$ is left invaraint under the action of these operators; $M_i|\Psi\rangle = N_i|\Psi\rangle = |\Psi\rangle$. An error operator E_k, which disrupts the code, is a tensor product of Pauli matrices, which anticommutes with some of M_i and N_i. Therefore the disrupted state has eigenvalue -1 for some operators in $\{M_i, N_i\}$ that anticommute with E_k. The set of the

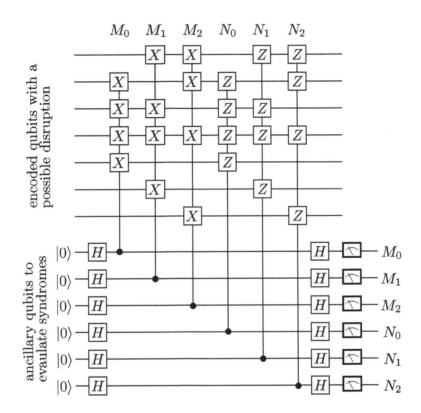

FIGURE 10.13

Error detecting circuit of the seven-qubit QECC. The unitary gates controlled by the ancillary qubits correspond to $\{M_i, N_i\}$, as denoted at the top of the circuit. The readout of the six ancillary qubits fixes the set of syndromes.

measurement outcomes of $\{M_i, N_i\}$ (syndrome) tells us which error operator acted on the encoded state. The Abelian (commuting) group $\{M_i, N_i\}$ is called the **stabilizer** [9, 10], while the subspace $C(S) = \{|\psi\rangle \in \mathbb{C}^{2^n} | \, M_i|\Psi\rangle = N_i|\Psi\rangle = |\Psi\rangle\}$ is called the **stabilizer code** associated with S.

10.4.2.3   Decoding

Once syndromes are detected and an appropriate correction is made, we have to decode the logical qubit so as to extract the input state $|\psi\rangle$. This is done by applying gates in Fig. 10.10 one by one in the reversed order and the initial state will be reproduced in the 0th qubit.

10.4.3   Gate Operations for Seven-Qubit QECC

It is shown in the next section that there exists more efficient encoding scheme, in which merely five qubits are required to encode a single qubit state. In spite of this, the seven-qubit QECC is advantageous over the five-qubit QECC in that gate operations for encoded qubits are easy to implement. Let us look at a few examples.

10.4.3.1   One-Qubit Gates

Let us start with simple examples. Suppose we introduce the $\tilde{X}$ gate which acts on $|0\rangle_L$ and $|1\rangle_L$ as

$$\tilde{X}|0\rangle_L = |1\rangle_L, \ \tilde{X}|1\rangle_L = |0\rangle_L. \tag{10.44}$$

This gate is implemented by $\bar{X} = X_0 X_1 X_2 X_3 X_4 X_5 X_6$ since

$$\bar{X}|0\rangle_L = \bar{X}\frac{1}{\sqrt{8}}(I + M_0)(I + M_1)(I + M_2)|0\rangle^{\otimes 7}$$

$$= \frac{1}{\sqrt{8}}(I + M_0)(I + M_1)(I + M_2)\bar{X}|0\rangle^{\otimes 7}$$

$$= |1\rangle_L, \tag{10.45}$$

where use has been made of the commutativity $[\bar{X}, M_i] = 0$. Application of $\bar{X}$ on Eq. (10.45) proves $\bar{X}|1\rangle_L = |0\rangle_L$.

Let us find $\tilde{Z}$, which satisfies

$$\tilde{Z}|0\rangle_L = |0\rangle_L, \ \tilde{Z}|1\rangle_L = -|1\rangle_L. \tag{10.46}$$

Again this is implemented by $\bar{Z} = Z_0 Z_1 Z_2 Z_3 Z_4 Z_5 Z_6$ since

$$\bar{Z}|0\rangle_L = \frac{1}{\sqrt{8}}(I + M_0)(I + M_1)(I + M_2)\bar{Z}|0\rangle^{\otimes 7}$$

$$= |0\rangle_L \tag{10.47}$$

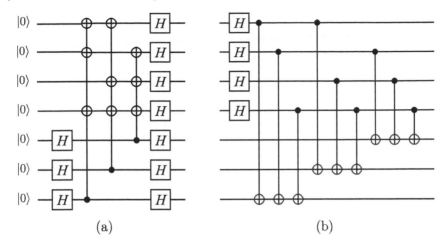

FIGURE 10.14
(a) Seven Hadamard gates applied on the logical qubit state $|x\rangle_L$. (b) A circuit equivalent with (a).

and

$$\bar{Z}|1\rangle_L = \frac{1}{\sqrt{8}}(I + M_0)(I + M_1)(I + M_2)\bar{Z}|1\rangle^{\otimes 7}$$
$$= -|1\rangle_L,\tag{10.48}$$

where we noted that $[\bar{Z}, M_i] = 0$ and $(-1)^7 = -1$.

Next, we consider the Haramard gate $\tilde{H}$ acting on the encoded states as

$$\tilde{U}_H|0\rangle_L = \frac{1}{\sqrt{2}}(|0\rangle_L + |1\rangle_L), \quad \tilde{U}_H|1\rangle_L = \frac{1}{\sqrt{2}}(|0\rangle_L - |1\rangle_L).\tag{10.49}$$

Surprisingly, this gate is also implemented by a tensor product of seven Hadamard gates,

$$\tilde{U}_H = U_H^{\otimes 7} = W_7,\tag{10.50}$$

where W_7 is the Walsh-Hadamard transform acting on seven qubits.

Let us consider the action of W_7 on $|0\rangle_L$ first. The circuit

$$W_7|0\rangle_L = W_7 \frac{1}{\sqrt{8}}(I + M_0)(I + M_1)(I + M_2)|0\rangle^{\otimes 7}$$

given in Fig. 10.14 (a) is put in the form given in (b) by making use of the trick introduced in Fig. 10.9 and $U_H^2 = I$. The resulting circuit is written as

$$W_7|0\rangle_L = [C_0(X_6 X_5)][C_1(X_6 X_4)][C_2(X_5 X_4)]$$
$$\times [C_3(X_6 X_5 X_4)]U_{H0}U_{H1}U_{H2}U_{H3}|0\rangle^{\otimes 7}$$
$$= [C_0(X_6 X_5)U_{H0}][C_1(X_6 X_4)U_{H1}][C_2(X_5 X_4)U_{H2}]$$
$$\times [C_3(X_6 X_5 X_4)U_{H3}]|0\rangle^{\otimes 7},\tag{10.51}$$

where
$$C_i U = |0_i\rangle\langle 0_i| \otimes I + |1_i\rangle\langle 1_i| \otimes U$$
is the controlled-U gate with the control bit i. (The vector $|x_i\rangle$ denotes the state $|x\rangle$ of the ith qubit.) By noting the identities $M_0 M_1 M_2 = X_3 X_4 X_5 X_6$ and (10.38), we obtain

$$
\begin{aligned}
W_7|0\rangle_L &= \frac{1}{4}(I + X_0 X_5 X_6)(I + X_1 X_4 X_6) \\
&\quad \times (I + X_2 X_4 X_5)(I + M_0 M_1 M_2)|0\rangle^{\otimes 7} \\
&= \frac{1}{4}(I + M_0\bar{X})(I + M_1\bar{X})(I + M_2\bar{X}) \\
&\quad \times (I + M_0 M_1 M_2)|0\rangle^{\otimes 7}.
\end{aligned}
\tag{10.52}
$$

After expansion and factorization of the last expression, we obtain

$$
\begin{aligned}
W_7|0\rangle_L &= \frac{1}{\sqrt{8}}(I + M_0)(I + M_1)(I + M_2)\frac{1}{\sqrt{2}}(I + \bar{X})|0\rangle^{\otimes 7} \\
&= \frac{1}{\sqrt{2}}(|0\rangle_L + |1\rangle_L).
\end{aligned}
\tag{10.53}
$$

The action of W_7 on $|1\rangle_L$ is easily obtained by applying W_7 on Eq. (10.53) as
$$W_7^2|0\rangle_L = |0\rangle_L = \frac{1}{\sqrt{2}}(W_7|0\rangle_L + W_7|1\rangle_L),$$
from which we obtain

$$W_7|1\rangle_L = \sqrt{2}|0\rangle_L - \frac{1}{\sqrt{2}}(|0\rangle_L + |1\rangle_L) = \frac{1}{\sqrt{2}}(|0\rangle_L - |1\rangle_L). \tag{10.54}$$

10.4.3.2 CNOT Gate

Another surprise is the CNOT gate acting on two logical qubits is also easily implmenented with the seven-qubit QECC. One simply introduces seven CNOT gates, $U_{\text{CNOT}}^{\otimes 7}$, connecting corresponding physical qubits in two logical qubits as shown in Fig. 10.15. Suppose the control bit is in the state

$$|0\rangle_L = \frac{1}{\sqrt{8}}\left(I + \sum_i M_i + \sum_{i<j} M_i M_j + M_0 M_1 M_2\right)|0\rangle^{\otimes 7}.$$

It follows from the linearity of $U_{\text{CNOT}}^{\otimes 7}$ that each term of the above expansion acts on $|0\rangle_L|x\rangle_L$ independently and outputs the same state since $|x\rangle_L$ is a simultaneous eigenvector of all M_i's with all the eigenvalues $+1$. Let us take $M_0 = X_4 X_3 X_2 X_1$ for example. We find

$$
\begin{aligned}
U_{\text{CNOT}}^{\otimes 7} M_0|0\rangle^{\otimes 7}|x\rangle_L &= U_{\text{CNOT}}^{\otimes 7}|0111100\rangle|x\rangle_L \\
&= |0111100\rangle \otimes X_4 X_3 X_2 X_1|x\rangle_L \\
&= M_0|0\rangle^{\otimes 7} \otimes M_0|x\rangle_L = M_0|0\rangle^{\otimes 7} \otimes |x\rangle_L.
\end{aligned}
$$

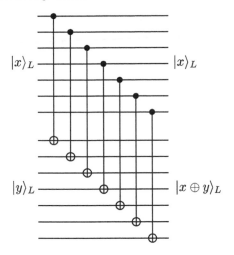

FIGURE 10.15

The CNOT gate acting on two logical qubits is simply a product of seven CNOT gates acting on pairs of physical qubits.

On the other hand, the target state $|x\rangle_L$ is flipped if the control bit is in the state

$$|1\rangle_L = \frac{1}{\sqrt{8}} \left(I + \sum_i M_i + \sum_{i<j} M_i M_j + M_0 M_1 M_2 \right) |1\rangle^{\otimes 7}.$$

Take the above example of M_0 again. We obtain for the component $M_0|1111111\rangle$, the output

$$\begin{aligned}
U^{\otimes 7}_{\text{CNOT}} M_0|1111111\rangle|x\rangle_L &= U^{\otimes 7}_{\text{CNOT}}|1000011\rangle|x\rangle_L \\
&= |1000011\rangle \otimes X_6 X_5 X_0|x\rangle_L \\
&= M_0|1111111\rangle \otimes \bar{X} M_0|x\rangle_L \\
&= M_0|1111111\rangle \otimes \bar{X}|x\rangle_L.
\end{aligned}$$

Clearly the similar output is obtained for other components in the expansion of $|1\rangle_L$.

In summary, we have proved that

$$U^{\otimes 7}_{\text{CNOT}} = |0\rangle_{LL}\langle 0| \otimes I + |1\rangle_{LL}\langle 1| \otimes \bar{X}. \tag{10.55}$$

Note that logical one-qubit gates are made of seven single-qubit gates acting on each physical qubit independently and the logical CNOT gates are made of seven CNOT gates acting on pairs of two qubits, one from the logical control bit and the other from the logical target bit. Since there are only one qubit involved in each logical qubit, any single qubit error during the gate operation may be corrected by the QECC. This desirable property of non-propagating error is called the **fault tolerance**, which is a key component in reliable quantum computing.

10.5 Five-Qubit QECC

DiVincenzo and Shor further reduced the number of qubits required for QECC [11].

Suppose n qubits are used to implement one logical qubit with the basis states $|0\rangle_L$ and $|1\rangle_L$. There are $3n$ operators X_i, Y_i and Z_i ($0 \leq i \leq n - 1$) and correspondingly $3n$ single-qubit errors. Each operator maps the basis $\{|0\rangle_L, |1\rangle_L\}$ to one of $\{X_i|0\rangle_L, X_i|1\rangle_L\}$, $\{Y_i|0\rangle_L, Y_i|1\rangle_L\}$ and $\{Z_i|0\rangle_L, Z_i|1\rangle_L\}$, each of which defines a two-dimensional subspace of 2^n-dimensional vector space associated with the n-qubit system. Therefore n must satisfy $2^n \geq 2(3n + 1)$, 1 for the nondisrupted basis, or equivalently

$$2^{n-1} \geq 3n + 1. \tag{10.56}$$

This inequality is saturated with $n = 5$. In other words, $n = 5$ is the optimal number to correct all types of single-qubit errors.

10.5.1 Encoding

Our exposition goes almost parallel to the previous seven-qubit encoding. Let us consider the algebra of the following operators,

$$\begin{aligned} M_0 &= X_2 X_3 Z_1 Z_4, \quad M_1 = X_3 X_4 Z_2 Z_0 \\ M_2 &= X_4 X_0 Z_3 Z_1 \quad M_3 = X_0 X_1 Z_4 Z_2. \end{aligned} \tag{10.57}$$

EXERCISE 10.11 (1) Show that $M_i^2 = I$. The eigenvalues of these operators are therefore ± 1.
(2) Show that $[M_i, M_j] = 0 \; \forall i, j$.
(3) Let $M_4 = X_1 X_2 Z_0 Z_3$. Show that $M_4 = M_0 M_1 M_2 M_3$. (M_4 is not an independent operator. The eigenvalues of M_0, M_1, M_2 and M_3 fix the eigenvalue of M_4 uniquely.)

Now we introduce the following encoding of logical qubit basis vectors,

$$\begin{aligned} |0\rangle_L &= \frac{1}{4}(I + M_0)(I + M_1)(I + M_2)(I + M_3)|00000\rangle \\ &= \frac{1}{4}\big[|00000\rangle + |11000\rangle + |01100\rangle + |00110\rangle + |00011\rangle + |10001\rangle \end{aligned}$$

$$-|10100\rangle - |01010\rangle - |00101\rangle - |10010\rangle - |01001\rangle$$
$$-|11110\rangle - |01111\rangle - |10111\rangle - |11011\rangle - |11101\rangle] \qquad (10.58)$$

and

$$|1\rangle_L = \frac{1}{4}(I + M_0)(I + M_1)(I + M_2)(I + M_3)|11111\rangle$$
$$= \frac{1}{4}[|11111\rangle + |00111\rangle + |10011\rangle + |11001\rangle + |11100\rangle + |01110\rangle$$
$$-|01011\rangle - |10101\rangle - |11010\rangle - |01101\rangle - |10110\rangle$$
$$-|00001\rangle - |10000\rangle - |01000\rangle - |00100\rangle - |00010\rangle]. \qquad (10.59)$$

It is easy to see that these vectors are orthonormal. Moreover they are simultaneous eigenvectors of $\{M_i\}$ with all the eigenvalues $+1$; $M_i|x\rangle_L = |x\rangle_L$. An input state $|\psi\rangle = a|0\rangle + b|1\rangle$ is encoded into the five-qubit state $|\Psi\rangle = a|0\rangle_L + b|1\rangle_L$.

Next, let us work out a quantum circuit which implements this encoding. We again use the identity

$$(|0\rangle\langle 0| \otimes I + |1\rangle\langle 1| \otimes U)(U_H \otimes I)|0\rangle|\Psi\rangle = \frac{1}{\sqrt{2}}(I + X \otimes U)|0\rangle|\Psi\rangle \qquad (10.60)$$

depicted in Fig. 10.9. Let us consider encoding $|0\rangle_L$ first. We rewrite

$$\frac{1}{\sqrt{2}}(I + M_3)|0\rangle^{\otimes 5} = \frac{1}{\sqrt{2}}(I + X_0 X_1 Z_2 Z_4)|0\rangle^{\otimes 5}$$
$$= \frac{1}{\sqrt{2}}(|0_1\rangle\langle 0_1| \otimes I + |1_1\rangle\langle 1_1| \otimes X_0 Z_2 Z_4)U_{H1}|0\rangle^{\otimes 5}$$
$$(10.61)$$

by making use of Eq. (10.60). Here U_{H1} is the Hadamard gate acting on qubit 1. Now $\frac{1}{\sqrt{2}}(I + M_3)|0\rangle^{\otimes 5}$ has been implemented as the controlled-$X_0 Z_2 Z_4$ gate with the control qubit 1.

Let us consider the action of $\frac{1}{\sqrt{2}}(I + M_2)$ next. We write the state $\frac{1}{\sqrt{2}}(I + M_3)|0\rangle^{\otimes 5}$ as $|0_4\rangle|\phi\rangle$, where we noted that the state of qubit 4 is $|0\rangle$ and wrote the state of the other qubits collectively as $|\phi\rangle$. It is also implemented as a controlled gate as

$$\frac{1}{2}(I + M_2)(I + M_3)|0\rangle^{\otimes 5} = \frac{1}{2}(I + M_2)|0_4\rangle|\phi\rangle$$
$$= \frac{1}{2}(I + X_0 X_4 Z_1 Z_3)|0_4\rangle|\phi\rangle$$
$$= (|0_4\rangle\langle 0_4| \otimes I + |1_4\rangle\langle 1_4| \otimes X_0 Z_1 Z_3)$$
$$\times U_{H4}|0_4\rangle|\phi\rangle. \qquad (10.62)$$

Note that the fourth qubit has been chosen to be the control bit since the initial control qubit state must be $|0\rangle$ for the identity (10.60) to be applicable.

We write the resulting state of this gate operation as $|0_3\rangle|\phi'\rangle$ by noting that the third qubit state is $|0\rangle$. Here $|\phi'\rangle$ collectively denotes the state of the other qubits.

Now we repeat the above substitution further for $\frac{1}{\sqrt{2}}(I + M_1)$ and $\frac{1}{\sqrt{2}}(I + M_0)$. The first operation is replaced by

$$\frac{1}{\sqrt{2}}(|0_3\rangle\langle 0_3| \otimes I + |1_3\rangle\langle 1_3| \otimes X_4 Z_0 Z_2)U_{H3}|0_3\rangle|\phi'\rangle \tag{10.63}$$

by taking the qubit 3 as the control bit. We write the state after this operation as $|0_2\rangle|\phi''\rangle$ by noting that the second qubit state is $|0\rangle$. Now the second operation is replaced by

$$\frac{1}{\sqrt{2}}(|0_2\rangle\langle 0_2| \otimes I + |1_2\rangle\langle 1_2| \otimes X_3 Z_1 Z_4)U_{H2}|0_2\rangle|\phi''\rangle. \tag{10.64}$$

Now we have shown how $|0\rangle_L$ is obtained from $|0\rangle^{\otimes 5}$ by applying four controlled gates.

We use the identity

$$M_3 M_2 M_0 = -X_1 X_2 X_3 X_4 Z_2 Z_3 \tag{10.65}$$

to generate $|1\rangle_L$ when the qubit 0 is in the state $|1\rangle$. We note that

$$|11111\rangle = -M_3 M_2 M_0|10000\rangle = M_3 M_2 M_0 Z_0|10000\rangle. \tag{10.66}$$

Suppose the encoding circuit for $|0\rangle_L$ acts on a state

$$Z_0(a|0\rangle + b|1\rangle)|0000\rangle = a|00000\rangle - b|10000\rangle.$$

Then the circuit outputs

$$\frac{1}{4}(I + M_0)(I + M_1)(I + M_2)(I + M_3)(a|0\rangle - b|1\rangle)|0\rangle^{\otimes 4}$$

$$= \frac{1}{4}(I + M_0)(I + M_1)(I + M_2)(I + M_3)(a|00000\rangle + bM_3 M_2 M_0|11111\rangle)$$

$$= a\frac{1}{4}(I + M_0)(I + M_1)(I + M_2)(I + M_3)|00000\rangle$$

$$+ bM_3 M_2 M_0 \frac{1}{4}(I + M_0)(I + M_1)(I + M_2)(I + M_3)|11111\rangle$$

$$= a|0\rangle_L + bM_3 M_2 M_0|1\rangle_L = a|0\rangle_L + b|1\rangle_L, \tag{10.67}$$

where we used the facts $[M_i, M_j] = 0$ and $|1\rangle_L$ is an simultaneous eigenvector of M_i with all the eigenvalues 1.

This shows that the circuit

$$U_{5\text{-encoding}} = \frac{1}{4}(I + M_0)(I + M_1)(I + M_2)(I + M_3)Z_0, \tag{10.68}$$

depicted in Fig. 10.16 is the correct encoding circuit which acts as

$$U_{5\text{-encoding}}(a|0\rangle + b|1\rangle)|0000\rangle = a|0\rangle_L + b|1\rangle_L. \tag{10.69}$$

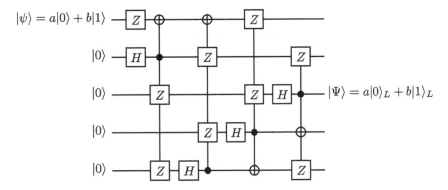

FIGURE 10.16
Encoding circuit for five-qubit QECC.

10.5.2 Error Syndrome Detection

The eigenvalues of four operators $\{M_i\}$ are used to discriminate possible single-qubit disruptions in codewords. Let us recall that $|0\rangle_L$, $|1\rangle_L$ and linear combinations thereof are the eigenvectors of M_i with all the eigenvalues $+1$. The eigenvalues of $\{M_i\}$ corresponding to single-qubit errors are [11, 3]

$$
\begin{array}{c|ccccccccccccccc}
& I & X_0 & X_1 & X_2 & X_3 & X_4 & Y_0 & Y_1 & Y_2 & Y_3 & Y_4 & Z_0 & Z_1 & Z_2 & Z_3 & Z_4 \\
M_0 & & * & & * & * & * & * & & & & & * & * & & & \\
M_1 & * & * & & & * & & * & * & * & & & & * & * & \\
M_2 & * & & * & & * & * & & * & * & * & & & & * & \\
M_3 & & * & & * & * & * & * & & * & * & * & & & & \\
\end{array}
\tag{10.70}
$$

where $*$ denotes the eigenvalue -1, while the eigenvalue $+1$ is denoted by an empty entry. The disrupted state $X_0(a|0\rangle_L + b|1\rangle_L)$, for example, produces the syndrome $(M_0, M_1, M_2, M_3) = (+1, -1, +1, +1)$.

The error detection circuit is constructed following the strategy employed in the seve-qubit QECC. Figure 10.17 shows the error detection circuit associated with our encoding scheme.

Correction of the disrupted encoded state can be done by consulting with Eq. (10.70).

10.5.2.1 Decoding

Once syndrome is determined and an appropriate correction is made, we have to decode the logical qubit so as to extract the input state $|\psi\rangle$. This is done by applying gates in Fig. 10.16 one by one in the reversed order and the initial state will be reproduced in the 0th qubit.

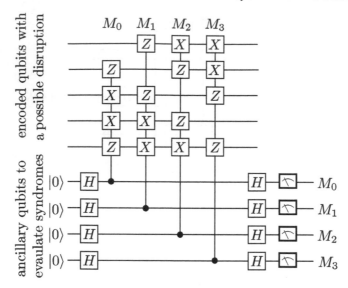

FIGURE 10.17
Error syndrome detecting circuit of the five-qubit QECC. The unitary gates controlled by the ancillary qubits correspond to $\{M_i\}$, as denoted at the top of the circuit. The readout of the four ancillary qubits detects the syndrome.

References

[1] M. A. Neilsen and I. L. Chuang, *Quantum Computation and Quantum Information*, Cambridge University Press (2000).

[2] A. M. Steane, eprint, quant-ph/0304016 (2003).

[3] N. D. Mermin, *Quantum Computer Science: An Introduction*, Cambridge University Press (2007).

[4] A. Hosoya, *Lectures on Quantum Computation* (in Japanese), Science Sha (1999).

[5] P. W. Shor, Phys. Rev. A **52**, 2493 (1995).

[6] F. J. MacWilliams and N. J. A. Sloane, *The Theory of Error-Correcting Codes*, North-Holland, Amsterdam (1977).

[7] A. M. Steane, Phys. Rev. Lett. **77**, 793 (1996).

[8] A. R. Calderbank and P. W. Shor, Phys. Rev. A **54**, 1098 (1996).

[9] D. Gottesman, Phys. Rev. A **54**, 1862 (1996).

[10] A. R. Calderbank *et al.*, Phys. Rev. Lett. **78**, 405 (1997).

[11] D. P. DiVincenzo and P. W. Shor, Phys. Rev. Lett. **77**, 3260 (1996).

Part II

Physical Realizations of Quantum Computing

11

DiVincenzo Criteria

11.1 Introduction

We have learned in Part I that information may be encoded and processed in a quantum-mechanical way. This new discipline called quantum information processing (QIP) is expected to solve a certain class of problems that current digital computers cannot solve in a practical time scale. Although a small-scale quantum information processor, such as quantum key distribution, is already available commercially, physical realization of large-scale quantum information processors is still beyond the scope of our currently available technology.

Classical information is encoded in a bit, which takes on values 0 and 1. We have seen in Part I that 0 and 1 in the classical information processing are replaced by the orthonormal vectors $|0\rangle$ and $|1\rangle$ of a two-dimensional complex vector space in the quantum information theory. Here information is encoded in a quantum bit (qubit), which takes the form $|\psi\rangle = \alpha|0\rangle + \beta|1\rangle$, where $|\alpha|^2 + |\beta|^2 = 1$. A quantum computer should have at least $10^2 \sim 10^3$ qubits to be able to execute algorithms that are more efficient than their classical counterparts. Although a quantum computer with several qubits is already available for some physical systems, actual construction of a working quantum computer is still a challenging task. DiVincenzo proposed necessary conditions, the so-called **DiVincenzo criteria**, that any physical system has to fulfill to be a candidate for a viable quantum computer [1]. In the next section, we outline these conditions as well as two additional criteria for networkability. The DiVincenzo criteria have been analyzed for several physical realizations, and the results of such analyses, as of the year 2004, are summarized in [2]. The following sections summarize the overview article by M Nakahara and M M Salomaa in [2].

11.2   DiVincenzo Criteria

In his influential article [1], DiVincenzo proposed five criteria that any physical system must satisfy to be a viable quantum computer. We summarize the relevant parts of these criteria, which may be helpful in reading subsequent chapters in Part II.

1. *A scalable physical system with well-characterized qubits.*

   To begin with, we need a quantum register made of many qubits to store information. Recall that a classical computer also requires memory to store information. The simplest way to realize a qubit physically is to use a two-level quantum system. For example, an electron, a spin 1/2 nucleus or two mutually orthogonal polarization states (horizontal and vertical, for example) of a single photon can be a qubit. We may also employ a two-dimensional subspace, such as the ground state and the first excited state, of a multi-dimensional Hilbert space, such as atomic energy levels. In the latter case, special care must be taken to avoid leakage of the state to the other part of the Hilbert space. In any case, the two states are identified as the basis vectors, $|0\rangle$ and $|1\rangle$, of the Hilbert space so that a general single qubit state takes the form $|\psi\rangle = \alpha|0\rangle + \beta|1\rangle$, where $|\alpha|^2 + |\beta|^2 = 1$. A multi-qubit state is expanded in terms of the tensor products of these basis vectors. Each qubit must be separately addressable. Moreover it should be scalable up to a large number of qubits. The two-dimensional vector space of a qubit may be extended to be three-dimensional (qutrit) or, more generally, d-dimensional (qudit).

   A system may be made of several different kinds of qubits. Qubits in an ion trap quantum computer, for instance, may be defined as: (1) hyperfine/Zeeman sublevels in the electronic ground state of ions (2) a ground state and an excited state of a weakly allowed optical transition and (3) normal mode of ion oscillation. A similar scenario is also proposed for Josephson junction qubits, in which two flux qubits are coupled through a quantized LC circuit. Simultaneous usage of several types of qubits may be the most promising way to achieve a viable quantum computer.

2. *The ability to initialize the state of the qubits to a simple fiducial state, such as $|00\ldots0\rangle$.*

Suppose you are not able to reset your (classical) computer. Then you will never trust the output of some computation even though processing is done correctly. Therefore initialization is an important part of both quantum and classical information processing.

In many realizations, initialization may be done simply by cooling to put the system in its ground state. Let ΔE be the difference between energies of the first excited state and the ground state. The system is in the ground state with a good precision at low temperatures satisfying $k_B T \ll \Delta E$. Alternatively, we may use projective measurement to project the system onto a desired state. In some cases, we observe the system to be in an undesired state upon such measurement. Then we may transform the system to the desired fiducial state by applying appropriate gates.

For some realizations, such as liquid state NMR, however, it is impossible to cool the system down to extremely low temperatures. In those cases, we are forced to use a thermally populated state as an initial state. This seemingly difficult problem may be amended by several methods if some computational resources are sacrificed. We then obtain an "effective" pure state, the so-called pseudopure state, which works as an initial state for most purposes.

A continuous fresh supply of qubits in a specified state, such as $|0\rangle$, is also an important requirement for successful quantum error correction as we have seen in Chapter 10.

3. *Long decoherence times, much longer than the gate operation time.*

The hardware of a classical computer lasts long, on the order of 10 years. It lasts so long that we often have a problem giving up a healthy computer when the operating system is superseded by a new one. Things are totally different for a quantum computer, which is fragile against external disturbance called decoherence, as we discussed in Chapter 9.

Decoherence is probably the hardest obstacle to building a viable quantum computer. Decoherence means many aspects of quantum state degradation due to interactions of the system with the environment and sets the maximum time available for quantum computation. Roughly speaking, this is the time required for a pure state

$$\rho_0 = (\alpha|0\rangle + \beta|1\rangle)(\alpha^*\langle 0| + \beta^*\langle 1|)$$

to "decay" into a mixed state of the form

$$\rho = |\alpha|^2|0\rangle\langle 0| + |\beta|^2|1\rangle\langle 1|.$$

Decoherence time itself is not very important. What matters is the ratio "decoherence time/gate operation time." For some realizations, decoherence time may be as short as $\sim \mu$s. This is not necessarily a big problem provided that the gate operation time, determined by the Rabi oscillation period and the qubit-coupling strength, for example, is much shorter than the decoherence time. If the typical gate operation time is $\sim$ ps, say, the system may execute $10^{12-6} = 10^6$ gate operations before the quantum state decays. We quote the number $\sim 10^5$ of gates required to factor 21 into 3 and 7 by using Shor's algorithm [3].

There are several ways to effectively prolong decoherence time. A closed-loop control method incorporates quantum error correcting codes (QECC) introduced in Chapter 10, while an open-loop control method incorporates noiseless subsystem [4] and decoherence free subspace (DFS) [5]. Both of these methods, however, require extra qubits.

4. A "universal" set of quantum gates.

Suppose you have a classical computer with a big memory. Now you have to manipulate the data encoded in the memory by applying various logic gates. You must be able to apply arbitrary logic operations on the memory bits to carry out useful information processing. It is known that the NAND gate is universal, i.e., any logic gates may be implemented with NAND gates.

Let $H(\gamma(t))$ be the Hamiltonian of an n-qubit system under consideration, where $\gamma(t)$ collectively denotes the control parameters in the Hamiltonian. The time-development operator of the system is

$$U[\gamma(t)] = \mathcal{T} \exp\left[-\frac{i}{\hbar} \int_0^T H(\gamma(t))dt\right] \in \mathrm{U}(2^n),$$

where $\mathcal{T}$ is the time-ordering operator. Our task is to find the set of control parameters $\gamma(t)$, which implements the desired gate U_{gate} as $U[\gamma(t)] = U_{\mathrm{gate}}$. Although this "inverse problem" seems to be difficult to solve, a theorem by Barenco *et al.*, presented in Chapter 4, guarantees that any $\mathrm{U}(2^n)$ gate may be decomposed into single-qubit gates $\in \mathrm{U}(2)$ and CNOT gates [6]. Therefore it suffices to find the control sequences to implement $\mathrm{U}(2)$ gates and a CNOT gate to construct an arbitrary gate. Note that a general unitary gate in $\mathrm{U}(2^n)$ is written as a product of an $\mathrm{SU}(2^n)$ gate and a physically irrelevant $\mathrm{U}(1)$-phase. Therefore we do not have to worry about the overall phase, and it suffices to concentrate on equivalent $SU(2^n)$ gates. This observation is noteworthy since the NMR Hamiltonian, for example, is traceless and is able to generate $SU(2^n)$ matrices only. Single-qubit gates are easily implemented if the one-qubit part of the Hamiltonian assumes two of the $\mathfrak{su}(2)$ generators by properly choosing the control parameters, where $\mathfrak{su}(2)$ stands for the Lie algebra of $SU(2)$. Implementation of a CNOT gate in any realization is considered to be a milestone in this respect. Note, however, that any two-qubit gates, which are neither a tensor product of two one-qubit gates nor a SWAP gate, work as a component of a universal set of gates [7].

Quantum circuit implementation requires fewer steps if multi-qubit gates acting on n (≥ 3) qubits are employed as modules [3].

5. *A qubit-specific measurement capability.*

The result of classical computation must be displayed on a screen or printed on a sheet of paper to readout the result. Although the readout process in a classical computer is regarded as too trivial a part of computation, it is a vital part in quantum computing.

The state after an execution of a quantum algorithm must be measured to extract the result of the computation. The measurement process depends heavily on the physical system under consideration. For most realizations, projective measurements are the primary method to extract the outcome of a computation. In liquid state NMR, in contrast, a projective measurement is impossible, and we have to resort to ensemble averaged measurements. This may cause a problem in some cases. Suppose the system is in the state $|\psi\rangle = (|00\rangle + |11\rangle)/\sqrt{2}$, for example. The outcome of a projective measurement of $|\psi\rangle$ is $|00\rangle$ with the probability $1/2$. The ensemble averaged measurement, on the other hand, yields *both* $|00\rangle$ and $|11\rangle$ with an equal weight. Another characteristic of ensemble mesasurement is that it is possible to measure noncommuting observables, such as σ_x and σ_y, simultaneously.

Measurement in general has no 100% efficiency due to decoherence, gate operation error and many more reasons. If this is the case, we have to repeat the same computation many times to achieve reasonably high reliability.

Moreover, we should be able to send and store quantum information to construct a quantum data processing network. This "networkability" requires following two additional criteria to be satisfied.

6. *The ability to interconvert stationary and flying qubits.*

> Some realizations are excellent in storing quantum information, while long-distance transmission of quantum information might require different physical resources. It may happen that some system has a Hamiltonian which is easily controllable and is advantageous in executing quantum algorithms. Compare this with a current digital computer, in which the CPU and the system memory are made of semiconductors while a hard disk drive is used as a mass storage device. Therefore a working quantum computer may involve several kinds of qubits and we are forced to introduce distributed quantum computing. Interconverting ability is also important in long-distance quantum teleportation using quantum repeaters.

7. *The ability to faithfully transmit flying qubits between specified locations.*

> Needless to say, this is an indispensable requirement for quantum communication such as quantum key distribution. This condition is also important in distributed quantum computing mentioned above.

The DiVincenzo criteria are not necessarily the gospel, and some conditions can be relaxed. For example, it is possible to replace unitary gates by irreversible non-unitary gates generated by measurements. This idea is already implemented in linear optics quantum computation [8]. An extreme in this approach must be the "one-way quantum computing," where conditional measurements send an initial "cluster state" to the final desired state [9].

There have also been active discussions concerning sufficiency of the criteria and what comes beyond the DiVincenzo criteria. Here is the list of some proposals:

1. Implementation of decoherence-free subsystems/subspaces

2. Implementation of quantum error correction

3. Fault-tolerant quantum computing and

4. Topologically protected qubits.

Gottesman's article "Requirements and Desiderata for Fault-Tolerant Quantum Computing: Beyond the DiVincenzo Criteria"[10] also discusses requirements for fault-tolerant quantum computing such as

1. Low gate error rates

2. Ability to perform operations in parallel

3. A way of remaining in, or returning to, the computational Hilbert space

4. A source of fresh initialized qubits during the computation and

5. Benign error scaling: error rates that do not increase as the computer gets larger, and no large-scale correlated errors.

It also lists "additional desiderata" for a practical quantum computer such as

1. Ability to perform gates between distant qubits

2. Fast and reliable measurement and classical computation

3. Little or no error correlation (unless the registers are linked by a gate)

4. Very low error rates

5. High parallelism

6. An ample supply of extra qubits and

7. Even lower error rates.

Many of the above conditions are necessary for quantum error corrections to work reasonably well.

11.3 Physical Realizations

There are numerous physical systems proposed as possible candidates for a viable quantum computer to date [11]. Here is the list of the candidates;

1. Liquid-state/solid-state NMR/ENDOR

2. Trapped ions

3. Neutral atoms in optical lattice

4. Cavity QED with atoms

5. Linear optics

6. Quantum dots (spin-based, charge-based)

7. Josephson junctions (charge, flux and current-biased qubits)

8. Electrons on liquid helium surface

and other unique realizations.

Subsequent chapters in this book give detailed accounts of some of these realizations in the light of the DiVincenzo criteria. [12] also introduces the above physical realizations. The ARDA QIST roadmap [11] evaluates each of these realizations from the same viewpoints. The roadmap is extremely valuable for the identification and quantification of progress in this multidisciplinary field.

References

[1] D. P. DiVincenzo, Fortschr. Phys. **48**, 771 (2000).

[2] M. Nakahara, S. Kanemitsu, M. M. Salomaa and S. Takagi (eds.), *Physical Realization of Quantum Computing: Are the DiVincenzo Criteria Fulfilled in 2004?* World Scientific, Singapore (2006).

[3] J. J. Vartiainen *et al.*, Phys. Rev. A **70**, 012319 (2004).

[4] E. Knill, R. Laflamme and L. Viola, Phys. Rev. Lett. **84**, 2525 (2000); P. Zanardi, Phys. Rev. A **63**, 012301 (2001); W. G. Ritter, Phys. Rev. A **72**, 012305 (2005).

[5] G. M. Palma, K. A. Suominen and A. K. Ekert, Proc. R. Soc. London A **452**, 567 (1996); L. M. Duan and G. C. Guo, Phys. Rev. Lett. **79**, 1953 (1997); P. Zanardi and M. Rasetti, Phys. Rev. Lett. **79**, 3306 (1997); D. A. Lidar, I. L. Chuang and K. B. Whaley, Phys. Rev. Lett. **81**, 2594 (1998); P. Zanardi, Phys. Rev. A **60**, R729 (1999); D. Bacon, D. A. Lidar and K. B. Whaley, Phys. Rev. A **60**, 1944 (1999).

[6] A. Barenco *et. al.*, Phys. Rev. A **52**, 3457 (1995).

[7] D. P. DiVincenzo, Phys. Rev. A **51**, 1015 (1995).

[8] E. Knill, R. Laflamme and G. J. Milburn, Nature **409**, 46 (2001).

[9] R. Raussendorf and H. J. Briegel, Phys. Rev. Lett. **86**, 5188 (2001).

[10] D. Gottesman, http://www.perimeterinstitute.ca/personal/dgottesman/FTreqs.ppt. See also D. Aharonov and M. Ben-Or, quant-ph/9906129 and J. Preskill, quant-ph/9712048.

[11] http://qist.lanl.gov/

[12] G. Chen *et al.*, *Quantum Computing Devices; Principles, Designs, and Analysis*, Chapman & Hill/CRC (2007).

12

NMR Quantum Computer

12.1 Introduction

The NMR quantum computer is one of the most established systems among many physical realizations of a quantum computer. In spite of its peculiar character associated with mixed states and lack of scalability, it still works as a prototypical quantum computer with at most 10 qubits. We should point out that it is the only quantum computer commercially available at the time of writing this book. For these reasons, an NMR quantum computer is introduced as one of the first examples of a physical system on which we can execute small-scale quantum algorithms. Qubits in this realization are nuclei with spin 1/2. Molecules with a certain number of such nuclei are employed as a quantum register. The system is made of a macroscopic number ($\sim 10^{20}$) of moleclues in thermal equilibrium, and we have to take care of these aspects in preparation of a state and measurements as explained in this chapter. Our exposition follows [1] and [2]. Other useful review is [3].

We employ the symbol

$$I_k = \frac{\sigma_k}{2} \quad (k = x, y, z) \tag{12.1}$$

throughout this chapter, which is a standard notation in the NMR community. We restrict ourselves within liquid state NMR in the present chapter.

12.2 NMR Spectrometer

NMR is a popular measurement instrument in physics, chemistry, pharmacology, medicine and many other areas. Recently it has been recognized that NMR is also the most convenient system on which we can execute quantum algorithms. In the present section, we introduce the relevant aspects of an NMR spectrometer so that the reader may construct his/her own NMR quantum computer.

12.2.1 Molecules

Molecules with a certain number of spin 1/2 nuclei are required to construct an NMR quantum computer. It is tempting to think that a spin 1 nucleus or a higher spin nucleus realizes a qutrit or a qudit more generally. It is known, however, that these higher spin nuclei have very short decoherence times and are not suitable as computational resources. Table 12.1 lists typical molecules used in NMR quantum computer experiments to date. It also lists relevant references. It should be noted that the nucleus of an ordinary carbon atom ^{12}C has no spin and it must be replaced by ^{13}C, whose nucleus has spin 1/2.

TABLE 12.1
Molecules often used in liquid state NMR quantum computation.
Their molecular structures are shown in Fig. 12.1.

Two-qubit molecules		
Molecule	Qubits	References
^{13}C-labelled chloroform	H and ^{13}C	[4], [5]
Partially deuterated cytosine	Two ^1H	[6], [7]
Three-qubit molecules		
Molecule	Qubits	References
^{13}C labelled carbons in alanine	Three ^{13}C	[8], [9]
Trichloroethylene	Two ^{13}C and one ^1H	[8], [10]
Bromotrifluoroethylene	Three ^{19}F	[11]
Five-qubit molecules		
Molecule	Qubits	Reference
Pentafluorobutadienl cyclopentadienyldicarbonyliron complex	Five ^{19}F	[12]
Seven-qubit molecules		
Molecule	Qubits	Reference
Perfluorobutadienyl iron complex with inner two carbons ^{13}C-labelled	Five ^{19}F and two ^{13}C	[13]

Molecules have to be put in a glass tube before they are loaded in the sample chamber of the equipment. Figure 12.2 shows a sample of carbon-13 labelled chloroform in a test tube. It is diluted in a solvent d-6 acetone.

12.2.2 NMR Spectrometer

A test tube containing molecules is placed in an NMR spectrometer. It is under a strong magnetic field B_0 on the order of 10 T, which defines well-defined spin-up and spin-down eigenstates of each nucleus. The energy difference between two spin states is $\hbar \gamma B_0 \equiv \hbar \omega_0$, where γ is the gyromagnetic ratio of

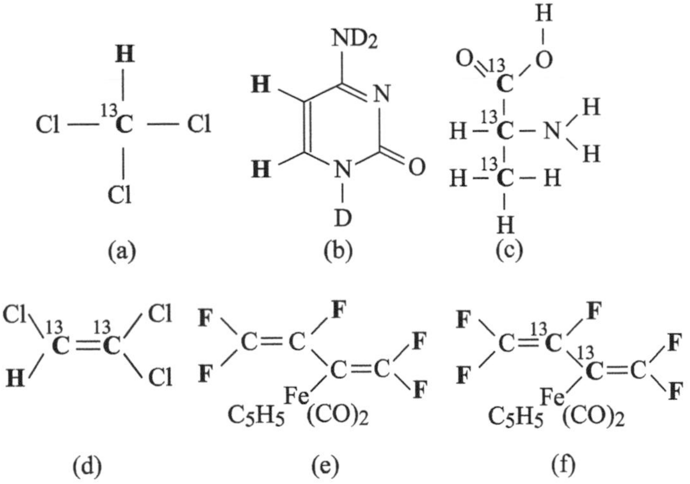

FIGURE 12.1

Structure of molecules listed in Table 12.1. Nuclei working as qubits are indicated in boldface. (a) Chloroform. (b) Partially deuterated cytosine. (c) ^{13}C labelled carbons in alanine. (d) Trichloroethylene. (e) Pentafluorobutadienl cyclopentadienyldicarbonyliron complex. (f) Perfluorobutadienyl iron complex with the inner two carbons ^{13}C-labelled.

the nucleus and ω_0 is called the **Larmor frequency**. The direction of B_0 is taken as the z-axis throughout this chapter.

A radio frequency (rf) magnetic field $B_1(t)$ perpendicular to B_0 is applied through a coil to control the spin state of a nucleus. It has been shown in §2.2 that a resonant rf field induces the Rabi oscillation which controls the spin state. It selectively accesses each spin by tuning its rf frequency ω_{rf} with the Larmor frequency of the target nucleus. The amplitude B_1, the frequency ω_{rf}, the phase ϕ and the pulse shape (square-well, Gaussian and so on) are controllable parameters. The same coil is also used to pick up signals from rotating spins through magnetic induction when measurement is done. See §12.6 for details of measurements. This measurement is not a projective measurement, as was remarked in the beginning of this chapter. The measurement outcome is an ensemble average of an observable and is called an **ensemble measurement**. It should be noted that ensemble measurement is non-demolishing. Moreover, quantum mechanically non-commuting variables may be measured simultaneously if ensemble measurement is employed. Taking advantage of this fact, Grover's search algorithm may be further accelerated if ensemble measurements are available [14]. We can arrange several coils when operating several nuclear species simultaneously. Each coil produces rf pulses for a particular nuclear species and receives induction signals from them. Figure 12.3 shows the schematic diagram of the setup.

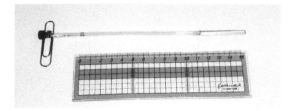

FIGURE 12.2
Trichloroethylene molecules in a test tube diluted in d-6 acetone. Courtesy of
Yasushi Kondo, Kinki University, Japan.

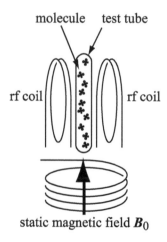

molecule   test tube

rf coil   rf coil

static magnetic field B_0

FIGURE 12.3
Test tube with a macroscopic number of molecules is placed in a strong static
field B_0 and an rf magnetic field $B_1(t)$ generated by a pair of rf coils.

Figure 12.4 shows the whole system of an NMR spectrometer. The parameters (amplitude, frequency, phase and pulse shape) of rf pulses to control
spins are specified by the sequencer in the spectrometer. A sequence of pulses,
namely a sequence of these parameters, is programmed beforehand according
to a quantum algorithm to be executed and fed to the host computer in the
beginning of an experiment.

Figure 12.5 shows the actual NMR spectrometer used to obtain the experimental data shown in this chapter.

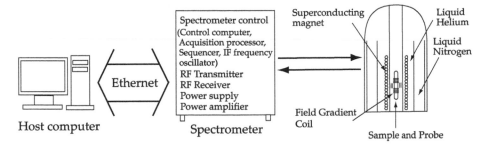

FIGURE 12.4

Schematic diagram of an NMR spectrometer. The test tube and the coils in Fig. 12.3 are in the right-most unit.

12.3 Hamiltonian

12.3.1 Single-Spin Hamiltonian

We are exclusively concerned with room-temperature liquid state NMR in this chapter. Due to rapid random motion of molecules in a liquid at room temperature, both rotational and translational intermolecular interactions are averaged to vanish, and each molecule may be regarded as being isolated from other molecules.

Let us consider a nucleus with spin $1/2$ in a strong static magnetic field $\boldsymbol{B}_0$. A typical value of B_0 employed for quantum computing is on the order of 10 T. Its direction is fixed along the z-axis as we mentioned previously.

The Hamiltonian of this nucleus is

$$H_0 = -\hbar\gamma \boldsymbol{B}_0 \cdot \boldsymbol{I} = -\hbar\omega_0 I_z, \qquad (12.2)$$

where γ is the nuclear gyromagnetic ratio and $\omega_0 = \gamma B_0$ is called the **Larmor frequency**. The eigenvalues of the Hamiltonian are $E_0 = -\hbar\omega_0/2$ and $E_1 = \hbar\omega_0/2$, and the corresponding eigenstates are

$$|0\rangle = \begin{pmatrix} 1 \\ 0 \end{pmatrix}, \quad |1\rangle = \begin{pmatrix} 0 \\ 1 \end{pmatrix}, \qquad (12.3)$$

respectively. Note that the state $|0\rangle$ ($|1\rangle$) denotes the spin up (down) state. Table 12.2 shows the Larmor frequencies of several nuclei which are often utilized in NMR quantum computing.

The spin state of a nucleus can be controlled by applying a radio frequency (rf) magnetic field in the xy-plane. Here we take its direction along the $-x$-axis as

$$\boldsymbol{B}_1(t) = -B_1(t)\cos(\omega_{\mathrm{rf}}t - \phi)\hat{\boldsymbol{x}}, \qquad (12.4)$$

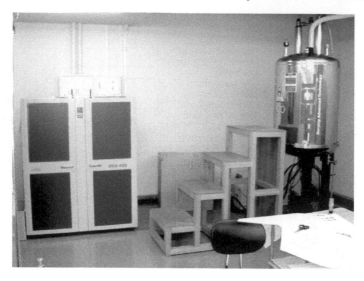

FIGURE 12.5
JEOL ECA-500 NMR spectrometer at Kinki University. The metal cylinder in the right is a superconducting magnet generating B_0. It also contains the test tube and the rf coils. The large box in the left is the spectrometer. The host computer is not shown in this picture.

TABLE 12.2
Larmor frequecies of typical nuclei
at $B_0 = 11.74$ T.

Nucleus	^{1}H	^{13}C	^{19}F	^{31}P
ω_0[MHz]	500	126	470	202

where $\hat{x}$ is the unit vector along the x-axis and $\omega_{\rm rf}$ and $-\phi$ are the angular frequency and the initial phase of the rf field, respectively. This field induces an extra term in the Hamiltonian of the form

$$H_{\rm rf} = 2\hbar\omega_1 \cos(\omega_{\rm rf}t - \phi)I_x, \tag{12.5}$$

where $2\hbar\omega_1 = \hbar\gamma B_1$. The factor 2 has been multiplied to make the corresponding Hamiltonian simpler. The total Hamiltonian in the laboratory frame (i.e., fixed coordinate axes) is therefore

$$H = H_0 + H_{\rm rf} = -\hbar\omega_0 I_z + 2\hbar\omega_1 \cos(\omega_{\rm rf}t - \phi)I_x. \tag{12.6}$$

The parameters ω_1, $\omega_{\rm rf}$ and ϕ are controllable as functions of time, while ω_0 (i.e., B_0) is fixed. It is always assumed in the following that the condition $\omega_0 \gg \omega_1$ is satisfied. Therefore a nuclear spin has two well-defined eigenstates

$|0\rangle = |\uparrow\rangle$, and $|1\rangle = |\downarrow\rangle$ and the rf field acts as a perturbation to control the spin states.

The ratio $k_B T / \hbar \omega_0$ is on the order of 8×10^4, at room temperature of $T \sim 300$ K, for $\omega_0 \sim 500$ MHz, which is the hydrogen Larmor frequency at $B_0 = 11$ T. Therefore, the liquid is in a thermal mixed state. For this reason, we use density matrices rather than wave functions to describe NMR quantum states. The one-spin density matrix of a thermal equilibrium state is

$$\rho(T) = \frac{e^{-H/k_B T}}{Z(T)}, \tag{12.7}$$

where T is the temperature and $Z(T) = \mathrm{tr}\, e^{-H/k_B T}$ is the partition function. The density matrix in the absence of an rf field is

$$\rho(T) = \frac{e^{\hbar \omega_0 I_z / k_B T}}{\mathrm{tr}\, e^{\hbar \omega_0 I_z / k_B T}} = \frac{1}{2}\left[I + \frac{\hbar \omega_0}{k_B T} I_z + O\left(\frac{\hbar \omega_0}{k_B T}\right)^2 \right]. \tag{12.8}$$

The dynamics of a density matrix is given by the Liouville-von Neumann equation

$$i\hbar \frac{d\rho}{dt} = [H, \rho]. \tag{12.9}$$

The Hamiltonian (12.6) has explicit time-dependence through coupling with an rf field. This is inconvenient in integrating the Liouville-von Neumann equation. This problem is solved if we change the frame of reference from the laboratory frame to a frame rotating with the Larmor frequency around the z-axis. Let

$$U_R = e^{-i\hbar \omega I_z t} \tag{12.10}$$

be a unitary transformation to a rotating frame with the angular velocity ω in general (we put $\omega = \omega_0$ later). Here we regard I_z as the generator of rotations around the z-axis. The density matrix is now transformed into

$$\rho_R = U_R \rho U_R^\dagger. \tag{12.11}$$

The Hamiltonian is also transformed to $\tilde{H}$, whose form is derived below. The Liouville-von Neumann equation in the rotating frame takes the same form as Eq. (12.9) and is given by

$$i\hbar \frac{d\rho_R}{dt} = [\tilde{H}, \rho_R]. \tag{12.12}$$

We substitute Eq. (12.10) into the above equation to obtain

$$\tilde{H} = U_R H U_R^\dagger - i\hbar U_R \frac{dU_R^\dagger}{dt}$$
$$= \hbar \begin{pmatrix} -(\omega_0 - \omega)/2 & \omega_1 e^{-i\omega t} \cos(\phi - \omega_{\mathrm{rf}} t) \\ \omega_1 e^{i\omega t} \cos(\phi - \omega_{\mathrm{rf}} t) & (\omega_0 - \omega)/2 \end{pmatrix}$$

$$= \frac{\hbar}{2} \begin{pmatrix} -\omega_0 + \omega \\ \omega_1 \left[e^{i[(\omega-\omega_{rf})t+\phi]} + e^{i[(\omega+\omega_{rf})t-\phi]} \right] \\ \qquad \omega_1 \left[e^{-i[(\omega-\omega_{rf})t+\phi]} + e^{-i[(\omega+\omega_{rf})t-\phi]} \right] \\ \omega_0 - \omega \end{pmatrix}. \quad (12.13)$$

Note that the main contribution $-\hbar\omega_0 I_z$ in the laboratory frame disappears under this transformation if we set $\omega = \omega_0$, which we will assume hereafter. Now we further simplify this Hamiltonian (12.13) by taking the "resonance condition" $\omega_{rf} = \omega_0$. Namely, we take ω_{rf} in resonance with the Larmor frequency ω_0 of the spin. Moreover, we note that the terms oscillating rapidly with the frequency $2\omega_0$ are averaged to vanish if we are interested in the time scale much longer than $1/\omega_0$. This approximation is known as the **rotating wave approximation**. We will see later that flipping a spin by angle π by making use of the Rabi oscillation takes time $\sim 1/\omega_1$, and this is much longer than $1/\omega_0$ due to the assumition $\omega_1 \ll \omega_0$. Therefore it is legitimate to replace the Hamiltonian (12.13) with a simpler time-independent Hamiltonian

$$\tilde{H} = \hbar\omega_1(\cos\phi I_x + \sin\phi I_y) = \hbar\omega_1 \begin{pmatrix} 0 & e^{-i\phi} \\ e^{i\phi} & 0 \end{pmatrix}. \quad (12.14)$$

Note that this Hamiltonian is traceless: $\mathrm{tr}\,\tilde{H} = 0$. A traceless Hamiltonian generates only elements of SU(2). In fact, let $\{\lambda_i\}$ be the set of eigenvalues of $\tilde{H}$ and let $U = e^{-i\tilde{H}t}$ be the time-evolution operator which $\tilde{H}$ generates. Suppose V is another unitary matrix which diagonalizes $\tilde{H}$, and hence U. Then we find $VUV^\dagger = \mathrm{diag}(e^{-i\lambda_j t})$ and

$$\det U = \det VUV^\dagger = \prod_j e^{-i\lambda_j t} = e^{-i\sum_j \lambda_j t} = e^{-i\,\mathrm{tr}\tilde{H}t} = 1.$$

Therefore, the one-qubit NMR Hamiltonian generates SU(2) gates only. Note, however, that this is by no means a restriction. Any $U \in$ U(2) may be mapped to $e^{i\alpha}U \in$ SU(2) by multiplying a proper phase factor. Since this extra overall phase is not observable, we may replace a U(2) gate U with an equivalent SU(2) gate $\tilde{U}$. This remains true for a multi-qubit unitary gate; see the next subsection.

12.3.2 Multi-Spin Hamiltonian

Molecules with n spins are required to execute n-qubit quantum algorithms. Let us consider a linear molecule in which each spin is coupled only to its nearest neighbor spins to simplify our argument. Although a more complicated spin network will be advantageous in saving the number of gates and the execution time, actual implementation requires more elaborated techniques in this case. We will take the natural unit in which $\hbar = 1$ hereafter to simplify mathematical expressions. It will be recovered whenever necessary.

Let us consider a molecule with two spins to begin with. We denote the Larmor frequency of the ith spin by $\omega_{0,i}$ ($i = 1, 2$). We assume there is a Heiseberg type interaction of the form

$$H_{\text{int}} = J \sum_{k=x,y,z} I_k \otimes I_k \qquad (12.15)$$

between spins, where J is the coupling strength. In fact there are other types of interaction including inter-molecular interaction. These interactions are averaged out, thanks to rapid translational and rotational motions of molecules at room temeparatures, and give no contribution to the Hamiltonian (12.15).

Suppose there are two oscillating magnetic fields along the $-x$-axis with frequency $\omega_{\text{rf},i}$ and amplitude $B_{1,i}$ ($i = 1, 2$). The Hamiltonian in the laboratory frame is

$$H = H_0 + H_{\text{rf},1} + H_{\text{rf},2}, \qquad (12.16)$$

where

$$H_0 = -\omega_{0,1} I_z \otimes I - \omega_{0,2} I \otimes I_z + J \sum_{k=x,y,z} I_k \otimes I_k, \qquad (12.17)$$

while

$$H_{\text{rf},1} = 2\omega_{1,1} \cos(\omega_{\text{rf},1} t - \phi_1)(I_x \otimes I + gI \otimes I_x) \qquad (12.18)$$

and

$$H_{\text{rf},2} = 2\omega_{1,2} \cos(\omega_{\text{rf},2} t - \phi_2)(g^{-1} I_x \otimes I + I \otimes I_x), \qquad (12.19)$$

where $2\omega_{1,i} = \gamma_i B_{1,i}$ and $g = \gamma_2/\gamma_1$ is the ratio of the gyromagnetic ratios of two nuclei. Here I is the unit matrix of dimension 2. The first (second) term in the parentheses in Eqs. (12.18) and (12.19) is the interaction Hamiltonian describing the coupling between the first (second) spin and the oscillating fields.

The transformation to a rotating frame of respective spin proceeds similarly to the single-spin case. Let us introduce the transformation

$$U_R = e^{-i\omega_{0,1} I_z t} \otimes e^{-i\omega_{0,2} I_z t}. \qquad (12.20)$$

The Hamitonian $\tilde{H}$ of the spins in respective rotating frames is defined as before as

$$\tilde{H} = \tilde{H}_0 + \tilde{H}_{\text{rf},1} + \tilde{H}_{\text{rf},2}, \qquad (12.21)$$

where

$$
\begin{aligned}
\tilde{H}_0 &= U_R H_0 U_R^\dagger - i U_R \frac{d}{dt} U_R^\dagger \\
&= J \left(e^{-i\omega_{0,1} I_z t} \otimes e^{-i\omega_{0,2} I_z t} \right) \sum_{k=x,y,z} I_k \otimes I_k \left(e^{i\omega_{0,1} I_z t} \otimes e^{i\omega_{0,2} I_z t} \right) \\
&= \pi J \begin{pmatrix} 0 & 0 & 0 & 0 \\ 0 & 0 & e^{i\Delta\omega_0 t} & 0 \\ 0 & e^{-i\Delta\omega_0 t} & 0 & 0 \\ 0 & 0 & 0 & 0 \end{pmatrix} + J I_z \otimes I_z.
\end{aligned} \qquad (12.22)
$$

Here $\Delta\omega_0 \equiv \omega_{0,2} - \omega_{0,1}$ is the difference in the Larmor frequencies of the spins. The matrix elements $e^{\pm i\Delta\omega_0 t}$ are averaged to vanish for the time scale τ satisfying $\Delta\omega_0\tau \gg 2\pi$. Table 12.3 shows relevant parameters for typical two-qubit

TABLE 12.3
Physical parameters of two-spin molecules, ^{13}C labelled chloroform and cytosine. The magnetic field is set to $B_0 = 11.74[\mathrm{T}]$.

	$\omega_{0,1}$	$\omega_{0,2}$	$\Delta\omega_0$	J
Chloroform	500 MHz	100 MHz	400 MHz	200 Hz
Cytosine	500 MHz	500 MHz	765 Hz	7.1 Hz

molecules, ^{13}C labelled chloroform and cytosine. ^{13}C labelled chloroform is a **heteronucleus molecule** whose qubits are hydrogen and ^{13}C nuclei. Cytosine is a **homonucleus molecule**, both qubits of which are hydrogen nuclei. It seems impossible at first glance to address a particular spin in the presence of other spins of the same species since they have the same resonance frequency. However, selective addressing is made possible through the so-called **chemical shift**. The Larmor frequency of a nucleus in a molecule depends not only on the nuclear species but also on its position in the molecule. The electron density at each nucleus varies according to the bonds around it, and therefore the effective magnetic field depends on where a particular nucleus sits in the molecule. This shift in the Larmor frequency is called the chemical shift and allows us to selectively address each nucleus of a properly designed molecule. We cannot employ methane (CH_4) as a four-qubit molecule since all the hydrogen nuclei sit in equivalent positions and therefore have the same chemical shift. Symmetry of the molecule must be broken to produce different chemical shifts.

The pulse width for one-qubit control is typically $\tau \sim 10$ μs for ^{13}C labelled chloroform for which $\Delta\omega_0\tau \sim 4000 \gg 1$. For cytosine, the one-qubit control pulse width τ cannot be too short. Let τ be the pulse width. Then its Fourier transform has a width $\sim 1/\tau$ in the frequency domain. Therefore selective addressing to each spin is impossible unless τ satisfies the condition $1/\tau \ll \Delta\omega_0$, In actual implementation, the pulse width τ is taken such that the condition

$$\Delta\omega_0\tau \gg 1 \gg J\tau \tag{12.23}$$

is satisfied. The second inequality must be satisfied for the effect of the J-coupling to be negligible during the one-qubit operation. Due to a large ratio $\Delta\omega_0/J \sim 10^2$ for cytosine, there always exists such τ which satisfies the condition (12.23). We have to resort to numerical optimization if one of the inequalities is not satified.

Now the interaction Hamiltonian takes a simple Ising form

$$\tilde{H}_0 = J I_z \otimes I_z \tag{12.24}$$

for both heteronucleus and homonucleus molecules, where a time scale $\tau \gg 1/\Delta\omega_0$ is assumed in the latter case. Disappearance of $I_x \otimes I_x$ and $I_y \otimes I_y$ is understood intuitively as follows. Suppose the rf fields are turned off. Then the i-th spin executes free precession with frequency $\omega_{0,i}$ around the z-axis. Since $\omega_{0,1}$ and $\omega_{0,2}$ differ by $\Delta\omega_0$, their x- and y-axes in the rotating frames rotate with relative angular frequencey $\Delta\omega_0$. Therefore, for a time scale τ such that $\Delta\omega_0\tau \gg 1$, the contribution from $I_x \otimes I_x$ and $I_y \otimes I_y$ is averaged out to vanish. The term $I_z \otimes I_z$ does not vanish since the z-axes in the rotating frame remain the same as the laboratory frame for both spins. Application of rf fields merely introduces slow motions of spins in the rotating frames and it does not alter this conclusion.

As for $H_{\text{rf},1}$, we obtain

$$\tilde{H}_{\text{rf},1} = U_R H_{\text{rf},1} U_R^\dagger$$
$$= \omega_{1,1}\Big[\left(e^{i(\omega_{\text{rf},1}t-\phi_1)} + e^{-i(\omega_{\text{rf},1}t-\phi_1)}\right)\left\{\left(e^{-i\omega_{0,1}I_z t}I_x e^{i\omega_{0,1}I_z t}\right) \otimes I\right\}$$
$$+ g\left(e^{i(\omega_{\text{rf},1}t-\phi_1)} + e^{-i(\omega_{\text{rf},1}t-\phi_1)}\right)\left\{I \otimes \left(e^{-i\omega_{0,2}I_z t}I_x e^{i\omega_{0,2}I_z t}\right)\right\}\Big].$$

Now we take the resonance condition $\omega_{\text{rf},i} = \omega_{0,i}$ $(i = 1, 2)$. Then $\tilde{H}_{\text{rf},1}$ is simplified as

$$\tilde{H}_{\text{rf},1} = \frac{\omega_{1,1}}{2}\Big[\begin{pmatrix} 0 & e^{-i\phi_1} \\ e^{i\phi_1} & 0 \end{pmatrix} \otimes I$$
$$+ gI \otimes \begin{pmatrix} 0 & e^{-i(\Delta\omega_0 t+\phi_1)} + e^{-i(\Omega_0 t-\phi_1)} \\ e^{i(\Delta\omega_0 t+\phi_1)} + e^{i(\Omega_0 t-\phi_1)} & 0 \end{pmatrix}\Big],$$

where $\Omega_0 \equiv \omega_{0,1} + \omega_{0,2}$. The second matrix vanishes for τ such that $\Omega\tau, \Delta\omega_0\tau \gg 1$, and finally we obtain

$$\tilde{H}_{\text{rf},1} = \omega_{1,1}\left[\cos\phi_1 I_x \otimes I + \sin\phi_1 I_y \otimes I\right]. \tag{12.25}$$

Similarly we prove that

$$\tilde{H}_{\text{rf},2} = \omega_{1,2}\left[\cos\phi_2 I \otimes I_x + \sin\phi_2 I \otimes I_y\right]. \tag{12.26}$$

In summary, the Hamiltonian for a two-qubit molecule in the rotating frames with respective Larmor frequency is

$$\tilde{H} = J I_z \otimes I_z + \omega_{1,1}\left[\cos\phi_1 I_x \otimes I + \sin\phi_1 I_y \otimes I\right]$$
$$+ \omega_{1,2}\left[\cos\phi_2 I \otimes I_x + \sin\phi_2 I \otimes I_y\right]. \tag{12.27}$$

From a control theoretical point of view, the first term is out of our control and is called the **drift term**, while the second and the third terms, altogether, are called the **control terms** since $\omega_{1,i}$ and ϕ_i are controllable.

Generalization of the above two-qubit Hamiltonian to an n-qubit Hamiltonian is straightforward. For a molecule with n spins coupled linearly, the Hamiltonian in the rotating frame of each spin with angluar frequency $\omega_{0,i}$ takes the form

$$\tilde{H} = \sum_{i=1}^{n-1} J_{i,i+1} I_{z,i} \otimes I_{z,i+1} + \sum_{i=1}^{n} \omega_{1i}(\cos\phi_i I_{x,i} + \sin\phi_i I_{y,i}), \qquad (12.28)$$

where $J_{i,i+1}$ stands for the coupling strength between spins i and $i+1$ and $I_{k,i} = I \otimes \ldots \otimes I_k \otimes \ldots \otimes I$ with I_k in the ith position. The resonance condition $\omega_{\mathrm{rf},i} = \omega_{0,i}$ and linear configuration of n spins are understood in deriving Eq. (12.28).

We will work exclusively with Hamiltonians in the rotating frame of each spin in the rest of this chapter.

12.4   Implementation of Gates and Algorithms

The Hamiltonians introduced in the previous section are employed to implement quantum gates. Here we consider one-, two-, and multi-qubit gates separately.

12.4.1   One-Qubit Gates in One-Qubit Molecule

The Hamiltonian

$$\tilde{H} = \omega_1(\cos\phi I_x + \sin\phi I_y)$$

contains only I_x and I_y as SU(2) generators. This is not a problem though since rotations generated by I_z can be implemented with $I_{x,y}$ generators as we see below. Let us define SU(2) gates which are often employed as building blocks of quantum circuits. Let $X, Y, Z, \bar{X}, \bar{Y}$ and $\bar{Z}$ be rotations by $\pi/2$ around $\tilde{x}$-, $\tilde{y}$-, $\tilde{z}$-, $-\tilde{x}$-, $-\tilde{y}$- and $-\tilde{z}$-axes respectively. Their explicit forms as SU(2) matrices are

$$X = e^{-i(\pi/2)I_x} = \frac{1}{\sqrt{2}}\begin{pmatrix} 1 & -i \\ -i & 1 \end{pmatrix}, \; Y = e^{-i(\pi/2)I_y} = \frac{1}{\sqrt{2}}\begin{pmatrix} 1 & -1 \\ 1 & 1 \end{pmatrix},$$

$$Z = e^{-i(\pi/2)I_z} = \frac{1}{\sqrt{2}}\begin{pmatrix} 1-i & 0 \\ 0 & 1+i \end{pmatrix}, \; \bar{X} = e^{i(\pi/2)I_x} = \frac{1}{\sqrt{2}}\begin{pmatrix} 1 & i \\ i & 1 \end{pmatrix}, \qquad (12.29)$$

$$\bar{Y} = e^{i(\pi/2)I_y} = \frac{1}{\sqrt{2}}\begin{pmatrix} 1 & 1 \\ -1 & 1 \end{pmatrix}, \; \bar{Z} = e^{i(\pi/2)I_z} = \frac{1}{\sqrt{2}}\begin{pmatrix} 1+i & 0 \\ 0 & 1-i \end{pmatrix}.$$

It is useful for later purposes to write down the explicit form of a gate $R(\theta, \phi)$, whose rotation angle is θ and phase angle is ϕ in the xy-plane,

$$R(\theta, \phi) = e^{-i\theta(\cos\phi I_x + \sin\phi I_y)} = \cos\frac{\theta}{2} I - 2i\sin\frac{\theta}{2}(\cos\phi I_x + \sin\phi I_y)$$

$$= \begin{pmatrix} \cos\dfrac{\theta}{2} & -i\sin\dfrac{\theta}{2}e^{-i\phi} \\ -i\sin\dfrac{\theta}{2}e^{i\phi} & \cos\dfrac{\theta}{2} \end{pmatrix}. \tag{12.30}$$

Let us consider implementing X, for example. We need to find parameters ϕ, ω_1 and τ such that

$$e^{-i\int_0^\tau \tilde{H}dt} = e^{-i\omega_1\tau(\cos\phi I_x + \sin\phi I_y)} = e^{-i\pi I_x/2}.$$

It is easily found that a pulse with phase $\phi = 0$, amplitude ω_1 and duration τ satisfying $\omega_1\tau = \pi/2$ does the job. We assume here the pulse shape is square and express it graphically as in Fig. 12.6. The parameter τ is called the **pulse width**. More sophisticated pulses are available, but we restrict ourselves within square pulses to simplify our calculation. Similarly $Y, \bar{X}, \bar{Y}$ are obtaind by applying pulses with $\phi = \pi/2, \pi$ and $-\pi/2$ and pulse duration $\tau = \pi/2\omega_1$, respectively. A typical value for ω_1 is ~ 100 kHz for heteronucleus molecules, and the above operation is implemented with the pulse width $\tau \sim 1/\omega_1 \sim 10$ μs.

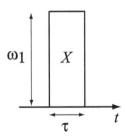

FIGURE 12.6
Square pulse. Amplitude of a continuous wave with frequency $\omega_{\rm rf}$ is modulated by this pulse. The amplitude corresponds to ω_1 and the pulse width to τ. They satisfy $\omega_1\tau = \pi/2$.

EXERCISE 12.1 Let $\phi = \pi/4$ in $\tilde{H}$ and write down the unitary matrix which $\tilde{H}$ generates when $\omega_1\tau = \pi/2$. Apply this unitary matrix to $|0\rangle$ and $|1\rangle$ and find the states obtained.

The Hamiltonian lacks the generator I_z. This does not imply Z and $\bar{Z}$ cannot be implemented with the Hamiltonian $\tilde{H}$. There are three ways to implement $U = e^{-i\alpha I_z}$ with $\tilde{H}$. The simplest one is to shift the clock of the NMR by a certain amount of time τ_z. The frame is rotating with the angular velocity ω_0 around the z-axis, and if we shift the clock of the NMR sequencer by τ_z, we will obtain a rotation equivalent with $U = e^{-i\omega_0 \tau_z I_z}$. Thus shifting the clock by $\tau_z = \alpha/\omega_0$ implements the gate $e^{-i\alpha I_z}$. The second one is to literally generate $U = e^{-i\alpha I_z}$ with $I_{x,y}$. By noting the identity $I_z = e^{-i(\pi/2)I_x} I_y e^{i(\pi/2)I_x}$ we immediately obtain

$$e^{-i\alpha I_z} = e^{-i(\pi/2)I_x} e^{-i\alpha I_y} e^{i(\pi/2)I_x}. \tag{12.31}$$

The third one is applicable only in the end of the computation and when the spin is in one of the eigenstates of I_z. It is clear that $e^{-i\alpha I_z}|j\rangle \sim |j\rangle$ if the phase is ignored since $\sigma_z|j\rangle = \pm|j\rangle$. Therefore if we can shift some of the I_z rotation matrices toward the very end of the algorithm, we may ignore them at all. Now we have shown that $\tilde{H}$ generates all SU(2) rotations.

The following relations are useful in designing pulse sequences for NMR quantum computing:

$$\begin{array}{llll}
XY\bar{X} = Z, & \bar{Y}XY = Z, & \bar{X}\bar{Y}X = Z, & Y\bar{X}\bar{Y} = Z \\
\bar{X}YX = \bar{Z}, & YX\bar{Y} = \bar{Z}, & X\bar{Y}\bar{X} = \bar{Z}, & \bar{Y}\bar{X}Y = \bar{Z} \\
XY = ZX, & XY = YZ, & \bar{Y}X = XZ, & Y\bar{X} = ZY \\
XZ = Z\bar{Y}, & \bar{Y}Z = Z\bar{X}, & \bar{X}Z = ZY, & YZ = ZX \\
XZZ = ZZ\bar{X}, & & YZZ = ZZ\bar{Y}. &
\end{array} \tag{12.32}$$

By making use of these relations, it becomes possible to replace Z and $\bar{Z}$ with other rotations. It also becomes possible to eliminate some of Z and $\bar{Z}$ by sending them to the both ends of a pulse sequence.

EXERCISE 12.2 Verify the above relations.

It is clear that the Hamiltonian $\tilde{H}$ is independent of t so far as ω_1 and ϕ are time-independent. In general, ω_1 and ϕ may change as functions of time. In actual experiments, they are often taken to be piecewise constant, for which case the time-evolution operator is given by

$$U = \mathcal{T} e^{-i\int_0^T \tilde{H}(t)dt} \equiv e^{-i\tilde{H}(t_n)\Delta t_n} e^{-i\tilde{H}(t_{n-1})\Delta t_{n-1}} \ldots e^{-i\tilde{H}(t_1)\Delta t_1}, \tag{12.33}$$

where $\mathcal{T}$ stands for the time-ordered product and

$$\tilde{H}(t_k) = \omega_1(t_k)\left[\cos\phi(t_k)I_x + \sin\phi(t_k)I_y\right]$$

is the Hamiltonian at the kth step whose temporal duration is Δt_k.

EXAMPLE 12.1 Let us consider implementing the Hadamard gate

$$U_{\mathrm{H}} = \frac{1}{\sqrt{2}}\begin{pmatrix} 1 & 1 \\ 1 & -1 \end{pmatrix}$$

with our Hamiltonian $\tilde{H}$. Since $\det U_H = -1$, we have to multiply i to U_H to make it an element of SU(2). (The factor $-i$ also does the job.) We are tempted to use Eq. (1.44) to find parameters ω_1, ϕ and τ such that

$$\tilde{H}\tau = -\frac{\pi}{\sqrt{2}}(I_x + I_z),$$

which certainly satisfies $e^{-i\tilde{H}\tau} = U_H$. However, this does not work since we do not have an I_z term in $\tilde{H}$. Therefore, we have to implement U_H using the formula (cf. Lemma 4.2)

$$
\begin{aligned}
&e^{-i\alpha I_x} e^{-i\beta I_y} e^{-i\gamma I_x} \\
&= \begin{pmatrix}
\cos\left(\frac{\beta}{2}\right)\cos\left(\frac{\alpha+\gamma}{2}\right) - i\sin\left(\frac{\beta}{2}\right)\sin\left(\frac{\alpha-\gamma}{2}\right) \\
\cos\left(\frac{\alpha-\gamma}{2}\right)\sin\left(\frac{\beta}{2}\right) - i\cos\left(\frac{\beta}{2}\right)\sin\left(\frac{\alpha+\gamma}{2}\right)
\end{pmatrix. \\
&\qquad \begin{pmatrix}
-\cos\left(\frac{\alpha-\gamma}{2}\right)\sin\left(\frac{\beta}{2}\right) - i\cos\left(\frac{\beta}{2}\right)\sin\left(\frac{\alpha+\gamma}{2}\right) \\
\cos\left(\frac{\beta}{2}\right)\cos\left(\frac{\alpha+\gamma}{2}\right) + i\sin\left(\frac{\beta}{2}\right)\sin\left(\frac{\alpha-\gamma}{2}\right)
\end{pmatrix}.
\end{aligned}
\tag{12.34}
$$

Comparison between U_H and the above expression immediately leads to the following solution:

$$\alpha = -\pi, \beta = \frac{\pi}{2}, \gamma = 0, \tag{12.35}$$

for example. Therefore U_H is implemented by two square pulses as

$$U_H = e^{i\omega_1\tau_2 I_x} e^{-i\omega_1\tau_1 I_y} = \bar{X}^2 Y, \tag{12.36}$$

where $\omega_1\tau_1 = \pi/2$ and $\omega_1\tau_2 = \pi$. The amplitude ω_1 need not be the same for the two pulses, but there is no reason to employ different amplitude either. The amplitude should be large to implement a gate with a shorter pulse width. However, a large amplitude pulse leads to overcurrent in the rf coil and eventually damages the coil. A typical pulse width for a π-pulse is on the order of 10 μs as mentioned before.

Using the symbols introduced above, this pulse sequence is conveniently expressed as

$$U_H : -Y - \bar{X}^2 - . \tag{12.37}$$

The time flows from left to right as before. We also describe the pulse sequence graphically as in Fig. 12.7.

EXERCISE 12.3 Implement the phase shift gate

$$U(\theta) = \begin{pmatrix} 1 & 0 \\ 0 & e^{-i\theta} \end{pmatrix} \tag{12.38}$$

using the Hamiltonian $\tilde{H}$. Here θ is a real constant. Note that $U \notin$ SU(2) and a phase must be multiplied to make it an element of SU(2).

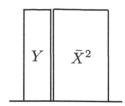

FIGURE 12.7
Control pulse sequence to implement the Hadamard gate H. $\bar{X}^2$ is a π-pulse around $-x$-axis. The time flows from left to right.

12.4.2 One-Qubit Operation in Two-Qubit Molecule: Bloch-Siegert Effect

Let us consider the effect of an off-resonance pulse on a qubit. We have to consider this effect when we have a multi-qubit molecule with several nuclei of the same species; addressing to one qubit may affect the other qubits of the same species since they have close resonance frequencies.

We first consider the effect of an off-resonance pulse on a one-qubit molecule. Let $\omega_{\text{rf}} = \omega_0 + \delta$, δ being the detuning parameter. Then we find from Eqs. (12.13) and (12.14) that

$$\tilde{H} = \delta I_z + \omega_1 \left(\cos \phi I_x + \sin \phi I_y \right) = \delta \left(\epsilon \cos \phi I_x + \epsilon \sin \phi I_y + I_z \right)$$
$$= \delta \sqrt{1 + \epsilon^2} \hat{\boldsymbol{n}} \cdot \boldsymbol{I}, \tag{12.39}$$

where $\epsilon = \omega_1/\delta$ and

$$\hat{\boldsymbol{n}} = \frac{1}{\sqrt{1 + \epsilon^2}} \left(\epsilon \cos \phi, \epsilon \sin \phi, 1 \right)^t \tag{12.40}$$

is a unit vector. The time-development operator is

$$U(t) = e^{-i\tilde{H}t} = e^{-i\delta\sqrt{1+\epsilon^2}\hat{\boldsymbol{n}} \cdot \boldsymbol{I}t}. \tag{12.41}$$

Suppose the detuning is large enough compared to ω_1 so that $|\epsilon| \ll 1$. Then it follows that $\hat{\boldsymbol{n}} \simeq (0, 0, 1)^t$, and we have an approximation

$$U(t) \simeq e^{-i\delta\sqrt{1+\epsilon^2}I_z t}. \tag{12.42}$$

In fact, the rotation axis $\hat{\boldsymbol{n}}$ is slightly tilted from the z-axis and the spin precesses around this axis, which remains near the z-axis. This observation justifies the negligence of components $I_{x,y}$ in Eq. (12.42). However, the effect of ϵ in the square root is not negligible if we are concerned with a long-term behavior of the spin, in which $t\delta\epsilon^2$ is sizeable. This effect is called the **Bloch-Siegert effect** [15], and this shift in the reference phase must be taken into account when designing pulse sequences which involve detuned rf fields. Suppose, for example, that $\epsilon = 10^{-1}$ and $\delta t = 20\pi$. Then we obtain $\delta\sqrt{1 + \epsilon^2}t - \delta t \simeq 0.31$ rad $\simeq 18°$, which is not negligible at all.

EXERCISE 12.4 Suppose a spin is in the state $| \uparrow \rangle$ at $t = 0$ and its time-development is driven by the operator (12.41). Find the spin wave function at later time $t > 0$. Find when the spin comes back to the initial state up to an overall phase.

Next we consider manipulating a single qubit in a two-qubit molecule. In case of a heteronucleus molecule, an rf field in resonance with one of the qubits has no effect on the other qubit. In this case, $\epsilon = \omega_1/\delta$ is typically on the order of 10^{-3}. For $\delta t = 20\pi$ as before, we obtain $\delta\sqrt{1 + \epsilon^2}t - \delta t \simeq 3 \times 10^{-5}$ rad $\simeq 1.8 \times 10^{-3}$ deg. If, in contrast, a homonucleus molecule is considered, we have to take a small amplitude pulse with $\omega_1 \ll \Delta\omega_0$, $\Delta\omega_0$ being the difference in the Larmor frequencies of two nuclei of the same species, for selective addressing to a particular qubit. This makes the pulse width τ longer, since $\omega_1\tau$ specifies the rotation angle. The effect of the J-coupling may not be negligible if $\tau \gtrsim 1/J$.

Let us consider the opposite limit in which $\delta \ll \omega_1$. This takes place when we apply a hard pulse (i.e., very short pulse) in resonance with one of the qubits, qubit 2, say, in a homonucleus molecule. Equation (12.39) with $\delta \ll \omega_1$ leads to a Hamiltonian $\tilde{H} \simeq \omega_1(\cos\phi I_x + \sin\phi I_y)$ acting on qubit 1. Therefore qubit 1 also gets rotated by the same amount as qubit 2. In other words, by applying a hard pulse in resonance with one of the qubits, both qubits are rotated similtaneously by the same angle. Therefore a gate $I \otimes U$, which is meant to act on the second qubit, works as $U \otimes U$ if it is implemented with a hard pulse.

12.4.3 Two-Qubit Gates

Any n-qubit gate may be implemented with single-qubit gates and the CNOT gates according to the universality theorem by Barenco *et al.* [16]. We have shown in the previous subsection how single-qubit gates are implemented. Let us consider the CNOT gate here. We recall that

$$U_{\text{CNOT}} = \begin{pmatrix} 1 & 0 & 0 & 0 \\ 0 & 1 & 0 & 0 \\ 0 & 0 & 0 & 1 \\ 0 & 0 & 1 & 0 \end{pmatrix}.$$

Note again that $\det U_{\text{CNOT}} = -1$ and we have to multiply U_{CNOT} by $e^{\pm i\pi/4}$, for example, to make it an element of SU(4). We must employ the J-coupling term to implement the CNOT gate since it cannot be decomposed into a tensor product of two SU(2) gates. A standard implementation of the CNOT

gate is [2]
$$U_{\text{CNOT}} = Z_1 \bar{Z}_2 X_2 U_J(\pi/J) Y_2, \tag{12.43}$$
where X_j is a $\pi/2$-rotation around the x-axis of the jth qubit while $\bar{X}_j$ is a $\pi/2$-rotation around the $-x$ axis of the jth qubit, for example. Explicitly,
$$X_1 = e^{-i\pi I_x/2} \otimes I, \quad \bar{X}_2 = I \otimes e^{-i\pi I_x/2},$$
for example. The matrix $U_J(\tau)$ is generated solely by the J-coupling term, without any rf pulses applied during period of time τ, as
$$U_J(\tau) = e^{-iJI_z \otimes I_z \tau} = \begin{pmatrix} e^{-iJ\tau/4} & 0 & 0 & 0 \\ 0 & e^{iJ\tau/4} & 0 & 0 \\ 0 & 0 & e^{iJ\tau/4} & 0 \\ 0 & 0 & 0 & e^{-iJ\tau/4} \end{pmatrix}. \tag{12.44}$$

Therefore
$$U_J(\pi/J) = e^{-i\pi I_z \otimes I_z} = \begin{pmatrix} e^{-i\pi/4} & 0 & 0 & 0 \\ 0 & e^{i\pi/4} & 0 & 0 \\ 0 & 0 & e^{i\pi/4} & 0 \\ 0 & 0 & 0 & e^{-i\pi/4} \end{pmatrix}. \tag{12.45}$$

Then it is easy to find that the LHS of Eq. (12.43) takes the form
$$Z_1 \bar{Z}_2 X_2 U_J(\pi/J) Y_2 = e^{-i\pi/4} \begin{pmatrix} 1 & 0 & 0 & 0 \\ 0 & 1 & 0 & 0 \\ 0 & 0 & 0 & 1 \\ 0 & 0 & 1 & 0 \end{pmatrix}$$
as promised.

EXERCISE 12.5 Implement the "inverted" CNOT gate
$$U_{\text{CNOT}'} = \begin{pmatrix} 1 & 0 & 0 & 0 \\ 0 & 0 & 0 & 1 \\ 0 & 0 & 1 & 0 \\ 0 & 1 & 0 & 0 \end{pmatrix}$$
with the two-qubit NMR Hamiltonian.

It is clear from the construction that the expansion such as Eq. (12.43) requires certain degrees of expertise in NMR pulse programming and/or trial and error to adjust all the matrix elements. We introduce in §12.5 a remarkable technique fully utilizing the theory of Lie algebras and Lie groups to obtain implementations of any two-qubit unitary gates. Although this technique is model independent, it is best suited for NMR quantum computing due to the reasons to be clarified below.

In principle, therefore, an NMR quantum computer is universal, and any $U(2^n)$ gate may be implemented by properly choosing the control parameters. One might wonder how a one- or two-qubit gate is embedded in a multi-qubit molecule. This is the subject of the next subsection.

12.4.4 Multi-Qubit Gates

Suppose we have a molecule with many qubits coupled linearly. Clearly we cannot turn off inter-qubit couplings even when we do not need them. This is rather inconvenient if we want to employ one-qubit gates and CNOT gates as building blocks of quantum algorithms. One-qubit operations are executed faster compared to $1/J$, and the effect of J-couplings is safely negligible. In contrast, a two-qubit operation involves a particular J-coupling, and we have to get rid of the time-evolution of the state due to other J-couplings. This "interaction on demand" is possible if a technique called **refocusing** is employed. Refocusing cancels unwanted inter-qubit couplings.

12.4.4.1 Three-Qubit Case

This is best understood from the following example of a three-qubit molecule. Suppose there are three spins 1, 2 and 3 in the molecule, each with the Larmor frequency $\omega_{0,i}$, ($i = 1, 2, 3$). The coupling strength between 1 and 2 is J_{12} and that between 2 and 3 is J_{23}. It is assumed that the spins are linear so that $J_{31} = 0$. The Hamiltonian in the rotating frame of each qubit is

$$\tilde{H} = J_{12} I_z \otimes I_z \otimes I + J_{23} I \otimes I_z \otimes I_z + \omega_{11}(\cos\phi_1 I_{1x} + \sin\phi_1 I_{1y})$$
$$+ \omega_{12}(\cos\phi_2 I_{2x} + \sin\phi_2 I_{2y}) + \omega_{13}(\cos\phi_3 I_{3x} + \sin\phi_3 I_{3y}), \quad (12.46)$$

where $I_{1x} = I_x \otimes I \otimes I$, for example.

Suppose we want to implement a gate

$$U_\alpha = \exp(-i\alpha I_z \otimes I_z \otimes I). \quad (12.47)$$

If it were not for the J_{23} coupling, we just need to turn off all the rf pulses and wait for a duration $\tau = \alpha/J_{12}$. In our case, however, the coupling between qubits 2 and 3 is also active, producing unwanted contribution

$$\exp\left(-iJ_{23}(\alpha/J_{12}) I \otimes I_z \otimes I_z\right),$$

which must be somehow nullified. The trick is to use the identity

$$e^{i\pi I_x} I_z e^{-i\pi I_x} = \bar{X}^2 I_z X^2 = -I_z.$$

To get rid of undesirable time evolution due to J_{23}, we apply the first π-pulse $e^{-i\pi I_{x3}}$ on the third qubit at $\tau/2$ and then allow the molecule to evolve freely for another duration $\tau/2$. The extra contribution cancels out by this flipping of the third qubit. Finally apply the second π-pulse $e^{i\pi I_{x3}}$ on the third qubit so that it comes back to its correct history.

More explicitly we verify that

$$U = \mathcal{T} \exp\left[-i \int_0^\tau \tilde{H}(t) dt\right]$$

$$
= (I \otimes I \otimes \bar{X}) \exp\left(-i\frac{\tau}{2}\tilde{H}_0\right)(I \otimes I \otimes X)\exp\left(-i\frac{\tau}{2}\tilde{H}_0\right)
$$

$$
= \exp\left[-i\frac{\tau}{2}(J_{12}I_z \otimes I_z \otimes I - J_{23}I \otimes I_z \otimes I_z)\right]
$$

$$
\times \exp\left[-i\frac{\tau}{2}(J_{12}I_z \otimes I_z \otimes I + J_{23}I \otimes I_z \otimes I_z)\right]
$$

$$
= \exp\left(-i\alpha I_z \otimes I_z \otimes I\right), \tag{12.48}
$$

where $\tilde{H}_0$ is the Hamiltonian (12.46) without rf pulses and use has been made of the identity

$$
[I_z \otimes I_z \otimes I, I \otimes I_z \otimes I_z] = 0.
$$

This result shows that we can eliminate the effect of J_{23} coupling by applying a pair of π-pulses to qubit 3. This technique is called **refocusing** or **decoupling**. Refocusing is also used to cancel field inhomogeneity and reduce transverse relaxation.

12.4.4.2   More-Qubit Case

Eliminating J-couplings in a molecule with more qubits requires some consideration. We look at the previous example of a three-qubit molecule from a slightly different viewpoint. We have seen the a pair of π-pulses introduces a change in the interaction Hamiltonian $I_z \otimes I_z \to -I_z \otimes I_z$. This change is equivalent with reversing the direction of time; $\tau I_z \otimes I_z \to -\tau I_z \otimes I_z$. Unwanted time-evolution has been cancelled by this reversed direction of time as

$$
e^{iJI_z \otimes I_z \tau/2}e^{-iJI_z \otimes I_z \tau/2} = I.
$$

Take a qubit and assign a number 1 if the time flows in a normal direction and -1 if it flows in a reversed direction. Then we obtain the following table for the three-qubit refocusing scheme;

	$\tau/2$	$\tau/2$
Qubit 1	1	1
Qubit 2	1	1
Qubit 3	1	-1

We notice the the inner product of the first and the second rows is nonvanishing, while that of the second and the third rows vanish. We infer that it is necessary to have a vanishing inner product of rows of the relevant qubits for refocusing of the J-coupling. If we want to get rid of all the J-coupling evolution for a duration τ, we may have

	$\tau/2$	$\tau/2$
Qubit 1	1	-1
Qubit 2	1	1
Qubit 3	1	-1

for example. Note that this pulse sequence applies only for a linear chain with nearest-neighbor couplings. Even in the presence of a coupling between qubit 1 and qubit 3, all the J-coupling evolutions are eliminated by taking

	$\tau/4$	$\tau/4$	$\tau/4$	$\tau/4$
Qubit 1	1	1	1	1
Qubit 2	1	-1	-1	1
Qubit 3	1	-1	1	-1

for example. Note that we need to divide τ into at least four intervals to cancel all J-couplings. This is easily extended to a four-qubit molecule with J-couplings between arbitary pairs as

	$\tau/4$	$\tau/4$	$\tau/4$	$\tau/4$
Qubit 1	1	1	1	1
Qubit 2	1	1	-1	-1
Qubit 3	1	-1	-1	1
Qubit 4	1	-1	1	-1

for example. This table defines a matrix whose elements are ± 1 as

$$H(4) = \begin{pmatrix} 1 & 1 & 1 & 1 \\ 1 & 1 & -1 & -1 \\ 1 & -1 & -1 & 1 \\ 1 & -1 & 1 & -1 \end{pmatrix}. \tag{12.49}$$

Note that this matrix has entries ± 1, and an arbitrary pair of rows has a vanishing inner product. Such a matrix is called a **Hadamard matrix** [17, 18, 19, 20]. Those Hadamard matrices whose first row and first column have entires 1 only are called normalized. Now we have learned it is possible to eliminate all the J-coupling evolutions by flipping qubits by a pair of π-pulses following the prescription given by an $n \times n$ Hadamard matrix $H(n)$.

Suppose we want to introduce a coupling between a pair of nearest neighbor qubits, i and $i + 1$, for example. Then we only need to replace the $i + 1$st row by the ith row in $H(n)$. For example, qubit 2–qubit 3 coupling in the four-qubit example above is introduced by taking the matrix

$$\begin{pmatrix} 1 & 1 & 1 & 1 \\ 1 & 1 & -1 & -1 \\ 1 & 1 & -1 & -1 \\ 1 & -1 & 1 & -1 \end{pmatrix}. \tag{12.50}$$

It is also possible to kill precession of each qubit by applying a pair of π-pulses. Precession of a qubit for a duration τ is nullified by a pair of π-pulses using the identity

$$\bar{X} e^{i\omega_0 I_z \tau/2} X e^{i\omega_0 I_z \tau/2} = e^{i\omega_0 I_z \tau/2} e^{-i\omega_0 I_z \tau/2} = I. \tag{12.51}$$

To eliminate one-qubit time-evolution due to precession, we need to have equal numbers of 1 and −1 in each row. We need to extend the $n \times n$ Hadamard matrix $H(n)$ to incorporate precession nullification in a refocusing program. Let us consider normalized Hadamard matrix $H(n+2)$ first. The first row is made of 1 only and must be removed. We also remove the second row and are left with an $n \times (n+2)$ matrix, in which the matrix elements of each row add up to vanish, and any two rows have a vanishing inner product. An example for a two-qubit molecule is obtained from $H(4)$ as

$$\begin{pmatrix} 1 & -1 & -1 & 1 \\ 1 & -1 & 1 & -1 \end{pmatrix}. \tag{12.52}$$

Precessions as well as J-coupling have been nullified with the above Hadamard matrix.

EXERCISE 12.6 Show, by explicitly evaluating the time-evolution matrix, that the above matrix (12.52) indeed nullifies time-evolution of the spins.

12.5   Time-Optimal Control of NMR Quantum Computer

We have implemented the CNOT gate in the end of §12.4.3. Although CNOT plays a particularly important role in the universality theorem, almost any two-qubit gate not in $SU(2) \otimes SU(2)$ does the job, an important exception being the SWAP gate.

In the present section, we consider a general strategy to implement any two-qubit gate. Our implementation is also optimal in terms of gate execution time. Let us start with some mathematical background materials.

12.5.1   A Brief Introduction to Lie Algebras and Lie Groups

It is assumed that the reader has some familiarity with the elementary theory of Lie algebras and Lie groups, such as SO(3) and SU(2). See [21] and [22], for example.

A Lie group G is a group equipped with a structure of an analytic manifold, where the group operations $G \times G \to G, G \to G$ defined by

$$xy \mapsto xy, \ x \mapsto x^{-1}, \tag{12.53}$$

respectively, are analytic with respect to local coordinates [22, 23].

Given a Lie group G, consider the tangent space $\mathfrak{g}$ of G at the unit element $I \in G$. In other words, $\mathfrak{g}$ is nothing but the vector space $T_I(G)$, which is constructed as follows [23]. Consider a curve $c : (a, b) \to G$ such that $c(0) = I$ and $c'(0) = X_c$, where it is assumed that $a < 0 < b$ and the curve belongs to C^1 class. For each choice of c, there exists a tangent vector X_c.* Suppose we take all the curves that pass I at $t = 0$ and consider the set of tangent vectors $\mathfrak{g} = \{X_c | c(0) = I, c \in C^1\}$. Then the set $\mathfrak{g}$ has a structure of a vector space, in which an addition and a scalar multiplication are well defined:

$$\forall X, Y \in \mathfrak{g}, \forall c_k \in \mathbb{R} \Rightarrow c_1 X + c_2 Y \in \mathfrak{g}. \tag{12.54}$$

Moreover, being a tangent vector space $\mathfrak{g}$ a Lie group G, the Lie bracket is well defined too:

$$X, Y \in \mathfrak{g} \to [X, Y] \in \mathfrak{g}, \tag{12.55}$$

where $[X, Y] = XY - YX$ is the Lie bracket of X and Y. The vector space $\mathfrak{g}$ is called the Lie algebra associated with a Lie group G. It is common to denote the Lie algebra of a Lie group G by the corresponding lower case German letter: the Lie algebra of $\mathrm{SU}(n)$ is denoted as $\mathfrak{su}(n)$, for example. Alternatively, the exponential map $\exp : \mathfrak{g} \to G$, $X \mapsto \exp X$ maps $\mathfrak{g}$ to a component G_0 of G, which contains the unit element I. By definition, this means that $G_0 = G$ for a simply connected Lie group G.

Let us work out an example $G = \mathrm{SU}(n)$. Consider a curve $c(t) : (a, b) \to \mathrm{SU}(n)$. It satisfies $c(t)^\dagger c(t) = I$ and $\det c(t) = 1$ for any $t \in (a, b)$. There exists a vector $X \in \mathfrak{g}$ such that $c(t) = \exp(Xt)$ in the vicinity of $t \sim 0$. The vector X satisfies the corresponding conditions

$$\det e^{Xt} = \exp(\mathrm{tr}\, X)t = 1, e^{Xt} e^{X^\dagger t} = e^{(X + X^\dagger)t} = I.$$

It is found from these conditions that

$$\mathrm{tr}\, X = 0 \quad \text{and} \quad X + X^\dagger = 0, \tag{12.56}$$

that is, X is traceless and skew-Hermitian. Conversely, any traceless skew-Hermitian matrix X defines $U = e^{Xt}$, which satisfies $\det U = 1$ and $U^\dagger U = I$. In summary

$$\mathfrak{su}(n) = \{X \in M(n, \mathbb{C}) | \mathrm{tr}\, X = 0, X + X^\dagger = 0\}. \tag{12.57}$$

The set $M(n, \mathbb{C})$ of $n \times n$ complex matrices has $2n^2$ real free parameters. The conditions $X = -X^\dagger$ reduces this down to n^2. In particular, the diagonal elements d_i of X must be pure imaginary. The condition $\mathrm{tr}\, X = 0$ introduces an additional condition $\sum_i d_i = 0$, which reduces the degrees of freedom to $n^2 - 1$ and hence dim $\mathfrak{su}(n) = n^2 - 1$. Let X_k ($1 \leq k \leq n^2 - 1$) be the generators of $\mathfrak{su}(n)$. Any element $U \in \mathrm{SU}(n)$ is then expressed as

$$U = \exp \left(\sum_{k=1}^{n^2-1} \alpha_k X_k \right). \tag{12.58}$$

---

*More formally, a tangent vector is defined as an equivalence class of curves that satisfies the conditions (12.53); see [23] for example.

For SU(2), for example, the vector space $\mathfrak{su}(2)$ is spanned by three traceless anti-Hermitian matrices, which we often take $i\sigma_k$ $(k = x, y, z)$.

We note that the condition $\det U = 1$ does not apply for $U \in \mathrm{U}(n)$, and accordingly the corresponding Lie algebra is

$$\mathfrak{u}(n) = \{X \in M(n, \mathbb{C}) | X + X^\dagger = 0\}, \tag{12.59}$$

for which $\dim \mathfrak{u}(n) = n^2$.

It is convenient to take the set of generators of $\mathrm{U}(2^n)$ as

$$I_{k_1} \otimes I_{k_2} \otimes \ldots \otimes I_{k_n}, \tag{12.60}$$

where $I_k \in \{I, I_x, I_y, I_z\}$. The generator $I \otimes I \otimes \ldots \otimes I$ must be excluded as a generator of $\mathfrak{su}(2^n)$ since it does not satisfy the traceless condition. In this way, we find there are $4^n - 1$ generators for $\mathfrak{su}(2^n)$.

EXAMPLE 12.2 Generators of $\mathfrak{su}(2^2)$ are

$$I_k \otimes I, \; I \otimes I_k, \; I_j \otimes I_k \quad (j, k = x, y, z).$$

Observe that there are $3 + 3 + 9 = 4^2 - 1$ generators.

12.5.2   Cartan Decomposition and Optimal Implementation of Two-Qubit Gates

We have seen in the preceeding sections that one-qubit operation takes a short time on the order $10 \, \mu$s for a heteronucleus molecule, while a two-qubit entangling operation takes time typically $\sim 1/J \sim 10$ ms. Therefore one-qubit operation time may be neglected in estimating the total execution time of a quantum algorithm [24]. Let us consider a molecule with two heteronucleus spins for definiteness, whose Hamiltonian, in the rotating frame with respective Larmor frequency, is given in Eq. (12.27), in which $\omega_{1,i}$ and ϕ_i are control parameters. Typically we have $\omega_{1,i} \gg J$, which justifies the above assumption of negligible one-qubit operation time compared to two-qubit operation time. This Hamiltonian generates a unitary matrix $U_{\mathrm{alg}} \in \mathrm{SU}(4)$ via the time-evolution equation

$$U_{\mathrm{alg}} = \mathcal{T} e^{-i \int_0^\tau \tilde{H}(t) dt}. \tag{12.61}$$

One may naively think that the path providing the shortest execution time corresponds to the shortest path connecting the unit matrix I (at $t = 0$) and U_{alg} at $t = T$. Note however that the one-qubit operation time is negligible and we may use one-qubit gates as many times as necessary. Thus we may identify $U_1, U_2 \in \mathrm{SU}(4)$ which differ by an element of $K \equiv \mathrm{SU}(2) \otimes \mathrm{SU}(2)$. This means that the relevant space for evaluating the time-optimal path is the coset space $\mathrm{SU}(4)/\mathrm{SU}(2) \otimes \mathrm{SU}(2)$ in which U_1 and $U_2 = KU_1$ are identified.

To find the time-optimal path connecting the unit matrix I and the matrix U_{alg}, therefore, amounts to finding the time-optimal path connecting cosets $[I]$ and $[U_{\text{alg}}]$, where $[U] \equiv \{kU | k \in K\}$. The Lie algebra $\mathfrak{su}(4)$ is decomposed as $\mathfrak{su}(4) = \mathfrak{k} \oplus \mathfrak{p}$ [24, 25, 26], where

$$\mathfrak{k} = \text{Span}(\{iI \otimes I_k, iI_k \otimes I\}), \quad (k = x, y, z), \tag{12.62}$$

$$\mathfrak{p} = \mathfrak{k}^{\perp} = \text{Span}(\{iI_j \otimes I_k\}), \quad (j, k = x, y, z). \tag{12.63}$$

They satisfy the commutation relations

$$[\mathfrak{k}, \mathfrak{k}] \subset \mathfrak{k}, \qquad [\mathfrak{p}, \mathfrak{k}] \subset \mathfrak{p}, \qquad [\mathfrak{p}, \mathfrak{p}] \subset \mathfrak{k}. \tag{12.64}$$

Decomposition of a Lie algebra $\mathfrak{g}$ into $\mathfrak{k}$ and $\mathfrak{p}$, satisfying the above commutation relations, is called a **Cartan decomposition**. The **Cartan subalgebra** $\mathfrak{h} = \text{Span}(\{iI_j \otimes I_j\}) \subset \mathfrak{p}$ plays an important role in our construction. A general theorem of Lie algebras proves that any element $U_{\text{alg}} \in \text{SU}(4)$ has a KP decomposition $U_{\text{alg}} = kp$ with $k \in K \equiv \exp \mathfrak{k}$ and $p \in P \equiv \exp \mathfrak{p}$. Moreover, any matrix $p \in P$ is rewritten in a conjugate form $p = k_1^{\dagger} h k_1$, where $k_1 \in K$ and h is an element of the **Cartan subgroup** H of $\text{SU}(4)$ defined as

$$H \equiv \exp \mathfrak{h} = \left\{ \exp\left(i \sum_{j=x,y,z} \alpha_j I_j \otimes I_j \right) \middle| \alpha_j \in \mathbb{R} \right\}. \tag{12.65}$$

Therefore we have a corresponding Cartan decomposition of a group element as $U_{\text{alg}} = kp = kk_1^{\dagger} h k_1 = k_2 h k_1$, where $k_i \in K$, $h \in H$ and $k_2 = kk_1^{\dagger}$. The quantum algorithm U_{alg} is now decomposed into one-qubit operations k_1, k_2 and a two-qubit entangling operation h. This decomposition determines an optimized pulse sequence of the NMR quantum computer as discussed in [24, 25, 26].

Cartan decomposition of an arbitrary $U \in \text{SU}(4)$ proceeds explicitly as follows. We take the magic basis [27] defined as

$$|\Psi_0\rangle = \frac{1}{\sqrt{2}}(|00\rangle + |11\rangle),$$

$$|\Psi_1\rangle = \frac{i}{\sqrt{2}}(|01\rangle + |10\rangle),$$

$$|\Psi_2\rangle = \frac{1}{\sqrt{2}}(|01\rangle - |10\rangle), \tag{12.66}$$

$$|\Psi_3\rangle = \frac{i}{\sqrt{2}}(|00\rangle - |11\rangle),$$

which is different from an ordinary Bell basis by phase. The transformation rule of a matrix U with respect to the standard binary basis $\{|00\rangle, |01\rangle, |10\rangle, |11\rangle\}$ into that with the magic basis $\{|\Psi_i\rangle\}$ is $U \to U_B \equiv$

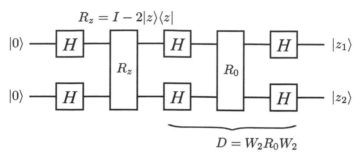

FIGURE 12.8
Implementation of the Grover database search algorithm for $n = 2$ qubits case. H is the Hadamard gate, $W_2 = U_H^{\otimes 2}$, while $R_z = I - 2|z\rangle\langle z|$.

$Q^\dagger U Q$, where

$$Q = \frac{1}{\sqrt{2}} \begin{pmatrix} 1 & 0 & 0 & i \\ 0 & i & 1 & 0 \\ 0 & i & -1 & 0 \\ 1 & 0 & 0 & -i \end{pmatrix}. \tag{12.67}$$

The matrix Q defines an isomorphism (1:1 linear map preserving the group product) between $K = \mathrm{SU}(2) \otimes \mathrm{SU}(2)$ and $\mathrm{SO}(4)$ and is used to classify two-qubit gates [25, 27]. In fact, it is easy to verify that $Q^\dagger k Q$ is an element of $\mathrm{SO}(4)$ for $k \in K$. Moreover, Q diagonalizes elements of the Cartan subgroup, viz $Q^\dagger h Q = \mathrm{diag}(e^{i\theta_0}, e^{i\theta_1}, e^{i\theta_2}, e^{i\theta_3})$ for $h \in H$. We find for $U = k_2 h k_1$ that

$$U_B = Q^\dagger U Q = Q^\dagger k_2 Q \cdot Q^\dagger h Q \cdot Q^\dagger k_1 Q = O_2 h_D O_1,$$

where $O_i \equiv Q^\dagger k_i Q$ is an element of $\mathrm{SO}(4)$ and $h_D \equiv Q^\dagger h Q$ is a diagonal matrix. From $U_B^t U_B = O_1^t h_D^2 O_1$, we notice that $U_B^t U_B$ is diagonalized by O_1 and its eigenvalues form the diagonal elements of h_D^2. Finally O_2 is found as $O_2 = U_B(h_D O_1)^{-1}$.

EXAMPLE 12.3 Let us consider implementing two-qubit Grover's database search algorithm U_z as a concrete example. The data are encoded in one of the basis vectors $|00\rangle, |01\rangle, |10\rangle, |11\rangle$, and the gate U_z picks out a particular binary basis vector $|z\rangle = |ij\rangle$ as a "target file" upon acting on $|00\rangle$ [33, 34]. Figure 12.8 shows the actual quantum cirucit implementation of the Grover algorithm. Here H is the Hadamard gate and $R_z = I - 2|z\rangle\langle z|$, $R_0 = I - 2|0\rangle\langle 0|$, cf Fig. 7.2. Here we do not explicitly give oracle cirucits R_z and R_0, but they are treated as black boxes.

Here we consider U_{10} which picks out the file $|10\rangle$ with a single step. The unitary matrix representing this algorithm takes the form

$$U_{10} = W_2 R_0 W_2 R_{10} W_2 = \begin{pmatrix} 0 & 1 & 0 & 0 \\ 0 & 0 & 0 & -1 \\ -1 & 0 & 0 & 0 \\ 0 & 0 & -1 & 0 \end{pmatrix}. \qquad (12.68)$$

We apply the above strategy and find the Cartan decomposition of $U_{10} = k_2 h k_1$ as

$$\begin{aligned} k_1 &= I_2 \otimes I_2, \\ h &= e^{i\pi(I_x \otimes I_x - I_y \otimes I_y)}, \qquad\qquad (12.69) \\ k_2 &= e^{-i(\pi/2)I_z} \otimes e^{i(\pi/\sqrt{2})(I_x + I_y)}. \end{aligned}$$

Actually, the decomposition is not unique, and we choose a solution that minimizes the execution time of an NMR quantum computer. To implement this decomposition with NMR, such terms as $e^{i\pi(I_x \otimes I_x)}$ must be rewritten in favor of the subset of generators of SU(4) contained in the Hamiltonian (12.27). We verify, for example, that

$$e^{i\pi(I_x \otimes I_x)}$$
$$= \left[e^{-i(\pi/2)I_y} \otimes e^{i(\pi/2)I_y} \right] \cdot e^{-i\pi(I_z \otimes I_z)} \cdot \left[e^{i(\pi/2)I_y} \otimes e^{-i(\pi/2)I_y} \right]. \quad (12.70)$$

Table 12.4 shows the pulse sequence to implement U_{10} with an NMR quantum computer. We call the hydrogen nucleus and the carbon nucleus qubit 1 and qubit 2, respectively. The time-optimal path requires the execution time of $1/J$, which happens to be the same as that for the conventional pulse sequence [35].

TABLE 12.4
Time-optimal pulse sequences for Grover's algorithm U_{10}. The number 1 (2) denotes the first (second) qubit. Here X ($\bar{X}$) and Y ($\bar{Y}$) denote $\pi/2$-pulse around x ($-x$) and y ($-y$) axis, respectively. The symbol Pi(θ) denotes a π-pulse around a direction $(\cos\theta, \sin\theta, 0)$ in the Bloch sphere. The symbol $(1/2J)$ indicates the length of the idle time, during which no external pulses are applied.

	Pulse sequence							Execution time
1:	X	(1/2J)	Xm	Y	(1/2J)	X	Ym	$1/J$
2:	X	(1/2J)	Xm	Ym	(1/2J)	Y	Pi($\pi/4$)	

EXERCISE 12.7 Find a Cartan decomposition of the controlled-Z gate

$$U_Z = |0\rangle\langle 0| \otimes I + |1\rangle\langle 1| \otimes \sigma_z$$

and implement the solution with an NMR pulse sequence which implements the decomposition.

EXERCISE 12.8 Find a Cartan decomposition of U_{CNOT}. Find an NMR pulse sequence which implements the decomposition.

12.6   Measurements

12.6.1   Introduction and Preliminary

We have concentrated on the NMR Hamiltonian and unitary gates so far and hardly discussed the states. We will be concerend with state measurement in this section. The NMR quantum computer is different from other candidates of quantum computer in that the qubit measurement is not a projective measurement but an ensemble measurement. Moreover, it has already been mentioned that the system works at room temperatures T, for which $\hbar\omega_0 \ll k_B T$. The system is, therefore, in an almost maximally mixed state.

Let ρ_{th} be a density matrix of a single nucleus without the rf field at thermal equilibrium;

$$\rho_{\text{th}} = \frac{e^{-H_0/k_B T}}{Z(T)}, \quad Z(T) = \text{tr}\, e^{-H_0/k_B T}. \tag{12.71}$$

It follows from the assumption $\hbar\omega_0 \ll k_B T$ that

$$\rho_{\text{th}} \simeq \frac{1}{2}\left(I + \frac{\hbar\omega_0}{k_B T} I_z\right), \tag{12.72}$$

in the basis $|\uparrow\rangle = |0\rangle$ and $|\downarrow\rangle = |1\rangle$. The density matrix in the rotating frame of the spin is

$$\tilde{\rho}_{\text{th}} = U\rho_{\text{th}}U^\dagger = \rho_{\text{th}}, \tag{12.73}$$

where $U = \exp(i\omega_0 I_z t)$.

It is depicted in Fig. 12.3 that an NMR spectrometer has a pair of rf coils, which also measure the transverse magnetization of the sample molecules through the induced electromotive force. Suppose the coils measure the x-component of the spin magnetization, which is proportional to $\langle I_x\rangle = \text{tr}\,(I_x\rho)$. Clearly $\langle I_x\rangle = 0$ for $\rho = \rho_{\text{th}}$ since the averaged magnetization points in the z-direction, which is free from precession. We must "tilt" the magnetization somehow to pick up the signal and measure the state of the molecules. Suppose an $\pi/2$-pulse along the x-axis $X = \exp(-i\pi I_x/2)$ has been applied at $t = 0$ in the rotating frame of the sample. The resulting state is

$$\tilde{\rho}_X = X\tilde{\rho}_{\text{th}}X^\dagger = \frac{1}{2}\left(I - \frac{\hbar\omega_0}{k_B T} I_y\right) \tag{12.74}$$

in the rotating frame. Now this state must be transformed back to the laboratory frame since the pickup coils are fixed in the laboratory. We obtain

$$\rho_X = U^\dagger \tilde{\rho}_X U = \frac{1}{2}\left(I - \frac{\hbar\omega_0}{k_B T}\left[\sin(\omega_0 t)I_x + \cos(\omega_0 t)I_y\right]\right). \tag{12.75}$$

M_x, the x-component of the magnetization of the molecules, is now evaluated as

$$M_x(t) \propto \langle I_x \rangle = \text{tr}\,(I_x \rho_X) = -\frac{\hbar\omega_0}{4k_B T}\sin(\omega_0 t). \tag{12.76}$$

Note that the prefactor $\hbar\omega_0/2k_B T$ originates from a tiny difference between the populations of the states $|\sigma_y = +1\rangle$ and $|\sigma_y = -1\rangle$ resulting after the $\pi/2$-pulse X is applied, reflecting an initial population difference between the states $|\uparrow\rangle$ and $|\downarrow\rangle$. The magnetic flux threading the coils oscillates as $\sim \sin\omega_0 t$, and the induced electromotive force as $\sim \cos\omega_0 t$.

12.6.2 One-Qubit Quantum State Tomography

Let

$$\tilde{\rho} = \frac{1}{2}I + c_x I_x + c_y I_y + c_z I_z \tag{12.77}$$

be a density matrix which resulted after some quantum gate operation. Here we temporarily assume that this qubit is not entangled with other qubits in the register. A more general case will be analyzed subsequently.

Our task is to measure c_k to identify the state $\tilde{\rho}$. Measurement of a density matrix is known as **quantum state tomography**. We first transform $\tilde{\rho}$ back to the laboratory frame,

$$\begin{aligned} \rho &= U^\dagger \tilde{\rho} U \\ &= \frac{1}{2}\begin{pmatrix} 1 + c_z & (c_x - ic_y)e^{i\omega_0 t} \\ (c_x + ic_y)e^{-i\omega_0 t} & 1 - c_z \end{pmatrix}. \end{aligned} \tag{12.78}$$

Measurement of the magnetization M_x yields

$$M_x(t) \propto \text{tr}\,(I_x \rho) = \frac{1}{2}\left(c_x \cos\omega_0 t + c_y \sin\omega_0\right). \tag{12.79}$$

The coefficients c_x and c_y are read out by making a Fourier transform of the received signal as will be shown in the next subsection.

We further need to evaluate c_z to identify ρ completely. The density matrix is "modified" to this end by operating an X pulse,

$$\tilde{\rho}_X = X\tilde{\rho}X^\dagger = \frac{1}{2}I + c_x I_x - c_z I_y + c_y I_z \tag{12.80}$$

in the rotating frame. The density matrix in the laboratory frame is

$$\rho_X = U^\dagger \tilde{\rho}_X U = \frac{1}{2}\begin{pmatrix} 1 + c_y & (c_x + ic_z)e^{i\omega_0 t} \\ (c_x - ic_z)e^{-i\omega_0 t} & 1 - c_y \end{pmatrix}. \tag{12.81}$$

Measurement of M_x yields

$$M_x^X(t) \propto \text{tr } (I_x \rho_X) = \frac{1}{2}(c_x \cos \omega_0 t - c_z \sin \omega_0 t). \qquad (12.82)$$

EXERCISE 12.9 Suppose $Y = \exp(-i(\pi/2)I_y)$ acts on $\tilde{\rho}$.
(1) Find $\tilde{\rho}_Y = Y \tilde{\rho} Y^\dagger$.
(2) Find $\rho_Y = U^\dagger \tilde{\rho}_Y U$.
(3) Show the expectation value of I_x with respect to the state ρ_Y is

$$\text{tr } (I_x \rho_Y) = \frac{1}{2}(c_z \cos \omega_0 t + c_y \sin \omega_0 t). \qquad (12.83)$$

The components $\cos \omega_0$ and $\sin \omega_0$ behave differently under Fourier transform, and their coefficients can be separated after Fourier-transforming the induced electromotive force. It is therefore possible to identify all of c_k from measurements of $\{M_x(t), M_x^X(t)\}$ or $\{M_x(t), M_x^Y(t)\}$ with calibration of the signal strengths with the thermal equilibrium state rotated by X, for example.

12.6.3   Free Induction Decay (FID)

It has been shown in the previous subsection that the measurement of $M_x(t)$ yields an outcome $\propto c_x \cos \omega_0 t + c_y \sin \omega_0 t = A \cos(\omega_0 + \alpha)$, where $A = \sqrt{c_x^2 + c_y^2}$ and $\tan \alpha = -c_y/c_x$.

In fact, precession of the magnetization does not last forever. Precession decays due to various reasons, such as field inhomogeneity in $\boldsymbol{B}_0$, for example, and actual signal is approximately given by

$$f(t) = Ae^{-t/T_2} \cos(\omega_0 t + \alpha), \qquad (12.84)$$

where T_2 is a parameter characterizing the decay time constant. Figure 12.9 shows the cosine-Fourier transform

$$\begin{aligned}
\tilde{f}_\alpha(\omega) &= \sqrt{\frac{2}{\pi}} \int_0^\infty dt Ae^{-t/T_2} \cos(\omega_0 t + \alpha) \cos \omega t \\
&= \frac{AT_2}{\sqrt{2\pi}} \left[\frac{\cos \alpha - (\omega - \omega_0)T_2 \sin \alpha}{(\omega - \omega_0)^2 T_2^2 + 1} + \frac{\cos \alpha - (\omega + \omega_0)T_2 \sin \alpha}{(\omega + \omega_0)^2 T_2^2 + 1} \right]
\end{aligned}$$

$$(12.85)$$

for $\alpha = 0, \pm\pi/4$ and $\pi/2$. The second term in the bottom equation is negligible in the vicinity of $\omega \sim \omega_0$.

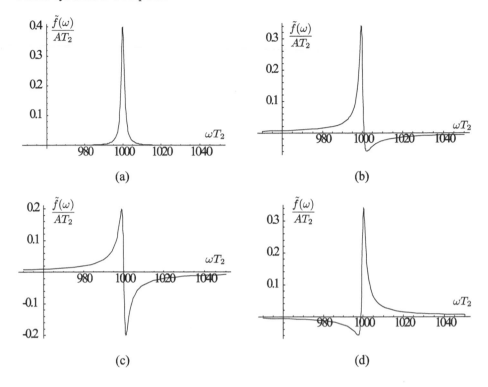

FIGURE 12.9

Cosine-Fourier transform of the FID signal. $\tilde{f}(\omega)/AT_2$ is plotted against ωT_2 for $\omega_0 T_2 = 10^3$ and (a) $\alpha = 0$, (b) $\alpha = \pi/4$, (c) $\alpha = \pi/2$ and (d) $\alpha = -\pi/4$.

12.6.4 Two-Qubit Tomography

The density matrix $\tilde{\rho}$ of two-qubit molecules in a rotating frame is parametrized as

$$\tilde{\rho} = \frac{I}{2} \otimes \frac{I}{2} + \begin{pmatrix} a_{00} & a_{01}+ib_{01} & a_{02}+ib_{02} & a_{03}+ib_{03} \\ a_{01}-ib_{01} & a_{11} & a_{12}+ib_{12} & a_{13}+ib_{13} \\ a_{02}-ib_{02} & a_{12}-ib_{12} & a_{22} & a_{23}+ib_{23} \\ a_{03}-ib_{03} & a_{13}-ib_{13} & a_{23}-ib_{23} & a_{33} \end{pmatrix}. \quad (12.86)$$

The first term $I/2 \otimes I/2$ is introduced for normalization and does not contribute to NMR signals. The number of independent parameters is $16-1 = 15$, because the constraint $\mathrm{tr}\tilde{\rho} = 1$ leads to a relation $\sum_i a_{ii} = 1$ among $\{a_{ii}\}$.

We identify the state $\tilde{\rho}$ completely if all a_{ij} and b_{ij} are measured by quantum state tomography. FID signal of each spin is measured by the NMR spectrometer at $t > 0$. Note, however, that the state $\tilde{\rho}$ has its own intrinsic time-development due to the interspin J-coupling. Let $U_J(t)$ be the time-development operator due to the J-coupling

$$U_J(t) = e^{-iJI_z \otimes I_z t}.$$

(12.87)

Then the density matrix in the rotating frame at $t > 0$ is

$$\tilde{\rho}(t) = U_J(t)\tilde{\rho}U_J^\dagger(t) \quad (t > 0).$$

(12.88)

The density matrix $\rho(t)$ in the laboratory frame is

$$\rho(t) = U_R^\dagger \tilde{\rho}(t) U_R,$$

(12.89)

where U_R is given by Eq. (12.20). The x-component of the magnetization in the laboratory frame is measured as

$$
\begin{aligned}
& \mathrm{tr}\,[(I_x \otimes I + I \otimes I_x)\rho(t)] \\
&= \frac{1}{2}\left[a_{01}\cos\left(\omega_{0,2} - \frac{J}{2}\right)t + a_{23}\cos\left(\omega_{0,2} + \frac{J}{2}\right)t \right. \\
&\quad \left. + a_{02}\cos\left(\omega_{0,1} - \frac{J}{2}\right)t + a_{13}\cos\left(\omega_{0,1} + \frac{J}{2}\right)t\right] \\
&\quad - \frac{1}{2}\left[b_{01}\sin\left(\omega_{0,2} - \frac{J}{2}\right)t + b_{23}\sin\left(\omega_{0,2} - \frac{J}{2}\right)t \right. \\
&\quad \left. + b_{02}\sin\left(\omega_{0,1} - \frac{J}{2}\right)t + b_{13}\sin\left(\omega_{0,1} - \frac{J}{2}\right)t\right].
\end{aligned}
$$

(12.90)

Therefore $a_{01}, a_{23}, a_{02}, a_{13}$ are measured from the cosine part of the spectrum and $b_{01}, b_{23}, b_{02}, b_{13}$ from the sine part. The cosine-Fourier transforms of these sine and cosine functions have different peaks at $\omega_{0,i} \pm J/2$ with different phases. Clearly the measurement of $\langle I_x \otimes I + I \otimes I_x \rangle$ is not sufficient to fix ρ. We have to deform ρ somehow so that other coefficients appear in the measured results. Of course, it is desirable that the number of the measurements is as small as possible. This is somewhat similar to describing a convex three-dimensional object by using photographs. A single-shot photograph is not sufficient to describe the object, and one has to rotate the object and take a couple of photographs to completely describe the object.

Suppose next that a $\pi/2$-pulse Y_1 along the y-axis of the rotating frame is applied to the spin 1 at $t = 0$. The density matrix after this operation is

$$Y_1 \tilde{\rho} Y_1^\dagger.$$

(12.91)

The resulting state in the laboratory frame is

$$\rho(t) = U_R^\dagger U_J(t) Y_1 \tilde{\rho} Y_1^\dagger U_J(t)^\dagger U_R.$$

(12.92)

The spectrometer measures the x-component of the magnetization of $\rho(t)$ as

$$\mathrm{tr}\,[(I_x \otimes I + I \otimes I_x)\rho(t)$$

$$\frac{1}{2}\left[\left(\frac{a_{01}}{2} - \frac{a_{03}}{2} - \frac{a_{12}}{2} + \frac{a_{23}}{2}\right)\cos\left(\omega_{0,2} - \frac{J}{2}\right)t\right.$$

$$+\left(\frac{a_{01}}{2} + \frac{a_{03}}{2} + \frac{a_{12}}{2} + \frac{a_{23}}{2}\right)\cos\left(\omega_{0,2} + \frac{J}{2}\right)t$$

$$\left.+\left(\frac{a_{00}}{2} - \frac{a_{22}}{2}\right)\cos\left(\omega_{0,1} - \frac{J}{2}\right)t + \left(\frac{a_{11}}{2} - \frac{a_{33}}{2}\right)\cos\left(\omega_{0,1} + \frac{J}{2}\right)t\right]$$

$$+\frac{1}{2}\left[-\left(\frac{b_{01}}{2} - \frac{b_{03}}{2} + \frac{b_{12}}{2} + \frac{b_{23}}{2}\right)\sin\left(\omega_{0,2} - \frac{J}{2}\right)t\right.$$

$$+b_{13}\sin\left(\omega_{0,2} + \frac{J}{2}\right)t - b_{02}\sin\left(\omega_{0,1} - \frac{J}{2}\right)t$$

$$\left.+\left(-\frac{b_{01}}{2} - \frac{b_{03}}{2} + \frac{b_{12}}{2} - \frac{b_{23}}{2}\right)\sin\left(\omega_{0,1} + \frac{J}{2}\right)t\right].$$

$$(12.93)$$

We extract from Eq. (12.93) new information on

$$a_{00} - a_{22}, \quad a_{11} - a_{33}, \quad a_{03} + a_{12}, \quad b_{03} - b_{12}.$$

Similarly, measurement after a Y_2-pulse provides

$$a_{00} - a_{11}, \quad a_{22} - a_{33}, \quad a_{03} + a_{12}, \quad b_{03} + b_{12},$$

while measurement after an X_1-pulse provides

$$b_{03} + b_{12}, \quad a_{00} - a_{22}, \quad a_{11} - a_{33}, \quad a_{03} - a_{12}.$$

Therefore, all a_{ij} and b_{ij} can be determined experimentally in four measurements. Although there are other combinations to measure $\{a_{ij}\}$ and $\{b_{ij}\}$, four is the minimal number of measurements required to fix $\rho(t)$ completely in general. This number may be reduced if $\tilde{\rho}$ has a more restricted form as we show in the next example.

EXAMPLE 12.4 Let us consider a simple case in which $\tilde{\rho} = \text{diag}(a_{00}, a_{11}, a_{22}, a_{33})$. Application of a Y_1-pulse yields $a_{00} - a_{22}$ and $a_{11} - a_{33}$, while a Y_2-pulse yeilds $a_{00} - a_{11}$ and $a_{22} - a_{33}$, from which all the diagonal components are found.

If, furthermore, it is known that only one of a_{ii} is nonvanishing, it can be identified by a single measurement with a Y_1-pulse, for example. It follows from Eq. (12.93) that the FID sigal is peaked at $\omega_{0,1} - J/2$ if either a_{00} or a_{22} is nonvanishing. The peak is positive if $a_{00} = 1$, while it is negative if $a_{22} = 1$. Similarly it is found from the signature of the peak at $\omega_{0,1} + J/2$ whether a_{11} or a_{33} is nonvanishing.

Figure 12.10 (a) summarizes the above statements, where J is taken as positive. Let $P(ij)$ $(i, j \in \{0, 1\})$ be the population of the state $|ij\rangle$. It is easy to see $P(00) > P(01), P(10) > P(11)$, where the order of $P(01)$ and

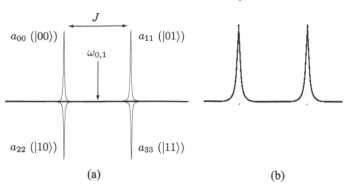

FIGURE 12.10

(a) Positions of the four peaks a_{ii}. The ket $|jk\rangle$ denotes the corresponding nuclear state. The positive peaks have slightly more amplitude than the negative peaks at thermal equilibrium. (b) Actual FID signal obtained with cytosine molecules. Courtesy of Y. Kondo, Kinki University, Japan.

$P(10)$ depends on whether $\omega_{0,1}$ is smaller or larger than $\omega_{0,2}$. If we apply the reading pulse Y_1 to a thermal equilibrium state $\rho \propto e^{-H/k_B T}$, we will obtain a superposition of the peaks as shown in Fig. 12.10 (b). It follows from the population differences $P(00) > P(10)$ and $P(01) > P(11)$ that both peak heights are positive. Figure 12.10 (b) is an actual FID signal from cytosine molecules whose qubits are two protons.

12.7 Preparation of Pseudopure State

Liquid state NMR at room temperature is different from other realizations in that the states are thermally populated. Therefore we must introduce various techniques to single out contributions from a fiducial initial state, $|00\ldots0\rangle$, for example. In the following subsections, we study these methods in detail. Note that the signal from a particular initial state reduces exponentially as a function of the number of qubits. This is one of the obstructions against scalablity of a liquid state NMR quantum computer. The other prominent future of NMR is that the measurement is not projective measurement but ensemble measurement with respect to a large number of molecules. Therefore it is possible to measure two noncummuting observables simultaneously.

We will be concerned mostly with two-qubit molecules in this section for definiteness. The Larmor frequencies of nuclei are denoted as $\omega_{0,i}$ and the coupling strength as J. Relevant references will be cited whenever generalization to more-qubit molecules is nontrivial.

An ensemble of two-qubit molecules at temperature T distributes according to the Bolzmann distribution

$$\rho(T) = \frac{e^{-H/k_BT}}{\mathrm{tr}\, e^{-H/k_BT}} \simeq \frac{1}{4}I \otimes I + \frac{\omega_{0,1}I_z \otimes I + \omega_{0,2}I \otimes I_z}{4k_BT}$$

$$= \frac{1}{4}I \otimes I + \frac{1}{8k_BT}\begin{pmatrix} \omega_{0,1} + \omega_{0,2} & 0 & 0 & 0 \\ 0 & \omega_{0,1} - \omega_{0,2} & 0 & 0 \\ 0 & 0 & -\omega_{0,1} + \omega_{0,2} & 0 \\ 0 & 0 & 0 & -\omega_{0,1} - \omega_{0,2} \end{pmatrix},$$

$$\tag{12.94}$$

where we used inequalities $\omega_{0,i}/k_BT \ll 1$ to reach the approximate form. For a hydrogen nucleus at $B_0 = 11.74$ T and $T = 300$ K, for example, we have $\hbar\omega_0 = 5.27 \times 10^{-26}$ J while $k_BT = 4.14 \times 10^{-21}$ J, justifying our assumption. Morevoer, $\omega_{0,i}$ are several orders of magnitude greater than J, and hence the contribution of J-coupling to $\rho(T)$ is virtually negligible.

If the molecule is homonucleus, in which $\Delta\omega_0 \equiv \omega_{0,2} - \omega_{0,1} \ll \omega_{0,i}$, the density matrix further simplifies as

$$\rho(T) = \frac{1}{4}I \otimes I + \frac{1}{8k_BT}\begin{pmatrix} \omega_{0,1} & 0 & 0 & 0 \\ 0 & 0 & 0 & 0 \\ 0 & 0 & 0 & 0 \\ 0 & 0 & 0 & -\omega_{0,1} \end{pmatrix}. \tag{12.95}$$

Our aim is to prepare an effective pure state out of the thermal equilibrium state. Let us write the equilibrium density matrix as

$$\rho(T) = \left(\frac{I}{2}\right)^{\otimes 2} + \Delta\rho. \tag{12.96}$$

The first term represents a uniformly mixed ensemble of all possible states, $|00\rangle, |01\rangle, |10\rangle$ and $|11\rangle$, while the second term denotes a tiny deviation from a uniformly mixed ensemble.

It has been shown in §12.6.3 that there is no FID signal if the spin-up component and spin-down component populate equal amounts. Therefore the uniform part $(I/2)^{\otimes n}$ has no contribution to the FID signal. Suppose we have a density matrix

$$\rho = \left(\frac{I}{2}\right)^{\otimes n} + \alpha\, \mathrm{diag}(1, 0, \ldots, 0). \tag{12.97}$$

It follows from the above observation that ρ effectively yields the FID signal from the pure state $\rho_{00} = |00\rangle\langle 00|$, albeit the amplitude is multiplied by the factor α. It is shown below that α is on the order of $\omega_{0,1}/2^n k_BT$, where n is the number of qubits. The effective pure state thus obtained is called the **pseudopure state**.

Molecules in the liquid state NMR are in a thermal equilibrium state, and it seems, at first sight, impossible to initialize the system in a fiducial pure state, such as $|00\rangle$. However, we may introduce an effective pure state if the density matrix is put in the form (12.97) by one way or another.

Note that a naive method to cool the molecules down to very low temperatures does not work since the sample is frozen and ceases to be a liquid. Then hidden interactions, which are averaged out in a liquid state, will be in action to render the Hamiltonian in a formidably complicated form. It should be also noted that nonunitary operations are required to produce an effective pure state out of a thermal distribution (12.94). This is because a thermal equilibrium density matrix has rank 4, while a pure state density matrix has rank 1, and any unitary transformation preserves the matrix rank; $\mathrm{rank}A = \mathrm{rank}(UAU^{\dagger})$.

We will introduce two methods to produce a pseudopure state in this section: temporal averaging method [28] and spatial averaging method [29, 30]. It is also possible to generate a pseudopure state using logical labelling. However, this method requires ancillary qubits and will not be treated here [31, 32].

12.7.1 Temporal Averaging

Molecules in NMR have a relatively long transverse relaxation time T_1, and a state out of thermal equilibrium maintains its distribution for a considerably long time. Then by preparing $2^n - 1$ different state populations, n being the number of qubits, we may execute a given quantum algorithm on each state population. It is possible to generate a pseudopure state by taking the average of the spectra over the different state populations. This is best illustrated for a two-qubit system.

Let us consider a two-qubit heteronuclear molecule for definiteness. Let $\omega_{0,1}$ and $\omega_{0,2}$ be the Larmor frequencies of the first and the second spins, respectively. The interqubit coupling may be ignored in evaluating the Bolzmann factor of the state population as was already pointed out. The thermal equilibrium density matrix is then given by Eq. (12.94), which we write as

$$\rho_0 \equiv \rho(T) = \mathrm{diag}(\rho_{00}, \rho_{11}, \rho_{22}, \rho_{33}). \qquad (12.98)$$

Now let us consider unitary gates which permutate the matrix elements $\rho_{11}, \rho_{22}, \rho_{33}$ cyclically leaving ρ_{00} fixed. This is done by the following permutation matrices

$$U_{\mathrm{cp}} = \begin{pmatrix} 1\,0\,0\,0 \\ 0\,0\,0\,1 \\ 0\,1\,0\,0 \\ 0\,0\,1\,0 \end{pmatrix}, \quad U_{\mathrm{cp}}^2 = \begin{pmatrix} 1\,0\,0\,0 \\ 0\,0\,1\,0 \\ 0\,0\,0\,1 \\ 0\,1\,0\,0 \end{pmatrix}. \qquad (12.99)$$

We note that these permutations are implemented with two CNOT gates as

$$U_{\mathrm{cp}} = U_{\mathrm{CNOT}}^{12} U_{\mathrm{CNOT}}^{21}, \quad U_{\mathrm{cp}}^2 = U_{\mathrm{CNOT}}^{21} U_{\mathrm{CNOT}}^{12}, \qquad (12.100)$$

where U_{CNOT}^{ij} stands for a CNOT gate with the control bit i and the target bit j.

The density matrix ρ_0 transforms under the action of these permutations as

$$\rho_1 \equiv U_{\mathrm{cp}}\rho_0 U_{\mathrm{cp}}^\dagger = \mathrm{diag}(\rho_{00}, \rho_{33}, \rho_{11}, \rho_{22}),$$
$$\rho_2 \equiv U_{\mathrm{cp}}^2 \rho_0 U_{\mathrm{cp}}^{2\dagger} = \mathrm{diag}(\rho_{00}, \rho_{22}, \rho_{33}, \rho_{11}),$$

(12.101)

respectively. Carrying out a given quantum algorithm with these three state populations and then averaging the three data is equivalent to executing the algorithm once for all with an effective initial state population

$$\rho_{\mathrm{ave}} = \frac{1}{3}(\rho_0 + \rho_1 + \rho_2) = \mathrm{diag}(\rho_{00}, s, s, s)$$
$$= sI_4 + \frac{4\rho_{00} - 1}{3}|00\rangle\langle 00|.$$

(12.102)

Here $s \equiv (\rho_{11} + \rho_{22} + \rho_{33})/3$ and we have used the fact $\sum_i \rho_{ii} = 1$. The spectral contribution from the first term, being proportional to I_4, cancels exactly, and we can pick up a contribution from the state $|00\rangle$ only. Since

$$\rho_{00} \simeq \frac{1 + \hbar(\omega_a + \omega_b)/k_B T}{4}$$

is very close to $1/4$ at room temperature, the coefficient $(4\rho_{00}-1)/3 \simeq 4\hbar(\omega_1 + \omega_2)/3k_B T$ is very small. Therefore the signal from the effective initial state $|00\rangle$ considerably reduces in this process.

Figure 12.11 shows the actual spectra employed for temporal averaging.

12.7.2 Spatial Averaging

Any NMR spectometer has a set of magnets which produces a field gradient along the static magnetic field $\boldsymbol{B}_0$. Molecules in a sample cell are under spatially varying field strength with which some components of a density matrix may be eliminated after averaging over the sample.

We confine ourselves to molecules with a small number of qubits for definiteness. Let us start with a single-qubit molecule. The Hamiltonian in a strong magnetic field along the z-axis takes the form

$$H_0(z) = -\omega_0(1 + b'z)I_z,$$

(12.103)

where b' is a parameter specifying the field gradient. Let us assume the sample extends in the region $|z| \leq L/2$, L being the sample extension along the z-axis. By transforming to the rotating frame with the unitary transformation (12.10) with $\omega = \omega_0$, we obtain a residual Hamiltonian

$$\tilde{H}_0 = -\omega_0 b' z I_z.$$

(12.104)

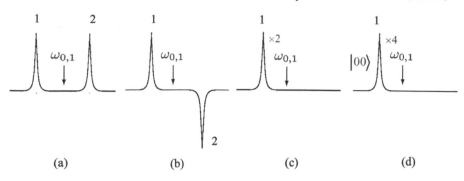

FIGURE 12.11

Spectrum of the qubit 1 obtained with each state population. Molecules employed are homonucleus cytosine, for which the density matrix elements satisfy $\rho_{11} = \rho_{22}$. (a) shows the spectrum obtained with the density matrix ρ_0. Peaks take positive values since $\rho_{00} > \rho_{22}$ and $\rho_{11} > \rho_{33}$. (b) is an FID spectrum obtained with ρ_1. Peak 1 is positive since $\rho_{00} > \rho_{11}$ while peak 2 is negative since $\rho_{33} < \rho_{22}$. (c) is an FID signal obtained with ρ_2 where peak 1 is positive since $\rho_{00} > \rho_{33}$, while peak 2 disappears since $\rho_{11} - \rho_{22} = 0$. These three contributions adds up to comprise the peak 1 in (d). This peak indicates that the pseudopure state $|00\rangle$ has been obtained after temporal averaging. Courtesy of Yasushi Kondo, Kinki University, Japan.

which generates the time-evolution operator

$$U_{\mathrm{FG}}(z,t) = e^{-i\int_0^t \tilde{H}_0 dt} = \exp\left(i\omega_0 b' z t I_z\right) \qquad (12.105)$$

after applying the field gradient for a duration t. Then the density matrix in the co-rotating frame develops as

$$
\begin{aligned}
\tilde{\rho}_{\mathrm{FG}}(z,t) &= U_{\mathrm{FG}}\tilde{\rho}U_{\mathrm{FG}}^{\dagger} \\
&= \frac{I}{2} + c_z I_z + \begin{pmatrix} 0 & (c_x - ic_y)e^{i\omega_0 b' z t} \\ (c_x + ic_y)e^{-i\omega_0 b' z t} & 0 \end{pmatrix},
\end{aligned}
$$

$$(12.106)$$

where we have expanded the initial density matrix as $\tilde{\rho} = I/2 + c_x I_x + c_y I_y + c_z I_z$). The FID signal consists of contributions from the whole sample. Therefore the effective density matrix is obtained by averaging $\tilde{\rho}_{\mathrm{FG}}(z,t)$ over z as

$$\tilde{\rho}_{\mathrm{FG}}(t) = \frac{1}{L}\int_{-L/2}^{L/2} \tilde{\rho}_{\mathrm{FG}}(z,t)dz. \qquad (12.107)$$

Suppose the condition $L\omega_0 b' t \gg 1$ is satisfied. Then the off-diagonal components in Eq. (12.106) average out to yield

$$\tilde{\rho}_{\mathrm{FG}}(t) = \frac{I}{2} + c_z I_z = \frac{1 - c_z}{2}I + c_z|0\rangle\langle 0|. \qquad (12.108)$$

The first term is the uniformly mixed state and does not contribute to the FID spectrum, while the second term behaves as an effective pure state $|0\rangle$.

Let us turn to a two-qubit molecule next. The Hamiltonian in the laboratory frame is

$$H_0 = -\omega_{0,1}(1 + b'z)I_z \otimes I - \omega_{0,2}(1 + b'z)I \otimes I_z + J \sum_{k=x,y,z} I_k \otimes I_k. \quad (12.109)$$

Transformation to the co-rotating frame produces the Hamiltonian

$$\tilde{H}_0 = -\omega_{0,1}b'zI_z \otimes I - \omega_{0,2}b'zI \otimes I_z + JI_z \otimes I_z, \quad (12.110)$$

which generates a local time-development operator

$$U_{\mathrm{FG}}(z,t) = \exp\left[i(\omega_{0,1}b'z + \omega_{0,2}b'z - JI_z \otimes I_z)t\right]. \quad (12.111)$$

Let $\tilde{\rho} = (\rho_{ij})$ be the initial density matrix in the rotating frame. After application of the field gradient for an interval t, we end up with a local density matrix

$$\tilde{\rho}_{\mathrm{FG}}(z,t) = U_{\mathrm{FG}}(z,t)\tilde{\rho}U_{\mathrm{FG}}^\dagger(z,t)$$

$$= \begin{pmatrix} \rho_{11} & * & * & * \\ * & \rho_{22} & e^{-i\Delta\omega_0 b'zt}\rho_{23} & * \\ * & e^{i\Delta\omega_0 b'zt}\rho_{32} & \rho_{33} & * \\ * & * & * & \rho_{44} \end{pmatrix}, \quad (12.112)$$

where $\Delta\omega_0 = \omega_{0,2} - \omega_{0,1}$ and $*$ stands for a matrix element which contains $e^{i2\omega_{0,i}b'zt}$ or $e^{i(\omega_{0,1}+\omega_{0,2})b'zt}$. Typical values of the parameters are $\omega_{0,i} \sim 10^9$ Hz, $b' \sim 10^{-2}$ T/m and $z \sim 10^{-2}$ m. The exponent with $t \sim 10^{-3}$ s is $\sim 10^2 \sim 16 \times 2\pi$. Therefore matrix elements denoted with $*$ vanish if the field gradient is applied for $t > 10^{-3}$ s and averaged over z.

We find $\Delta\omega_0 \sim \omega_{0,i}$ for a heteronucleus molecule, in which case the 23- and 32-matrix elements also vanish after averaging over z and we obtain

$$\tilde{\rho}_{\mathrm{FG}}(t) = \frac{1}{L} \int_{-L/2}^{L/2} \tilde{\rho}_{\mathrm{FG}}(z,t)dz$$

$$= \mathrm{diag}(\rho_{11}, \rho_{22}, \rho_{33}, \rho_{44}). \quad (12.113)$$

For a homonucleus molecule, in contrast, $\Delta\omega_0$ may be as small as 10^3 Hz, and the phase multiplying ρ_{23} and ρ_{32} is ~ 1 for sufficiently small t. Therefore we obtain

$$\tilde{\rho}_{\mathrm{FG}}(t) = \begin{pmatrix} \rho_{11} & 0 & 0 & 0 \\ 0 & \rho_{22} & \rho_{23} & 0 \\ 0 & \rho_{32} & \rho_{33} & 0 \\ 0 & 0 & 0 & \rho_{44} \end{pmatrix}. \quad (12.114)$$

Now we are ready to consider a pseudopure state generated with the field gradient technique. Suppose we have a thermal equilibrium density matrix

$$\tilde{\rho}(T) = \frac{e^{-H/k_B T}}{Z(T)}$$

$$= \frac{\hbar}{8k_B T} \text{diag}(\omega_{0,1} + \omega_{0,2}, \omega_{0,1} - \omega_{0,2}, -\omega_{0,1} + \omega_{0,2}, -\omega_{0,1} - \omega_{0,2}).$$

$$= \frac{\hbar \omega_{0,1}}{4k_B T} I_z \otimes I + \frac{\hbar \omega_{0,2}}{4k_B T} I \otimes I_z, \tag{12.115}$$

where we have dropped the irrelevant term $I/4$ in $\tilde{\rho}(T)$. Define an angle η by $\cos\eta = \omega_{0,2}/2\omega_{0,1}$, assuming $\omega_{0,2} \leq 2\omega_{0,1}$, and operate an η-pulse with $\phi_2 = \pi/2$ on qubit 2 to yield

$$U_{\pi/2,2}(\eta)\tilde{\rho}(T)U_{\pi/2,2}(\eta)^\dagger$$

$$= \frac{\hbar}{8k_B T} \begin{pmatrix} 3\omega_{0,2}/2 & \omega_{0,1}\sin\eta & 0 & 0 \\ \omega_{0,1}\sin\eta & \omega_{0,2}/2 & 0 & 0 \\ 0 & 0 & -\omega_{0,2}/2 & \omega_{0,1}\sin\eta \\ 0 & 0 & \omega_{0,1}\sin\eta & -3\omega_{0,2}/2 \end{pmatrix}. \tag{12.116}$$

We then get rid of off-diagonal matrix elements by applying the field gradient to obtain

$$\tilde{\rho}_1 = \frac{\hbar \omega_{0,2}}{4k_B T}(2I_z \otimes I + I \otimes I_z). \tag{12.117}$$

We need to eliminate ρ_{22}, ρ_{33} and ρ_{44} to obtain the pseudopure state $|00\rangle\langle00|$, which is done by applying (1) a $\pi/4$-pulse along the x-axis on qubit 1, (2) free evolution without pulses for $t = \pi/4J$, (3) a $\pi/4$-pulse along $-y$-axis on qubit 1 and finally (4) the field gradient again. A staightforward but tedious calculation shows that the density matrix obtained by applying operations (1) through (3) is

$$\tilde{\rho}_2 = \frac{\hbar \omega_{0,2}}{8k_B T} \begin{pmatrix} 3 & 0 & 0 & 0 \\ 0 & -1 & 0 & -2 \\ 0 & 0 & -1 & 0 \\ 0 & -2 & 0 & -1 \end{pmatrix}. \tag{12.118}$$

Finally the field gradient (4) eliminates the off-diagonal components to yield

$$\tilde{\rho}_3 = \frac{\hbar \omega_{0,1}}{2k_B T}(I_z \otimes I_z + \frac{1}{2}I_z \otimes I + \frac{1}{2}I \otimes I_z)$$

$$= \frac{\hbar \omega_{0,1}}{2k_B T}|00\rangle\langle00|, \tag{12.119}$$

where a term proportional to the unit matrix has been dropped.

EXERCISE 12.10 (1) Elaborate on the above steps (1) through (4) to derive Eq. (12.119).
(2) It has been assumed in the above analysis that $\omega_{0,i}$ satisfies the condition $\omega_{0,2} \leq 2\omega_{0,1}$. Repeat the above analysis by assuming $\omega_{0,2} > 2\omega_{0,1}$.

It should be noted that off-diagonal components ρ_{23} and ρ_{32} do not appear in the density matrices throughout the above calculation; see Eqs. (12.116) and (12.118). Therefore the above prescription to generate a pseudopure state using the spatial averaging is also applicable to homonucleus molecules provided that η is set to $\cos^{-1}(1/2) = \pi/3$.

See [29, 30] for generation of pseudopure states for general n-qubit systems by the spatial averaging method.

12.8 DiVincenzo Criteria

DiVincenzo criteria for an NMR quantum computer are evaluated as follows.

1. A scalable physical system with well-characterized qubits:

   Spin $1/2$ nuclei in a molecule are used as qubits. They cannot be cooled down to ultralow temperature since a molecule must be solved in a liquid to simplify nucleus-nucleus interaction. Selective addressing to each spin is possible by taking advantage of Larmor frequency differences. Chemical shifts among the same nuclear spices make it possible to access spins selectively even in a homonucleus molecule. However, selective addressing becomes harder and harder as the number of the same nuclei grows. The initialization outlined in §12.7 is also difficult for a large number of spins, and the estimated upper bound in the number of qubits in an NMR quantum computer is ~ 10.

2. The ability to initialize the state of the qubits to a simple fiducial state, such as $|00\ldots0\rangle$:

   Molecules in a liquid solvent at room temperature are in a thermal equilibrium state, which is quite close to the uniform mixture of all possible spin states. Since any unitary transformation cannot map a mixed state to a pure state, we need to employ nonunitary operations, such as temporal averaging, spatial averaging or logical labelling, to prepare a pseudopure state $|00\ldots0\rangle$, for example. The number of steps (the number of pulses, say) required to prepare the pseudopure state diverges exponentially as a function of the number of qubits n. The number of steps cannot be too large for an NMR quantum computer because of a finite decoherence time. The maximum number of qubits is estimated to be ~ 10 from this viewpoint too.

3. Long decoherence times, much longer than the gate operation time:

   Decoherence time depends on the molecule employed as a quantum computer. It may be as large as $10^2 \sim 10^3$ s. Single-qubit gate operation time can be as short as $\sim 10^{-5}$ s, while two-qubit gate operation time, making use of the J-couplings, takes $\sim 10^{-2} \sim 10^{-1}$ s. It has been shown in [36] that a faithful implementation of Shor's algorithm providing the factorization $21 = 3 \times 7$ requires approximately 10^5 gate operations, among which $\sim 10^4$ are two-qubit gate operations. Therefore we need at least $\sim 10^{-2} \times 10^4 = 10^2$ s decoherence time to execute this modest factorization.

4. A "univeral" set of quantum gates:

   One-qubit operations are implemented with rf pulses, by making use of the Rabi oscillations. Two-qubit operations are realized by using the J-coupling between nuclei. Some important two-qubit gates, such as the CNOT gate and the SWAP gate, are realized. In fact, a simplified version of Shor's factorization algorithm has already been demonstrated [13].

5. A qubit-specific measurement capability:

   Measurement of qubit states with the free induction decay (FID) is a well-established measurement technique in NMR, having several decades of history. It is also possible to measure the density matrix itself (quantum state tomography) and the unitary gate (quantum process tomography) within the current technology. However, the signal to noise ratio scales as ne^{-an}, $a \sim 1$ being a constant, and the readout becomes more and more difficult as the number of qubits grows.

In addition to the difficulties listed above, thermal density matrix at room temperature is not entangled, and it is often criticized that NMR is not a true quantum computer. However, it works as a simulator to a real quantum computer on which we can execute quantum algorithms. Several important techniques have been developed from these standpoints in the past. It should also be addressed that NMR is the only quantum computer which is commercially available. An NMR quantum computer is expected to remain an important tool to develop various techniques necessary to materialize a real working quantum computer to come.

References

[1] Y. Kondo, M. Nakahara and S. Tanimura, in *Physical Realizations of Quantum Computing: Are the DiVincenzo Criteria Fulfilled in 2004?*, ed. M. Nakahara *et al.*, World Scientific, Singapore (2006).

[2] L. M. K. Vandersypen and I. L. Chuang, Rev. Mod. Phys. **76**, 1037 (2004).

[3] J. Jones, in *Quantum Entanglement and Information Processing: Lecture Notes of the les Houches Summer School 2003*, eds. J.-M. Raimond, J. Dalibard and D. Esteve, Elsevier Science and Technology (2004).

[4] E. Knill, I. L. Chuang and R. Laflamme, Phys. Rev. A **57**, 3348 (1998).

[5] I. L. Chuang *et al.*, Nature **393**, 143 (1998).

[6] J. A. Jones, M. Mosca and R. H. Hansen, Nature **393**, 344 (1998).

[7] J. A. Jones and M. Mosca, Phys. Rev. Lett. **83**, 1050 (1999).

[8] D. G. Cory *et al.*, Phys. Rev. Lett. **81**, 2152 (1998).

[9] M. D. Price *et al.*, Phys. Rev. A **60**, 2777 (1999).

[10] B. R. Laflamme *et al.*, Phil. Trans. R. Soc. Lond. A **356**, 1941 (1998).

[11] L. M. K. Vandersypen *et al.*, Phys. Rev. Lett. **83**, 3085 (1999).

[12] L. M. K. Vandersypen *et al.*, Phys. Rev. Lett. **85**, 5452 (2000).

[13] L. M. K. Vandersypen *et al.*, Nature **414**, 883 (2001).

[14] R. Brüschweiler, Phys. Rev. Lett. **85**, 4815 (2000).

[15] F. Bloch and A. Siegert, Phys. Rev. **57**, 522 (1940).

[16] A. Barenco *et al.*, Phys. Rev. A **52**, 3457 (1995).

[17] J. J. Sylvester, Phil. Mag. **34**, 461 (1867).

[18] J. Hadamard, Bull. Sci. Math. **17**, 240 (1893).

[19] J. A. Jones and E. Knill, J. Magn. Reson., Ser. A **141**, 322 (1999).

[20] D. W. Leung *et al.*, Phys. Rev. A **61**, 042310 (2000).

[21] W. Rossmann, *Lie Groups*, Oxford Univ. Press, New York (2002).

[22] S. Helgason, *Differential Geometry, Lie Groups and Symmetric Spaces*, Academic Press, New York (1978).

[23] M. Nakahara, *Geometry, Topology and Physics* (2nd ed.), Taylor and Francis, Boca Raton, Ann Arbor, London, Tokyo (2003).

[24] N. Khaneja, R. Brockett and S. J. Glaser, Phys. Rev. A **63**, 032308 (2001).

[25] J. Zhang *et al.*, Phys. Rev. A **67**, 042313 (2003).

[26] M. Nakahara *et al.*, Phys. Lett. A **350**, 27 (2006).

[27] Y. Makhlin, Quant. Info. Proc. **1**, 243 (2002).

[28] E. Knill, I. L. Chuang and R. Laflamme, Phys. Rev. A **81**, 5672 (1998).

[29] U. Sakaguchi, H. Ozawa and T. Fukumi, Phys. Rev. A **61**, 042313 (2000).

[30] D. G. Cory, A. F. Fahmy and T. F. Havel, Proc. Natl. Acad. Sci. USA **94**, 1634 (1997).

[31] N. A. Gershenfeld and I. L. Chuang, Science **275**, 350 (1997).

[32] L. M. K. Vandersypen *et al.*, Phys. Rev. Lett. **83**, 3085 (1999).

[33] L. K. Grover, in *Proceedings of the 28th Annual ACM Symposium on the Theory of Computation*, 212, ACM Press, New York (1996).

[34] L. K. Grover, Phys. Rev. Lett. **79**, 325 (1997).

[35] I. L. Chuang, N. Gershenfeld and M. Kubince, Phys. Rev. Lett. **80**, 3408 (1998).

[36] J. J. Vartiainen *et al.*, Phys. Rev. A **70**, 012319 (2004).

13

Trapped Ions

Cirac and Zoller proposed in their seminal paper [1] published in 1995 to use ions trapped in an external potential as qubits. Trapped ions are expected to be scalable to a large number of qubits if the trap is segmented into several subtraps. Currently the largest trapped ion register comprises eight qubits, which is the largest among all proposed physical realizations.

Excellent review articles are [2] and [3], which we follow in this chapter.

13.1 Introduction

It has become possible to cool atoms and ions down to submicro Kelvins thanks to progress in laser cooling technology. Let us consider an ion with two energy levels, the ground state $|g\rangle$ with energy $\hbar\omega_g$ and an excited state $|e\rangle$ with $\hbar\omega_e$. Let $\hbar\omega_0 = \hbar\omega_e - \hbar\omega_g$ be the difference between their energy levels. Suppose an ion, with optical absorption curve Fig. 13.1 (a), is moving along the x-axis with speed v as shown in Fig. 13.1 (b). There are two counterpropagating laser beams along $\pm x$-directions with the frequency ω_L, which is red-detuned from ω_0;* see Fig. 13.1 (a). Then the laser beam propagating in the $-x$-direction is blue-shifted for the ion so that the frequency is shifted to $\omega_L' > \omega_L$, while the laser beam in the x-direction is red-shifted to $\omega_L'' < \omega_L$. These frequencies are also shown in Fig. 13.1 (a), which shows that the photons with ω_L' are more likely to be absorbed by the ion than those with ω_L''. In contrast, the spontaneous emission is isotropically distributed. As a result, the ion absorbs more head-on photons than follow-up photons and finally stops moving along the x-axis. If three pairs of laser beams along the $\pm x$, $\pm y$ and $\pm z$ directions are applied to the atom, it will give up all the kinetic energy eventually, set aside possible thermal motions. Neutral atom gas and ion gas at submicro Kelvins will be obtained with this **Doppler cooling** technique [4]. Evaporative cooling is used for further cooling a gas down to nano Kelvin regime [5].

---

*A laser is "red-detuned" if its frequency ω_L is slightly lower than a resonance frequency ω_0. It is "blue-detuned" if its frequency is slightly higer than ω_0.

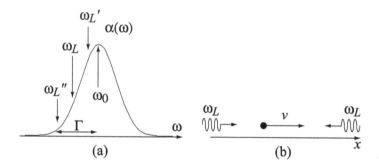

FIGURE 13.1

(a) The absorption coefficient of an ion (atom) in the vicinity of some resonance frequency ω_0. The width of the resonance is denoted as Γ. The atom
is irradiated by a red-detuned laser beam with the frequency ω_L in the laboratory frame. (b) An ion (atom) moving along the x-axis with velocity v.
Two red-detuned laser beams with frequency ω_L are applied along the $\pm x$-
directions. The laser frequency ω_L is blue-shifted to a frequency ω'_L closer to
ω_0 for an ion moving toward the laser beam, while it is red-shifted to ω''_L for
an atom moving away from the laser beam. It follows from the absorption
coefficient in (a) that photons having head-on collision with the ion are more
likely to be absorbed than the photons following-up with the ion. Spontaneous
emissions of the absorbed photons are isotropic. As a result, the ion gives up
its linear momentum.

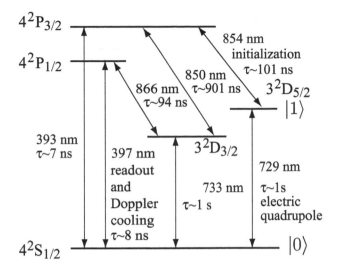

FIGURE 13.2
Relevant energy levels of a ^{40}Ca$^+$ ion. The numbers in nm denote the wave
lengths of the transition frequency. The first two letters denote the orbit
which is occupied by the last electron (4s for the ground state, for example).
The next two letters denote L_J, where L is the orbital angular momentum of
the last electron, while J is the quantum number of $J = L + S$.

In the following exposition, ions cooled in this way and stored in a linear
trap, as well as the vibrational normal modes of the ion crystal in the trap,
are used to encode information.

An eight-qubit entangled state has been realized with an ion trap [6]. At
the writing of this book, it seems the ion trap is one of the most promising
candidates of a scalable quantum computer. The definition of qubits, how to
confine them in a trap, and control of one and two qubits are outlined in the
present chapter.

13.2   Electronic States of Ions as Qubits

Let us start with an example of a ^{40}Ca$^+$ ion to make our discussion concrete.
^{40}Ca atom has electron configuration $1s^2 2s^2 2p^6 3s^2 3p^6 4s^2$. Note that the 3d
orbitals are empty in the ground state. We are interested in an ion ^{40}Ca$^+$,
whose electron configuration in its ground state is $1s^2 2s^2 2p^6 3s^2 3p^6 4s$. The
orbitals except for 4s form a closed shell. Figure 13.2 shows the energy levels
of a ^{40}Ca$^+$ ion. The states $4\,^2S_{1/2}$ and $3\,^2D_{5/2}$, abbreviated simply as $S_{1/2}$ and

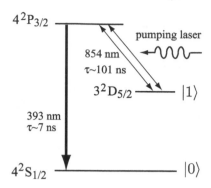

FIGURE 13.3

Initialization of ion qubit state by optical pumping. Those ions in the state $|1\rangle$ are pumped to the $P_{3/2}$ state by a laser beam in resonance with the transition frequency $D_{5/2} \to P_{3/2}$. Excited ions decay to $|0\rangle$ with the lifetime of $\tau \sim 7$ ns, while a small number of ions decay back to $|1\rangle$. They are excited again to $P_{3/2}$ and decay mainly to $|1\rangle$. Repeating this cycle many times eventually transforms all the qubits in $|1\rangle$ to $|0\rangle$.

$D_{5/2}$ hereafter, denote the qubit states $|0\rangle$ and $|1\rangle$, respectively. The transition between $S_{1/2}$ and $D_{5/2}$ is dipole-forbidden, and the metastable state $D_{5/2}$ has a lifetime ~ 1 s. Transition from $D_{5/2}$ to $S_{1/2}$ is made possible via the 4p $P_{3/2}$ state. The $D_{5/2}$ state is first excited to the 4p $P_{3/2}$ state by an 854 nm laser, and then it decays to the $S_{1/2}$ state spontaneously.

It is assumed that ions are well-separated in space from each other so that selective addressing to each ion is possible with a well-focused laser beam. A linear chain of N ions works as a register. The chain also supports $3N$ normal modes, which may be employed as bus qubits connecting separate ions. The above mentioned transition is also employed for sideband cooling of the linear chain normal modes to the ground state, which we will learn more about later.

The transition $S_{1/2} \leftrightarrow$ 4p $P_{1/2}$, with $\lambda = 397$ nm and width $\Gamma \simeq 20$ MHz, is used for Doppler cooling, in which a red detuned laser with ~ 10 MHz detuning is applied. This transition is also used for readout. Let the qubit under measurement be in the state $|0\rangle = |S_{1/2}\rangle$. Then irradiation of a laser beam of $\lambda = 397$ nm leads to resonance fluorescence. On the other hand, the laser beam has no effect if the qubit is in the state $|1\rangle = |D_{5/2}\rangle$, and no fluorescence is observed. In case the qubit state is a superposition $|\psi\rangle = a_0|0\rangle + a_1|1\rangle$, it is possible to read out the ratio $|a_0|^2/|a_1|^2$ by repeating the measurement many times and counting the number of times fluorescence was observed.

Initialization of the ioninc qubit state is achieved by depleting the ions in the $D_{5/2}$ level ($|1\rangle$) by **optical pumping** as shown in Fig. 13.3. This is done by making use of the transition $D_{5/2} \to P_{3/2}$ with an 854 nm pumping laser. The ions in the state $P_{3/2}$ subsequently decay mainly to the $S_{1/2}$ state ($|0\rangle$). Ions which decay to the $D_{5/2}$ state will be excited to $P_{3/2}$ again and the occupancy of $P_{3/2}$ will eventually vanish. A sharply focused laser beam initializes an individual ion, while a broad laser beam initializes all the ions.

Initialization of the phonon modes is achieved by sideband cooling; see §13.4.2.

13.3 Ions in Paul Trap

13.3.1 Trapping Potential

There are several variants of ion traps available. In the present section, we will analyze the linear Paul trap for concreteness [7, 8]. Earnshaw's theorem claims that a point charge cannot be maintained in stable stationary equilibrium in a static electric field. This is most easily seen from the Laplace equation $\Delta\Phi = 0$, where Φ is the potential energy acting on a charge. Φ may be expanded around a local extremum, which is chosen to be the origin, as $\Phi = ax^2 + by^2 + cz^2$. Substituting this into the Laplace equation leads to $a + b + c = 0$, from which we know at least one of a, b and c must be negative, and the potential takes a local maximum at the origin along the corresponding direction.

If we want to confine a collection of charges, therefore, we have to find a way out of this theorem. A possible loophole is to introduce a time-dependent electric field. The Paul trap shown in Fig. 13.4 (a) is made of four bars and a set of endcap electrodes. Figure 13.5 shows an actual ion trap.

Let us look at the potential produced by the four bars, which attains radial confinement of ions. Let a be the distance between the center of the trap and that of the axes of the bars as shown in Fig. 13.4 (b). We take the trap axis as the z-axis as depicted in Fig. 13.4 (a). The potential produced by the bar A is

$$V_A(x, y) = \frac{V_0}{2} \cos(\Omega t) \left[\ln \frac{\sqrt{x^2 + (y-a)^2}}{d} + 1 \right]$$

$$= \frac{V_0}{2} \cos(\Omega t) \left[\ln \frac{\sqrt{x^2 + (y-a)^2}}{a} + \ln \frac{a}{d} + 1 \right],$$

where $(V_0/2) \cos \Omega t$ is the applied voltage and d is the radius of the electrode bar. By adding all the contributions from the four bars we obtain

$$V_\perp(x, y) = \frac{V_0}{4} \cos(\Omega t) \left[\ln \frac{x^2 + (y-a)^2}{a^2} + \ln \frac{x^2 + (y+a)^2}{a^2} \right.$$

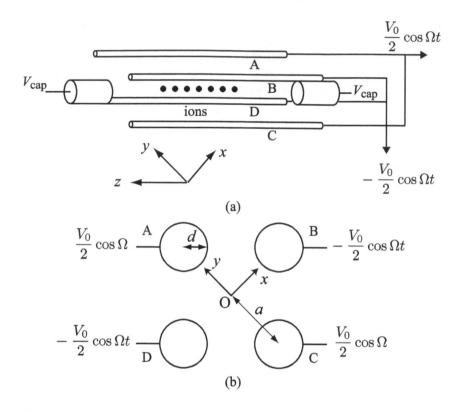

(a)

(b)

FIGURE 13.4

(a) Schematic diagram of a Paul trap. (b) Oscillating voltage $(V_0/2)\cos\Omega t$ is applied to bars A and C while $-(V_0/2)\cos\Omega t$ is applied to bars B and D for radial confinement. A voltage V_{cap} is applied to a pair of endcaps in (a) for axial confinement.

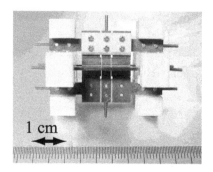

FIGURE 13.5

Actual ion trap. Courtesy of Shinji Urabe, Osaka University, Japan.

$$-\ln\frac{(x-a)^2+y^2}{a^2}-\ln\frac{(x+a)^2+y^2}{a^2}\Bigg]$$

$$\simeq V_0\cos(\Omega t)\frac{x^2-y^2}{a^2}, \tag{13.1}$$

where we assumed $|x|,|y|\ll a$ in the expansion. Moreover, we have ignored the mirror image contribution due to the presence of other bars by assuming $d\ll a$. Including the mirror images to the potential introduces a multiplicative factor to $V_\perp(x,y)$, but its functional form remains the same.

The four bars have been shown to produce a quadrupole oscillating field (13.1) around the origin. We need to study the equation of motion of an ion to show that it is radially confined under this potential. Let Q be the charge and M be the mass of the ion. Then the radial equations of motion are

$$M\frac{d^2x}{dt^2}=-\frac{2QV_0}{a^2}\cos(\Omega t)x \tag{13.2}$$

$$M\frac{d^2y}{dt^2}=\frac{2QV_0}{a^2}\cos(\Omega t)y. \tag{13.3}$$

By dividing the above equations by M and introducing the parameter

$$q=\frac{4QV_0}{M\Omega^2a^2}, \tag{13.4}$$

we obtain

$$\frac{d^2x}{dt^2}+q\frac{\Omega^2}{2}\cos(\Omega t)x=0 \tag{13.5}$$

$$\frac{d^2y}{dt^2}-q\frac{\Omega^2}{2}\cos(\Omega t)y=0. \tag{13.6}$$

These equations are called the **Mathieu equations**, and their mathematical properites are well known. General motion of an ion in the xy-plane is separated into a slow secular harmonic motion, called **guiding center motion**, around the origin with the angular frequencies

$$\omega_{\perp 0}\equiv\frac{\Omega q}{2\sqrt{2}} \tag{13.7}$$

and fast oscillation, called the **micromotion**, with the frequency Ω. To show this, we separate the coordinate x as $x=x_0+\delta x$, where x_0 denotes secular motion, while δx describes the micromotion. If this separation is substituted into Eq. (13.2), we obtain

$$\ddot{x}_0+\delta\ddot{x}=-q\frac{\Omega^2}{2}\cos(\Omega t)(x_0+\delta x). \tag{13.8}$$

We assume here that $|\ddot{x}_0|\ll|\delta\ddot{x}|$ and $|x_0|\gg|\delta x|$. Then δx satisfies

$$\delta\ddot{x}=-q\frac{\Omega^2}{2}\cos(\Omega t)x_0.$$

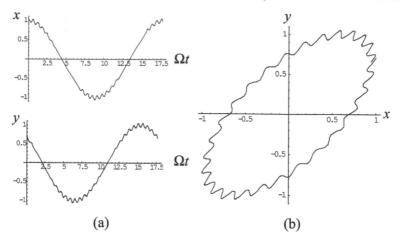

(a)   (b)

FIGURE 13.6
Motion of an ion in a trap with $q = 0.1$. (a) The x- and y-coordinates of an ion as functions of Ωt. (b) The trajectory in the xy-plane.

Assume that x_0 oscillates slowly compared to δx and substitute the ansatz $\delta x \propto \cos(\Omega t)$ in the above equation to find

$$\delta x = \frac{q}{2} \cos(\Omega t) x_0. \tag{13.9}$$

Substituting this into Eq. (13.8), we find x_0 satisfies the equation of motion

$$\ddot{x}_0 = -q\frac{\Omega^2}{2}\cos(\Omega t)\delta x = -\frac{q^2}{4}\Omega^2 \cos^2(\Omega t) x_0. \tag{13.10}$$

Assume $x_0 \propto \cos(\omega_{\perp 0} t)$ and $\omega_{\perp 0} \ll \Omega$. Then we may replace $\cos^2(\Omega t)$ in the above equation by $\langle \cos^2(\Omega t)\rangle = 1/2$ to obtain

$$\ddot{x}_0 = -\frac{q^2 \Omega^2}{8} x_0, \tag{13.11}$$

from which we find

$$\omega_{\perp 0} = \frac{\Omega q}{2\sqrt{2}}. \tag{13.12}$$

Figure 13.6 shows the trajectory of an ion in the trap.

Now we introduce the axial-confinement potential V_{cap} to the endcap electrodes. Let L be the distance between the two endcaps, whose tips are at $z = \pm L/2$. Then the z-dependence of the potential in the vicinity of the z-axis takes the form $c_1[z^2 - (x^2 + y^2)/2] + c_2$ in the lowest order approximation in z. Here the x- and y-dependences are added to satisfy the Laplace equation. From the condition $c_1(L/2)^2 + c_2 = V_{\text{cap}}$, we may put the potential produced by the endcaps as

$$V_z(x, y, z) = c_1 \left[z^2 - \frac{1}{2}(x^2 + y^2) - \left(\frac{L}{2}\right)^2 \right] + V_{\text{cap}}.$$

The coefficient c_1 is a constant fixed by the geometry of the trap. For later convenience, we write $c_1 = 4V_{\text{cap}}\alpha/L^2$, where α is a dimensionless constant. We then find

$$V_z(x, y, z) = \frac{4V_{\text{cap}}\alpha}{L^2} \left[z^2 - \frac{1}{2}(x^2 + y^2) \right], \tag{13.13}$$

where constant terms are dropped.

By adding the radial and the axial potentials, we finally obtain the total potential

$$V(x, y, z) = V_0 \cos(\Omega t) \frac{x^2 - y^2}{a^2} + \frac{4V_{\text{cap}}\alpha}{L^2} \left[z^2 - \frac{1}{2}(x^2 + y^2) \right]. \tag{13.14}$$

The equations of motion are derived from Eq. (13.14) as

$$M \frac{d^2 x}{dt^2} - Q \left[\frac{4V_{\text{cap}}\alpha}{L^2} - \frac{2V_0}{a^2} \cos(\Omega t) \right] x = 0 \tag{13.15}$$

$$M \frac{d^2 y}{dt^2} - Q \left[\frac{4V_{\text{cap}}\alpha}{L^2} + \frac{2V_0}{a^2} \cos(\Omega t) \right] y = 0 \tag{13.16}$$

$$M \frac{d^2 z}{dt^2} + Q \frac{8V_{\text{cap}}\alpha}{L^2} z = 0. \tag{13.17}$$

By dividing the above equations by M and introducing the parameters

$$b = \frac{Q\alpha V_{\text{cap}}}{ML^2\Omega^2}, \quad q = \frac{2QC_0}{Mr_0^2\Omega^2},$$

we obtain the equations of motion

$$\frac{d^2 x}{dt^2} + \frac{\Omega^2}{4} \left[b - 2q \cos(\Omega t) \right] x = 0, \tag{13.18}$$

$$\frac{d^2 y}{dt^2} + \frac{\Omega^2}{4} \left[b + 2q \cos(\Omega t) \right] y = 0, \tag{13.19}$$

$$\frac{d^2 z}{dt^2} + \frac{b\Omega^2}{2} z = 0. \tag{13.20}$$

The x- and y-components are described again by the Mathieu equations. Equations (13.5) and (13.5) are special cases of (13.18) and (13.19), respectively, in which b is set to 0. The equation of motion along the z-axis is nothing but that of a simple harmonic oscillator with the angular frequency

$$\omega_z = \sqrt{\frac{b}{2}} \Omega. \tag{13.21}$$

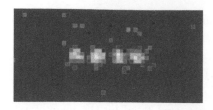

FIGURE 13.7
Four ions in a trap. Photo courtesy of Shinji Urabe, Osaka University, Japan.

The transverse frequencies $\omega_{x,y}$ are still given by $\omega_{\perp 0}$ in Eq. (13.12) since we assume $b \ll q$.

Figure 13.7 shows the image of four ions trapped in the apparatus shown in Fig. 13.5.

13.3.2   Lattice Formation

Suppose there are N ions of the same species along the z-axis of the Paul trap. Let z_n be the z coordinate of the nth ion. We also assume, as before, that the ions are trapped tightly in the radial (xy) direction. The potential energy of this system is

$$U = \frac{1}{2}M\omega_z^2 \sum_{j=1}^{N} z_j^2 + \frac{Q^2}{8\pi\epsilon_0} \sum_{j,k=1 \ (j\neq k)}^{N} \frac{1}{|z_k - z_j|}, \qquad (13.22)$$

where we have put $x_j = y_j = 0$ ($1 \le j \le N$). The equilibrium positions of the ions are fixed by solving the potential minimization equations

$$\frac{\partial U}{\partial z_i} = 0 \quad (1 \le i \le N). \qquad (13.23)$$

A few examples are in order. Let $N = 2$. The condition (13.23) yields

$$M\omega_z^2 z_1 + \frac{Q^2}{4\pi\epsilon_0(z_2 - z_1)^2} = M\omega_z^2 z_2 - \frac{Q^2}{4\pi\epsilon_0(z_2 - z_1)^2} = 0,$$

where we assumed $z_1 < 0 < z_2$. The above equations have a solution

$$z_2 = -z_1 = \left(\frac{1}{4}\right)^{1/3} l \simeq 0.630\, l, \qquad l^3 = \left(\frac{Q^2}{4\pi\epsilon_0 M\omega_z^2}\right). \qquad (13.24)$$

The parameter l characterizes the interionic distance.

Let us consider the case $N = 3$, for which we take $z_1 < z_2 < z_3$. Equation (13.23) yields

$$z_1 + l^3 \left[\frac{1}{(z_2 - z_1)^2} + \frac{1}{(z_3 - z_1)^2}\right] = 0$$

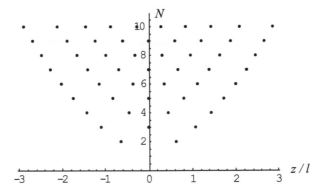

FIGURE 13.8
Equilibrium positions of ions for $2 \leq N \leq 10$.

$$z_2 + l^3 \left[-\frac{1}{(z_2 - z_1)^2} + \frac{1}{(z_3 - z_1)^2} \right] = 0$$

$$z_3 - l^3 \left[\frac{1}{(z_3 - z_2)^2} + \frac{1}{(z_3 - z_1)^2} \right] = 0.$$

By substituting a symmetry ansatz $z_3 = -z_1, z_2 = 0$, we immediately find the solution

$$z_3 = -z_1 = \left(\frac{5}{4} \right)^{1/3} l \simeq 1.077\, l, \quad z_2 = 0. \tag{13.25}$$

We have to resort to numerical solutions for general N. We obtain the results

$$N = 4: z_4 = -z_1 = 1.437\, l, \quad z_3 = -z_2 = 0.454\, l$$
$$N = 5: z_5 = -z_1 = 1.743\, l, \quad z_3 = -z_2 = 0.822\, l, \quad z_3 = 0 \tag{13.26}$$
$$N = 6: z_6 = -z_1 = 2.012\, l, \quad z_5 = -z_2 = 1.136\, l,$$
$$z_4 = -z_3 = 0.370\, l,$$

where we have arranged the coordinates in such a way that $z_i < z_j$ for $i < j$. It is found from the above solutions and Fig. 13.8 that the interionic distance for a fixed ion number N is least at the center, and this distance is a monotonically decreasing function of N. A numerical estimate shows that the interionic distance $\Delta z(N)$ at the trap center scales as

$$\Delta z(N) \simeq \frac{2.018\, l}{N^{0.59}} \tag{13.27}$$

with a good precision [9].

13.3.3   Normal Modes

It will be shown below that normal modes of the one-dimensional crystal play
a very important role in an ion trap quantum computer. Low-frequency modes
act as a qubit bus connecting ionic qubits.

Suppose the ions oscillate with small amplitudes around their equilibrium
positions. Let us expand the potential energy U around the equilibrium position of each ion as

$$U_{jj} \equiv \frac{\partial^2 U}{\partial z_j^2} = M\omega_z^2 + \frac{Q^2}{4\pi\epsilon_0} \sum_{k=1(k\neq j)}^{N} \frac{1}{|z_k - z_j|^3}$$

$$= M\omega_z^2 \left(1 + l^3 \sum_{k=1(k\neq j)}^{N} \frac{1}{|z_k - z_j|^3} \right), \tag{13.28}$$

$$U_{jk} \equiv \frac{\partial^2 U}{\partial z_j \partial z_k} = -\frac{Q^2}{4\pi\epsilon_0} \frac{1}{|z_k - z_j|^3} = -M\omega_z^2 \frac{l^3}{|z_k - z_j|^3}. \tag{13.29}$$

Let z_j^0 be the equilibrium position of the jth ion and expand z_j as $z_j = z_j^0 + \delta z_j$.
The Lagrangian is then expanded as

$$L = \frac{M}{2} \sum_j \dot{z}_j^2 - \frac{M}{2}\omega_z^2 \sum_j z_j^2 - \frac{Q^2}{8\pi\epsilon_0} \sum_{j,k(j\neq k)} \frac{1}{|z_k - z_j|}$$

$$\simeq \frac{M}{2} \sum_j \dot{\delta z}_j^2 - \frac{1}{2} \sum_{j,k=1}^{N} U_{jk}\delta z_j \delta z_k. \tag{13.30}$$

The Euler-Lagrange equation derived from this Lagrangian is

$$\frac{d}{dt}\frac{\partial L}{\partial \dot{\delta z}_j} - \frac{\partial L}{\partial z_j} = M\ddot{\delta z}_j + \sum_{k\neq j} U_{jk}\delta z_k. \tag{13.31}$$

Substituting the ansatz $\delta z_j = A_j e^{-i\nu t}$ into the above equation, we obtain
$-M\nu^2 A_j + \sum_k U_{jk}A_k = 0$ for $1 \leq j \leq N$ or

$$\det(U_{jk} - M\nu^2 \delta_{jk}) = 0. \tag{13.32}$$

Let us introduce a dimensionless eigenvalue λ by $\nu^2 = \omega_z^2 \lambda$. By inspecting the
characteristic equation

$$\sum_k U_{jk}u_k^{(p)} = M\omega_z^2 \lambda_p u_j^{(p)}, \tag{13.33}$$

we find that

$$u^{(1)} = \frac{1}{\sqrt{N}}(1, 1, \ldots, 1)^t \quad (\lambda_1 = 1),$$

$$u^{(2)} = \frac{1}{\sqrt{\sum_j z_j^2}}(z_1, z_2, \ldots, z_N)^t \quad (\lambda_2 = 3) \tag{13.34}$$

are normalized eigenvectors. The first mode with the frequency ω_z has no nodes. It is the motion of the ion lattice as a whole and is called the **center of mass mode** (COM). The second mode, with a single node, with the frequency $\sqrt{3}\omega_z$ is called the **breathing mode**. Note that the eigenvalues as well as the form of the eigenvectors are independent of N. All the other modes have higher energies, and their explicit forms must be obtained numerically. It follows from orthogonality between eigenvectors of a symmetric matrix U_{jk} that

$$\sum_{k=1}^{N} u_k^{(p)} = \sqrt{N} u^{(p)} \cdot u^{(1)} = 0 \quad (2 \leq p \leq N). \tag{13.35}$$

All the modes are evaluated explicitly for a small number of ions. For example,

$$u^{(1)} = \frac{1}{\sqrt{2}}(1,1)^t \qquad \lambda_1 = 1$$
$$u^{(2)} = \frac{1}{\sqrt{2}}(-1,1)^t \qquad \lambda_2 = 3 \tag{13.36}$$

for $N = 2$ and

$$u^{(1)} = \frac{1}{\sqrt{3}}(1,1,1)^t \qquad \lambda_1 = 1$$

$$u^{(2)} = \frac{1}{\sqrt{2}}(-1,0,1)^t \qquad \lambda_2 = 3 \tag{13.37}$$

$$u^{(3)} = \frac{1}{\sqrt{6}}(1,-2,1)^t \qquad \lambda_3 = \frac{29}{5}$$

for $N = 3$. See [9] for normal modes up to $N = 10$. Modes $u^{(1)}$ and $u^{(2)}$ are the only relevant modes we are concerned with at low enough temperatures.

EXERCISE 13.1 Verify that $u^{(3)}$ in Eq. (13.37) for $N = 3$ is indeed an eigenvector with a proper eigenvalue $\lambda_3 = 29/5$.

Now the displacement δz_j from the equilibrium position z_j^0 is expanded in terms of the normal modes as

$$\delta z_j(t) = \sum_{p=1}^{N} Q_p(t) u_j^{(p)} \quad (1 \leq j \leq N). \tag{13.38}$$

Substitution of this expansion into the Lagrangian (13.30) yields decoupled Lagrangian

$$L = \frac{M}{2} \sum_{p=1}^{N} \left(\dot{Q}_p^2 - \omega_z^2 \lambda_p Q_p^2 \right). \tag{13.39}$$

The coodinate z_j of the jth ion is

$$z_j = z_j^0 + \text{Re}\left[\sum_{p=1}^{N} C_p u_j^{(p)} e^{-i\omega_z \sqrt{\lambda_p} t}\right], \qquad (13.40)$$

where we put $Q_p = C_p e^{-i\omega_z \sqrt{\lambda_p} t}$. Suppose only the lowest mode $u^{(1)}$ (COM mode) is excited. The coordinate then simplifies as

$$z_j = z_j^0 + \text{Re}\left[\frac{C_1}{\sqrt{N}} e^{-i\omega_z t}\right]. \qquad (13.41)$$

These modes are treated as a quantized harmonic oscillator in the following.

13.4   Ion Qubit

13.4.1   One-Spin Hamiltonian

Let us consider $^{40}\text{Ca}^+$ ions for definiteness in which states $4^2\text{S}_{1/2}$ and $3^2\text{D}_{5/2}$ form the qubit basis vectors. These states are coupled through quadrupole interaction, whose Hamiltonian is

$$H_{\text{int}} = \frac{e}{2} \sum_{a,b=x,y,z} r_a r_b \frac{\partial E_a}{\partial R_b}(\sigma_+ + \sigma_-), \quad \sigma_\pm = \frac{1}{2}(\sigma_x \pm i\sigma_y), \qquad (13.42)$$

where $\boldsymbol{R}$ is the position of the ion, $\boldsymbol{r}$ is the position of the valence electron with respect to the nucleus and $\boldsymbol{E}$ is the electric field of the laser beam propagating with the wave vector $\boldsymbol{k}$, which is written as

$$\boldsymbol{E} = \boldsymbol{E}_0 \sin(\boldsymbol{k} \cdot \boldsymbol{x} - \omega_L t - \phi) \qquad (13.43)$$

assuming it is linearly polarized.

The Rabi oscillation frequency in this case is

$$\hbar\Omega = \frac{e}{2}\langle 3^2\text{D}_{5/2}|(\boldsymbol{k} \cdot \boldsymbol{r})(\boldsymbol{E}_0 \cdot \boldsymbol{r})|4^2\text{S}_{1/2}\rangle. \qquad (13.44)$$

We assume the laser beam is focused so that it irradiates the jth ion only and has no effect on the other ions in the chain. The interaction between the laser and the jth ion at $\boldsymbol{R}_j = (0, 0, z_j)$ is

$$H_{\text{int}} = \hbar\Omega \cos(kz_j \cos\theta - \omega_L t - \phi)(\sigma_+ + \sigma_-), \qquad (13.45)$$

where θ is the angle between $\boldsymbol{k}$ and the z-axis and we assume θ is independent of j, i.e., the distance between the laser and an ion is much larger than the ion lattice size.

Let us consider the pth normal mode of the lattice vibration. The quantized Hamiltonian and the normal coordinate are

$$H_{\mathrm{h}} = \hbar \nu_p a_p^\dagger a_p, \quad Q_p = \sqrt{\frac{\hbar}{2M\nu_p}} (a_p^\dagger + a_p), \tag{13.46}$$

where $\nu_p = \sqrt{\lambda_p} \omega_z$. The correspondence between the normal coordinate Q_p and the actual displacement z_j of the jth ion is

$$\delta z_j = u_j^{(p)} Q_p. \tag{13.47}$$

We drop the index p from Q_p, a_p and ν_p hereafter to simplify the notation. Now the interaction Hamiltonian is put in the form

$$H_{\mathrm{int}} = \hbar \Omega \cos \left[\eta_j \left(a + a^\dagger \right) - \omega_L t - \phi_j \right] (\sigma_+ + \sigma_-), \tag{13.48}$$

where

$$\eta_j = u_j k \sqrt{\frac{\hbar}{2M\nu}} \cos \theta,$$

and ϕ_j is the phase of the laser beam at the position of the jth ion. The dimensionless quantity

$$\eta \equiv k \sqrt{\frac{\hbar}{2M\nu}} \tag{13.49}$$

is called the **Lamb-Dicke parameter**. Note that $\sqrt{\hbar/2M\nu}$ is on the order of $\sqrt{\langle \delta z_j^2 \rangle}$ in the pth normal mode, and hence the Lamb-Dicke parameter η denotes the approximate ratio of $\sqrt{\langle \delta z_j^2 \rangle}$ and the laser wave length when N is sufficiently small.

Let us introduce the interaction picture in which the main part of the Hamiltonian is $H_{\mathrm{h}} + H_{\mathrm{ion}}$, where

$$H_{\mathrm{ion}} = -\frac{\hbar}{2} \omega_0 \sigma_z \tag{13.50}$$

and $\hbar \omega_0 = \hbar \omega_D - \hbar \omega_S$ is the difference between the energies of the excited state D and the ground state S. Transformation to the interaction picture is realized by the unitary matrix

$$U(t) = \exp \left[-\frac{i}{\hbar} (H_{\mathrm{h}} + H_{\mathrm{ion}}) t \right]. \tag{13.51}$$

The interaction Hamiltonian in this picture is

$$\begin{aligned}
H_{\mathrm{int}}'(t) &= U(t)^\dagger H_{\mathrm{int}} U(t) \\
&= \frac{\hbar}{2} \Omega \left[\sigma_+ e^{i\eta(a e^{-i\nu t} + a^\dagger e^{i\nu t})} e^{i(\phi - \delta t)} + \text{Hermitian conjugate} \right],
\end{aligned} \tag{13.52}$$

where $\delta = \omega_L - \omega_0$ and we have dropped rapidly oscillating terms proportional to $e^{\pm i(\omega_L + \omega_0)}$ (**rotating wave approximation**). We have set $u_j = 1$ and $\cos\theta = 1$ in the Hamiltonian to simplify our argument.

It is possible to couple a lattice vibrational mode to a particular ionic state by controlling the detuning δ. There appear terms of the form $\sigma_+ a^k a^{\dagger l}$ and $\sigma_- a^l a^{\dagger k}$ in the interaction Hamiltonian H'_{int} if the exponential function containing a and $a^\dagger$ is expanded. These terms introduce coupling between the states $|0\rangle|n\rangle$ and $|1\rangle|n+s\rangle$ when $\delta \sim s\nu$, where $s = k - l$. Here $|n\rangle$ is the eigenvector of the operator $a^\dagger a$ with the phonon number $n \in \{0, 1, 2, \ldots\}$.

The laser frequency ω_L is detuned toward the blue side from the resonance fequency δ in case $s > 0$. Accordingly this transition is called the **blue sideband transition**. If it is shifted toward the red side ($s < 0$), on the other hand, the corresponding transition is called the **red sideband transition**. The laser changes the internal electronic states while leaving the vibration mode unchanged when $s = 0$. This transition is called the **carrier transition**.

Suppose the quantum number of the harmonic mode is n. Then the size of the wave function scales as $\sim \sqrt{2n+1}a_0$, where $a_0 \simeq \sqrt{\hbar/2M\nu}$. Consider the **Lamb-Dicke limit** $\eta^2(2n+1) \ll 1$, in which the laser field may be treated as slowly varying over the range of ion's oscillatory motion. The interaction Hamiltonian H'_{int} is expanded in terms of η, in this case, and reduces to

$$H_{LD} = \frac{\hbar}{2}\Omega\sigma_+ \left[1 + i\eta(ae^{-i\nu t} + a^\dagger e^{i\nu t})\right] e^{i(\phi - \delta t)} + \text{Hermitian conjugate.} \tag{13.53}$$

Suppose ω_L is in tune with ω_0 (carrier transition) so that $\delta = 0$. Then H_{LD} further simplifies, by dropping terms oscillating with time, to a time-independent form as

$$H_{CT} = \frac{\hbar}{2}\Omega(\sigma_+ e^{i\phi} + \sigma_- e^{-i\phi}). \tag{13.54}$$

This Hamiltonian introduces the transitions $|0\rangle|n\rangle \leftrightarrow |1\rangle|n\rangle$ with the Rabi frequency Ω.

Let us consider the red sideband transition, for which $\delta < 0$, next. We take $\delta = -\nu$ as an example, for which $s = -1$. Then H_{LD} becomes

$$H_{RST} = i\frac{\hbar}{2}\eta\Omega \left(a\sigma_+ e^{i\phi} - a^\dagger \sigma_- e^{-i\phi}\right), \tag{13.55}$$

again by dropping time-dependent terms. This Hamiltonian gives rise to the transitions $|0\rangle|n\rangle \leftrightarrow |1\rangle|n-1\rangle$ with Rabi frequency $\Omega_{n,n-1} = \Omega\eta\sqrt{n}$ as shown in the next section.

Similarly blue-sideband transtion with $\delta = \nu$ ($s = +1$) simplifies the Hamiltonian to

$$H_{BST} = i\frac{\hbar}{2}\eta\Omega \left(a^\dagger \sigma_+ e^{i\phi} - a\sigma_- e^{-i\phi}\right), \tag{13.56}$$

where only terms with no time dependence have been kept. It introduces the transitions $|0\rangle|n\rangle \leftrightarrow |1\rangle|n+1\rangle$ with Rabi frequency $\Omega_{n,n+1} = \Omega\eta\sqrt{n+1}$; see the next section.

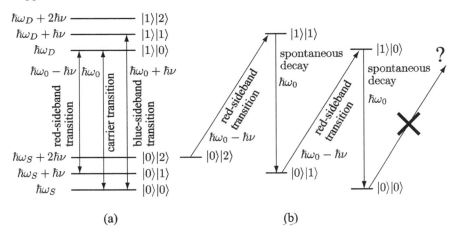

(a) (b)

FIGURE 13.9

(a) Ionic and vibrational energy levels. A state $|m\rangle|n\rangle$ stands for the tensor product of an ionic energy eigenstate $|m\rangle$ and phonon eigenstate $|n\rangle$, where $|m = 0\rangle = |S_{1/2}\rangle$ and $|m = 1\rangle = |D_{5/2}\rangle$, respectively, and $n = 0, 1, 2, \ldots$ is the phonon number eigenvalue. The energy difference between the two ionic states is $\hbar\omega_0 = \hbar\omega_D - \hbar\omega_S$. Each ionic eigenstate has a tower of phonon eigenstates, separately by equal distance $\hbar\nu$. The phonon energy quantum $\hbar\nu$ ($\ll \hbar\omega_0$) is exaggerated in the figure. The transition $|0\rangle|n\rangle \leftrightarrow |1\rangle|n + s\rangle$ ($s \in \mathbb{Z}$) is called a carrier transition if $s = 0$, a red-sideband transition if $s < 0$ and a blue-sideband transition if $s > 0$. (b) Sideband cooling mechanism. An ion is irradiated by a laser beam of the frequency $\omega_0 - \nu$. An excited phonon state $|0\rangle|2\rangle$, for example, absorbs a photon and is excited to $|1\rangle|1\rangle$. The excited state then decays, by emitting a photon with frequencey ω_0, to $|0\rangle|1\rangle$. The ion is excited to $|1\rangle|0\rangle$ in the next cycle and decays into the phonon vacuum state $|0\rangle|0\rangle$. It is impossible to excite this ion state any more since there is no state to which $|0\rangle|0\rangle$ is excited by a red-sideband laser beam.

13.4.2 Sideband Cooling

It is possible to cool an ionic system by employing the Doppler cooling such that the average number of residual phonons is approximately 15. The number of phonons is further reduced to ~ 0 if sideband cooling is introduced [10]. As a result, it is possible to employ phonon eigenstates $|0\rangle$, a state with no phonons, and $|1\rangle$, a state with a single phonon, to construct a phonon qubit.

Let Γ be the natural line width of the excited state and ν be the angular frequency of the phonon. They must satisfy the inequality $\nu \gg \Gamma$ for the phonon levels to be resolved and sideband cooling to work; see Fig. 13.9. The transition between the ground state $S_{1/2}$ and the excited state $D_{5/2}$ with a very long natural lifetime ~ 1 s is employed to satisfy this condition in a Ca^+ ion. The line width of this transition is $\Gamma \sim 20$ MHz, and it is easy to confine

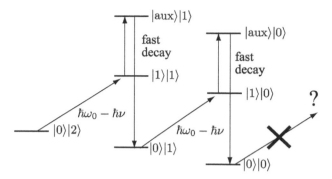

FIGURE 13.10
Efficient sideband cooling of a vibrational mode. $D_{5/2}$ state with a long life-
time is further excited to an auxiliary state $|aux\rangle$ which decays fast.

a lattice in a trap with $\nu \gg \Gamma$. The resonance transition between the states
$|S\rangle|n\rangle$ and $|D\rangle|n + s\rangle$, i.e., between the phonon number eigenstates $|n\rangle$ and
$|n + s\rangle$ is realized if we take $\delta = \nu s$. Processes with $s \neq 0$ are called **sideband
resonance transitions** to discriminate them from the transition with $s = 0$.
The process $|S\rangle|n\rangle \rightarrow |D\rangle|n - 1\rangle$ is driven if the laser is set to red-detuned
frequency with $s = -1$. Subsequently the ionic state decays from $|D\rangle$ to $|S\rangle$
by spontaneous photon emission. The matrix element of the phonon number
changing process is proportional to η, which is very small in the Lamb-Dicke
limit. Therefore the phonon number preserving process $|D\rangle|n-1\rangle \rightarrow |S\rangle|n-1\rangle$
is the dominant one in this limit. As a result the phonon number reduces as
$|S\rangle|n\rangle$ to $|S\rangle|n - 1\rangle$ in the end of these processes. The state with vanishing
phonon number is reached if this process is repeated many times.

The very long lifetime ~ 1 s of the $D_{5/2}$ state is an obstacle against efficient
cooling in the above process. To avoid this difficulty, the ion in the state $D_{5/2}$
is excited to $P_{3/2}$, which decays to $S_{1/2}$ with a very short life time ~ 8 ns,
leading to efficient cooling of the vibrational mode as shown in Fig. 13.10.

13.5   Quantum Gates

13.5.1   One-Qubit Gates

13.5.1.1   Carrier Transition

The time-development operator for the carrier transition is

$$U_{CT}(t) = e^{-\frac{i}{\hbar}H_{CT}t} = \exp\left[-i\frac{\Omega t}{2}\left(\sigma_+ e^{i\phi} + \sigma_- e^{-i\phi}\right)\right]$$

$$= \sum_{k=0}^{\infty} \frac{1}{(2k)!} \left(-i\frac{\Omega t}{2}\right)^{2k} \left[(\sigma_+\sigma_-)^k + (\sigma_-\sigma_+)^k\right]$$

$$+ \sum_{k=0}^{\infty} \frac{1}{(2k+1)!} \left(-i\frac{\Omega t}{2}\right)^{2k+1} \left[(\sigma_+\sigma_-)^k\sigma_+e^{i\phi} + (\sigma_-\sigma_+)^k\sigma_-e^{-i\phi}\right].$$

$$(13.57)$$

We implicitly assume that the harmonic oscillator mode stays in an eigenstate and no transition to different phonon number states takes place, which implies that only operators of the form $|n\rangle\langle n|$ appear in the expansion of $U_{\text{CT}}(t)$. Now $U_{\text{CT}}(t)$ takes the form

$$\sum_{n=0}^{\infty} \cos \frac{\Omega t}{2} \left[|0\rangle\langle 0| \otimes |n\rangle\langle n| + |1\rangle\langle 1| \otimes |n\rangle\langle n|\right]$$

$$-i \sum_{n=0}^{\infty} \sin \frac{\Omega t}{2} \left[|1\rangle\langle 0| \otimes |n\rangle\langle n|e^{i\phi} + |0\rangle\langle 1| \otimes |n\rangle\langle n|e^{-i\phi}\right]. \quad (13.58)$$

However, it follows from these observations that any unitary rotation of the form

$$R(\theta, \phi) = \cos\left(\frac{\theta}{2}\right) I - i\sin\left(\frac{\theta}{2}\right) \begin{pmatrix} 0 & e^{-i\phi} \\ e^{i\phi} & 0 \end{pmatrix} = e^{-i(\theta/2)(\cos\phi\sigma_x + \sin\phi\sigma_y)}$$

$$(13.59)$$

in the subspace spanned by $\{|0\rangle|n\rangle, |1\rangle|n\rangle\}$ is realized by irradiating an ion with a laser beam in the carrier frequency for a duration t, where $\theta = \Omega t$. Note that any SU(2) rotation may be implemented with σ_x and σ_y rotations.

13.5.1.2 Red Sideband Transition

Let us consider the red sideband transition next. The time development operator is

$$U_{\text{RST}} = e^{-iH_{\text{RST}}t} = \exp\left[-i\frac{\eta\Omega t}{2}\left(ia\sigma_+e^{i\phi} - ia^\dagger\sigma_-e^{-i\phi}\right)\right]$$

$$= \sum_{k=0}^{\infty} \frac{1}{(2k)!} \left(-i\frac{\eta\Omega t}{2}\right)^{2k} \left[(a\sigma_+a^\dagger\sigma_-)^k + (a^\dagger\sigma_-a\sigma_+)^k\right]$$

$$+ \sum_{k=0}^{\infty} \frac{1}{(2k+1)!} \left(-i\frac{\eta\Omega t}{2}\right)^{2k+1}$$

$$\times \left[(a\sigma_+a^\dagger\sigma_-)^k ia\sigma_+e^{i\phi} - (a^\dagger\sigma_-a\sigma_+)^k ia^\dagger\sigma_-e^{-i\phi}\right]. \quad (13.60)$$

The transformation $U_{\text{RST}}(t)$ takes the form

$$\sum_{n=0}^{\infty} \cos \frac{\Omega_n t}{2} \left[|0\rangle\langle 0| \otimes |n\rangle\langle n| + |1\rangle\langle 1| \otimes |n-1\rangle\langle n-1|\right]$$

$$+ \sum_{n=0}^{\infty} \sin \frac{\Omega_n t}{2} \left[|1\rangle\langle 0| \otimes |n-1\rangle\langle n| e^{i\phi} - |0\rangle\langle 1| \otimes |n\rangle\langle n-1| e^{-i\phi} \right],$$

$$(13.61)$$

where $\Omega_n = \sqrt{n}\eta\Omega$. Let us restrict ourselves within the subspace spanned by $\{|0\rangle|n\rangle, |1\rangle|n-1\rangle\}$. Then the matrix U_{RST} generates a rotation

$$R_n^-(\theta, \phi) = \cos \frac{\sqrt{n}\theta}{2} I + \sin \frac{\sqrt{n}\theta}{2} \begin{pmatrix} 0 & -e^{-i\phi} \\ e^{i\phi} & 0 \end{pmatrix}, \qquad (13.62)$$

where $\theta = \eta\Omega t$.

13.5.1.3   Blue-Sideband Transition

Finally let us consider the blue-sideband transition. The time development operator is

$$U_{\mathrm{RST}} = e^{-iH_{\mathrm{BST}}t} = \exp\left[-i\frac{\eta\Omega t}{2} \left(ia^\dagger \sigma_- e^{i\phi} - ia\sigma_- e^{-i\phi} \right) \right]$$

$$= \sum_{k=0}^{\infty} \frac{1}{(2k)!} \left(-i\frac{\eta\Omega t}{2} \right)^{2k} \left[\left(a^\dagger \sigma_+ a\sigma_- \right)^k + \left(a\sigma_- a^\dagger \sigma_+ \right)^k \right]$$

$$+ \sum_{k=0}^{\infty} \frac{1}{(2k+1)!} \left(-i\frac{\eta\Omega t}{2} \right)^{2k+1}$$

$$\times \left[\left(a^\dagger \sigma_+ a\sigma_- \right)^k ia^\dagger \sigma_+ e^{i\phi} - \left(a\sigma_- a^\dagger \sigma_+ \right)^k ia\sigma_- e^{-i\phi} \right]. \quad (13.63)$$

The transformation $U_{\mathrm{BST}}(t)$ takes the form

$$\sum_{n=0}^{\infty} \cos \frac{\Omega_n t}{2} \left[|0\rangle\langle 0| \otimes |n-1\rangle\langle n-1| + |1\rangle\langle 1| \otimes |n\rangle\langle n| \right]$$

$$+ \sum_{n=0}^{\infty} \sin \frac{\Omega_n t}{2} \left[|1\rangle\langle 0| \otimes |n\rangle\langle n-1| e^{i\phi} - |0\rangle\langle 1| \otimes |n-1\rangle\langle n| e^{-i\phi} \right].$$

$$(13.64)$$

It generates a rotation

$$R_n^+(\theta, \phi) = \cos \frac{\sqrt{n}\theta}{2} I + \sin \frac{\sqrt{n}\theta}{2} \begin{pmatrix} 0 & -e^{-i\phi} \\ e^{i\phi} & 0 \end{pmatrix} \qquad (13.65)$$

in the subspace spanned by $\{|0\rangle|n-1\rangle, |1\rangle|n\rangle\}$, where $\theta = \eta\Omega t$.

13.5.2   CNOT Gate

The CNOT gate proposed by Cirac and Zoller [1] has been realized by Schmidt-Kaler *et al.* [11]. The pulse sequence the latter employed was slightly

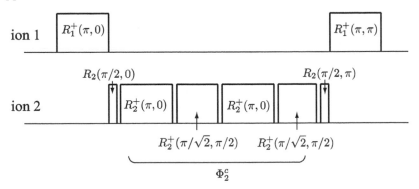

FIGURE 13.11
Pulse sequence to implement the CNOT gate experimentally [11]. The first and the last pulses are applied to ion 1, while six pulses are applied to ion 2. Note that the Rabi frequencies for $R(\theta, \phi)$ and $R^+(\theta, \phi)$ are different, and hence the pulse widths are also different.

different from one in the origial proposal in [1]. It is given as

$$R_1^+(\pi, \pi) R_2(\pi/2, \pi) \Phi_2^c R_2(\pi/2, 0) R_1^+(\pi, 0), \tag{13.66}$$

where the subscript $j \in \{1, 2\}$ in R_j refers to which ion the pulse acts on and Φ_2^c is a composite single ion phase gate

$$\Phi_2^c = R_2^+(\pi/\sqrt{2}, \pi/2) R_2^+(\pi, 0) R_2^+(\pi/\sqrt{2}, \pi/2) R_2^+(\pi, 0) \tag{13.67}$$

acting on ion 2. Here $R_j(\theta, \phi)$ and $R_j^+(\theta, \phi)$ are rotation operators associated with carrier and blue-sideband transitions, respectively. Figure 13.11 shows the pulse sequence to implement the CNOT gate experimentally. Let us see how the basis vectors

$$\{|0\rangle_1|0\rangle_2|0\rangle, |0\rangle_1|1\rangle_2|0\rangle, |1\rangle_1|0\rangle_2|0\rangle, |1\rangle_1|1\rangle_2|0\rangle\}$$

transform under the above gate operations, where $|n\rangle$ denotes the phonon states, while $|0\rangle_j$ and $|1\rangle_j$ ($j \in \{1, 2\}$) denote the jth ion states S and D, respectively. We find

$$\begin{array}{ccc}
R_1^+(\pi, 0) & & R_2(\pi/2, 0) \\
|0\rangle_1|0\rangle_2|0\rangle & \rightarrow & |1\rangle_1|0\rangle_2|1\rangle & \rightarrow & |1\rangle_1 \frac{1}{\sqrt{2}}(|0\rangle_2 - i|1\rangle_2)|1\rangle \\
|0\rangle_1|1\rangle_2|0\rangle & \rightarrow & |1\rangle_1|1\rangle_2|1\rangle & \rightarrow & |1\rangle_1 \frac{1}{\sqrt{2}}(|1\rangle_2 - i|0\rangle_2)|1\rangle \quad (13.68) \\
|1\rangle_1|0\rangle_2|0\rangle & \rightarrow & |1\rangle_1|0\rangle_2|0\rangle & \rightarrow & |1\rangle_1 \frac{1}{\sqrt{2}}(|0\rangle_2 - i|1\rangle_2)|0\rangle \\
|1\rangle_1|1\rangle_2|0\rangle & \rightarrow & |1\rangle_1|1\rangle_2|0\rangle & \rightarrow & |1\rangle_1 \frac{1}{\sqrt{2}}(|1\rangle_2 - i|0\rangle_2)|0\rangle.
\end{array}$$

The composite single ion phase gate Φ_2^c is applied next to the second ion. There are second-ion states $|0\rangle_2|0\rangle$, $|0\rangle_2|1\rangle$, $|1\rangle_2|0\rangle$, $|1\rangle_2|1\rangle$ participating in Eq. (13.68). The state $|1\rangle_2|0\rangle$ does not change under the operation of Φ_2^c. The states $|0\rangle_2|0\rangle$ and $|1\rangle_2|1\rangle$ correspond to the subspace with $n = 1$ in Eq. (13.65), where Φ_2^c is explicitly given as $\Phi_2^c = -I$, where I is the unit matrix of order 2. The state $|0\rangle_2|1\rangle$ and its companion state $|1\rangle_2|2\rangle$ correspond to the $n = 2$ subspace in Eq. (13.65), where Φ_2^c is again given by $-I$. Therefore the states $\{|0\rangle_2|0\rangle, |0\rangle_2|1\rangle, |1\rangle_2|0\rangle, |1\rangle_2|1\rangle\}$ are transformed into the states $\{-|0\rangle_2|0\rangle, -|0\rangle_2|1\rangle, |1\rangle_2|0\rangle, -|1\rangle_2|1\rangle\}$ under the action of Φ_2^c. As a result, the output states in Eq. (13.68) are further trasformed as

$$
\begin{array}{cccccc}
\Phi_2^c & & R_2(\pi/2, \pi) & & R_1^+(\pi, \pi) & \\
\to |1\rangle_1 \tfrac{1}{\sqrt{2}}(-|0\rangle_2 + i|1\rangle_2)|1\rangle & \to & -|1\rangle_1|0\rangle_2|1\rangle & \to & |0\rangle_1|0\rangle_2|0\rangle & \\
\to |1\rangle_1 \tfrac{1}{\sqrt{2}}(-|1\rangle_2 + i|0\rangle_2)|1\rangle & \to & -|1\rangle_1|1\rangle_2|1\rangle & \to & |0\rangle_1|1\rangle_2|0\rangle & \\
\to |1\rangle_1 \tfrac{1}{\sqrt{2}}(-|0\rangle_2 - i|1\rangle_2)|0\rangle & \to & -i|1\rangle_1|1\rangle_2|0\rangle & \to & -i|1\rangle_1|1\rangle_2|0\rangle & \\
\to |1\rangle_1 \tfrac{1}{\sqrt{2}}(|1\rangle_2 + i|0\rangle_2)|0\rangle & \to & i|1\rangle_1|0\rangle_2|0\rangle & \to & i|1\rangle_1|0\rangle_2|0\rangle. &
\end{array}
$$
$$(13.69)$$

In summary, the two-qubit gate obtained by the operation of the pulse sequences (13.68) and (13.69) is

$$
U = \begin{pmatrix}
1 & 0 & 0 & 0 \\
0 & 1 & 0 & 0 \\
0 & 0 & 0 & i \\
0 & 0 & -i & 0
\end{pmatrix},
$$
$$(13.70)$$

which differs from the CNOT gate by a phase $\pm i$ in two matrix elements.

The CNOT gate is obtained if extra single qubit rotation gates are applied on qubit 2 as

$$R_2(\pi/2, -\pi/2)R_2(\pi/2, 0)U R_2(\pi/2, \pi)R_2(\pi/2, \pi/2). \qquad (13.71)$$

EXERCISE 13.2 Show that the pulse sequence (13.71) in fact implements the CNOT gate.

13.6 Readout

Readout of an ion qubit state is conducted by the **electron shelving method** [12, 13, 14], depicted in Fig. 13.12. The ion lattice is irradiated by a laser beam with the frequency in resonance with the transition $S_{1/2} \to P_{1/2}$. Suppose an

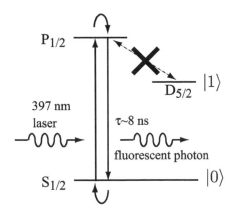

FIGURE 13.12
Electron shelving method for ionic state readout. An ion in the state $|0\rangle$ absorbs the resonant laser photon to be excited to $P_{1/2}$, which decays fast to $|0\rangle$ by emitting a photon. This process is repeated many times, and fluorescent light is observed from those ions in $|0\rangle$. Ions in the $|1\rangle$ state remain dark during laser irradiation.

ion is in the state $S_{1/2}$ ($|0\rangle$). The ion absorbs a photon and makes transition to $P_{1/2}$. Due to a short lifetime ~ 7 ns, this state decays to the initial state $|0\rangle$ by emitting a photon. The ion absorbs a photon from a laser beam again and makes transition to $P_{1/2}$ again, followed by a decay to $|0\rangle$ with an emitted photon. This cycle is repeated while the laser beam is turned on. If, in contrast, the ion is in the state $|1\rangle$, transition to $P_{1/2}$ is forbidden and it does not change the qubit state. Therefore, it does not emit any photons while the laser is turned on. In this way, the photon count from each ion tells whether the ion was initially in the state $|0\rangle$ or $|1\rangle$. This readout method is called the electron shelving method.

13.7 DiVincenzo Criteria

DiVincenzo criteria for an ion trap quantum computer are evaluated as follows.

1. A scalable physical system with well-characterized qubits:

   Two internal energy levels of an ion serve as a qubit. Selective addressing is possible by using a focused laser beam directed toward each ion. A linear trap is expected to accommodate up to ~ 100 ions. Putting more ions leads to instability of the system. Eight ions in a trap is the maximum number of qubits ever realized among all physical realizations

[6]. Distant ions may interact with each other through the vibrational modes of the ionic lattice. This saves lots of gates when we need to operate a two-qubit gate on two distant ionic qubits. The number of qubits may be further increased by segmenting the trap into several subtraps [15]. Ions may be shuttled from one subtrap to the other for interaction if necessary.

2. The ability to initialize the state of the qubits to a simple fiducial state, such as $|00\ldots0\rangle$:

   Intialization of the ionic internal state is done by optical pumping to an excited state, from which ions decay fast into the ground state (Fig. 13.3). Lattice vibration is initialized by the optical sideband cooling method (§13.4.2).

3. Long decoherence times, much longer than the gate operation time:

   Decoherence in ionic qubits and lattice vibrations must be discussed separately. For ionic qubits, coherence times of 10^{0-1} s (Ca) and 10^{2-3} s (Be) have been reported. Coherence time is limited by unwanted transitions between qubit states or transition to states orthogonal to the qubit states. Phonon qubits (lattice vibrations) decohere due to several reasons. The most important one is noise in the trapping potential caused by voltage fluctuation, for example. Collision of the residual gas may lead to decoherence. Micromotion and lattice anharmonicity also cause deviation from an ideal harmonic oscillation. Mode-mode coupling between differenct normal modes is introduced through the lattice nonlinearity.

4. A "univeral" set of quantum gates:

   Single-qubit gates are realized by focused laser beams acting on individual qubits. Lattice vibrational modes and ionic modes are entangled by making use of sideband transitions. The same technique makes it possible to entangle distant qubits through the vibrational mode. The CNOT gate acting on two qubits and controlled-NOT gate with many control qubits are already demonstrated [11].

5. A qubit-specific measurement capability:

   The states of ions after the execution of a quantum algorithm are read out by the electron shelving method (§13.6).

References

[1] J. I. Cirac and P. Zoller, Phys. Rev. Lett., **74**, 4091 (1995).

[2] M. Šašura and V. Bužek, J. Mod. Opt., **49**, 1593 (2002).

[3] M. Riebe, University of Innsbruck Thesis (2005). Available at http://heart-c704.uibk.ac.at/publications/dissertation/riebe_diss.pdf

[4] S. Chu, Rev. Mod. Phys. **70**, 685 (1998); C. N. Cohen-Tannoudji, Rev. Mod. Phys. **70**, 707 (1998); W. D. Phillips, Rev. Mod. Phys. **70**, 721 (1998).

[5] E. A. Cornell and C. E. Wieman, Rev. Mod. Phys. **74**, 875 (2002); W. Ketterle, Rev. Mod. Phys. **72**, 1131 (2002).

[6] H. Häffner *et al.*, Nature **438**, 643 (2005).

[7] H. G. Dehmelt, Advances in atomic and molecular physics, **3**, 53 (1967).

[8] W. Paul, Rev. Mod. Phys., **62**, 531 (1990).

[9] D. F. V. James, Appl. Phys. B **66**, 181 (1998).

[10] D. J. Wineland and W. M. Itano, Phys. Rev. A **20**, 1521 (1979).

[11] F. Schmidt-Kaler *et al.*, Nature **422**, 408 (2003).

[12] J. C. Bergquist *et. al.*, Phys. Rev. Lett. **57**, 1699 (1986).

[13] TH. Sauter *et al.*, Phys. Rev. Lett. **57**, 1696 (1986).

[14] W. Nagourney, J. Sandberg and H. Dehmelt, Phys. Rev. Lett. **56**, 2797 (1986).

[15] D. Kielpinski, C. Monroe and D. J. Wineland, Nature **417**, 709 (2002).

14

Quantum Computing with Neutral Atoms

14.1   Introduction

It is possible to trap neutral atoms in a periodic potential of various dimensions created by laser beams so that each potential minimum traps exactly a single atom. Then each atom works as a qubit, which has the following advantages over other realizations.

1. Neutral atoms trapped in an optical potential produced by laser beams interact with the environment very weakly. Therefore it is expected that the decoherence time of this system is considerably long.

2. Single-qubit gates acting on an isolated atom as well as two-qubit operations are possible with current technology.

3. A large qubit system can be realized by trapping atoms in an optical lattice. It is possible, even in a many-qubit system, to make the interatomic interaction very small when all the atoms are in their ground state.

4. Interatomic interaction can be introduced on demand.

Alkali (metal) atoms with simple atomic structure are used as qubits in many experiments.

14.2   Trapping Neutral Atoms

14.2.1   Alkali Metal Atoms

The ground state of an alkali metal atom has closed electron shells and an additional electron in $S_{1/2}$-orbit. Figure 14.1 shows the energy levels of an alkali metal atom. Besides the ground state $n^2S_{1/2}$, excited states $n^2P_{1/2}$ and $n^2P_{3/2}$ also play important roles in the following.

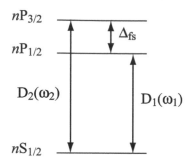

FIGURE 14.1
Energy level scheme of an alkali metal atom. The parameter Δ_{fs} denotes the fine structure splitting energy, while D_k denotes the D_k line and ω_k, its resonance frequency.

Hyperfine structure becomes relevant in the decription of an atomic state at ultra-low temperature. Let S be the electron spin operator and I be the nuclear spin operator. The **hyperfine interaction** Hamiltonian is

$$H_{\mathrm{hf}} = \lambda \sum_k S_k \otimes I_k, \qquad (14.1)$$

which is often abbriviated as $\lambda S \cdot I$. The coupling constant λ is a positive number for alkali metal atoms. Let us define the **hyperfine spin** operator

$$F = S \otimes I + I \otimes I. \qquad (14.2)$$

The unit matrix I and the nuclear spin operator I, especailly its quantum number I, should not be confused. Which is meant by I should be clear from the context. We again introduce a less strict form $F = S + I$. By taking the quantum number $S = 1/2$, we find that F spin has a quantum number $F = I \pm 1/2$. Conisder the identity

$$S \cdot I = \frac{1}{2}\left[F(F+1) - I(I+1) - 3/4\right] \qquad (14.3)$$

derived from $F^2 = F(F+1) = I(I+1) + 3/4 + 2S \cdot I$. We find from Eq. (14.3) and $\lambda > 0$ that the state with $F = I - 1/2$ has a lower energy than the other state with $F = I + 1/2$. Atoms ^{23}Na, ^{87}Rb and ^{133}Cs are often employed in quantum computing experiments. Na and Rb have nuclear spin $I = 3/2$ and hence there are a three-fold degenerate $F = 1$ state and a five-fold degenerate $F = 2$ state. Cs has $I = 7/2$ and has $F = 3$ and $F = 4$.

14.2.2 Magneto-Optical Trap (MOT)

MOT (magneto-optical trap) is one of the methods to trap neutral atoms spatially at ultra-low temperatures. We outline the principle of trapping below

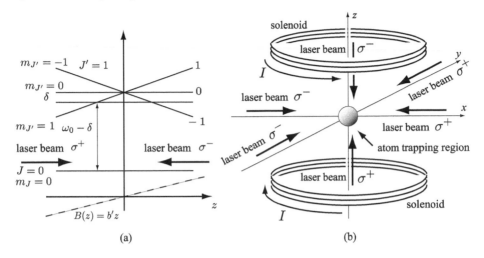

FIGURE 14.2

Magneto-optical trap (MOT). (a) Excited states with $J' = 1$ Zeeman-split under magnetic field gradient linearly varying in z. If an atom moves toward the $-z$ direction from the center of the trap, it absorbs σ^+ polarized photons with a positive momentum along the z-axis, leading to the transition $m_J = 0 \rightarrow m_{J'} = 1$. As a result, the atom is pushed back to the center. (b) Six polarized laser beams and anti-Helmholtz coils to generate quadrupole magnetic field for 3-d MOT. Note that the polarizations of the beams are different in the x- and y- directions from the z-direction. Adopted from [1].

by taking a one-dimensional model as an example to simplify our discussion.

Let us consider a fictitious atom with energy levels shown in Fig. 14.2 (a). The ground state is assumed to be a singlet ($J = 0$), while the excited state is a triplet ($J = 1$). We will denote the quantum numbers of the triplet state with J' to avoid confusion. The quantum number m_J of a triplet state is denoted $m_{J'}$, for example. Suppose a space-dependent magnetic field with $B_z(z) = (dB_z/dz)z$ is applied to the atom. The energy of the excited state $|J'm_{J'}\rangle$ splits under this field as in Fig. 14.2 (a) with

$$\Delta E = m_{J'} g_{J'} \mu_B \left(\frac{dB_z}{dz}\right) z,$$

where $g_{J'}$ is the Lanade g-factor and $m_{J'}$ is the eigenvalue of the z-component of $\boldsymbol{J}$. It is assumed in the figure that $dB_z/dz > 0$ and $g_J > 0$. Suppose a circularly polarized (σ^-) laser light, red-detuned from the excitation energy, is applied to the atom from the postive z-direction and another circularly polarized (σ^+) laser light, also red-detuned, is applied from the negative z-direction. It was shown in §13.1, where Doppler cooling was discussed, that an atom moving toward the positive z-direction in the area $z > 0$ absorbs mainly photons with negative momentum and hence is under force with the negative z-component. Similarly an atom moving toward the negative z-direction in the area $z < 0$ absorbs mainly photons with positive momentum and hence is under force with the positive z-component. Atoms at around $z = 0$ are confined in this way. Note that the trapping region shrinks in size when the field gradient dB_z/dz is increased while the laser frequency is kept fixed.

A quadrupole magnetic field is used to confine atoms in a three-dimensional space. The components of the field are given by

$$\boldsymbol{B}(\boldsymbol{x}) = \frac{dB_z}{dz}\left(-\frac{x}{2}, -\frac{y}{2}, z\right),\tag{14.4}$$

for example and laser beams along the $\pm x$- and $\pm y$-axes are also applied to confine atoms along the x- and y-directions. The minus sign in the x- and y-components of $\boldsymbol{B}$ accounts for the opposite polarizations of laser beams along these axes in Fig. 14.2 (b).

The fine structure states $nS_{1/2}$ and $nP_{3/2}$ of alkali metal atoms are often employed as the ground state and the excited state, respectively, in an experiment. Although the actual scenario is slightly more complicated than the above simplified model, the underlying mechanism is the same, and trapping potential is formed by MOT [2].

Atoms trapped in MOT absorb laser beams and always radiate fluoresence light, which may be used for qubit readout.

14.2.3   Optical Dipole Trap

Widely detuned laser beams are used to confine neutral atoms in an optical trap with small atomic loss. Let

$$\boldsymbol{E}(\boldsymbol{x}, t) = \mathrm{Re}\left(\boldsymbol{E}_0(\boldsymbol{x})e^{-i\omega_L t}\right)\tag{14.5}$$

be the electric field of a laser beam, where ω_L is the laser frequency. We assume here that the beam is linearly polarized, and hence $\boldsymbol{E}_0$ is a real function. The laser beam produces a potential acting on atoms in the ground state through the **ac Stark shift**, also known as the **light shift**. Let $\boldsymbol{d}$ be the dipole operator of the atom. The atom-radiation interaction Hamiltonian is

$$H_i = -\frac{1}{2}\left(\boldsymbol{E}_0(\boldsymbol{x})\cdot\boldsymbol{d}\right)\left(e^{-i\omega_L t} + e^{i\omega_L t}\right).\tag{14.6}$$

Let us calculate the light shift assuming that an atom is sitting at a point $\boldsymbol{x}$. Let $|g\rangle$ and $|e\rangle$ be the ground state and an excited state, respectively, and let $|\psi\rangle = c_g|g\rangle + c_e|e\rangle$ be the atomic state in the presence of a laser beam. The coefficients c_g and c_e satisfy the Schrödinger equations

$$i\frac{\partial c_g}{\partial t} = -\frac{\Omega_{eg}}{2}e^{i\omega_L t}c_e\tag{14.7}$$

$$i\frac{\partial c_e}{\partial t} = \left(\omega_{eg} - i\frac{\Gamma_e}{2}\right)c_e - \frac{\Omega_{eg}}{2}e^{-i\omega_L t}c_g.\tag{14.8}$$

Here $\omega_{eg} = (E_e - E_g)/\hbar$ is the transition frequency of the laser in resonance with the energy difference between the ground state $|g\rangle$ with the energy E_g and the excited state $|e\rangle$ with E_e, while Γ_e is the natural line width of the excited state. The Rabi frequency is denoted as $\Omega_{eg} = \langle e|\boldsymbol{d}|g\rangle \cdot \boldsymbol{E}_0/\hbar$.

Now we redefine c_e by $c_e \to c_e e^{-i\omega_L t}$ to obtain

$$i\frac{\partial c_g}{\partial t} = -\frac{\Omega_{eg}}{2} c_e \tag{14.9}$$

$$i\frac{\partial c_e}{\partial t} = \left(-\Delta_{eg} - i\frac{\Gamma_e}{2}\right) c_e - \frac{\Omega_{eg}}{2} c_g, \tag{14.10}$$

where $\Delta_{eg} = \omega_L - \omega_{eg}$ is the detuning of the laser frequency from the resonance frequency ω_{eg}. Suppose the detuning is sufficiently large so that $|\Delta_{eg}| \gg \Omega_{eg}$ is satisfied and hence the time derivative in Eq. (14.10) is negligible. Then we obtain

$$c_e = -\frac{\Omega_{eg}}{2\left(\Delta_{eg} + i\Gamma_e/2\right)} c_g$$

and, substituting this result into Eq. (14.9), we find

$$i\frac{\partial c_g}{\partial t} = \frac{\Omega_{eg}^2}{4\left(\Delta_{eg} + i\Gamma_e/2\right)} c_g. \tag{14.11}$$

This result shows that the ground state energy has shifted in the presence of a laser beam. This light shift

$$V = \frac{\hbar\Omega_{eg}^2}{4(\Delta_{eg} + i\Gamma_e/2)} = \frac{\hbar\Omega_{eg}^2}{4(\omega_L - \omega_{eg} + i\Gamma_e/2)} \tag{14.12}$$

may be looked upon as a potential acting on an atom. Considering the space dependence of the electric field, we obtain

$$V(\boldsymbol{x}) = \frac{\hbar\Omega_{eg}^2(\boldsymbol{x})}{4\left(\Delta_{eg} + i\Gamma_e/2\right)}. \tag{14.13}$$

A large electric field region with large Ω_{eg} acts as a repulsive potential with a blue-detuned laser ($\Delta_{eg} > 0$), while it works as an attractive potential with a red-detuned laser ($\Delta_{eg} < 0$).

Atom loss associated with the potential (14.13) is due to spontaneous photon scattering. It comes from the imaginary part of the potential and is given by

$$\frac{\Gamma_e}{2} \frac{\Omega_{eg}^2}{(2\Delta_{eg})^2} \tag{14.14}$$

when $|\Delta_{eg}| \gg \Gamma_e$. It follows from Eqs. (14.13) and (14.14) that the loss scales as $|E_0|^2\Gamma_e/\Delta_{eg}^2$, whereas the potential depth scales as $|E_0|^2/\Delta_{eg}$. Therefore a far off-resonant intense laser beam is able to produce a deep enough potential with sufficiently small atomic loss.

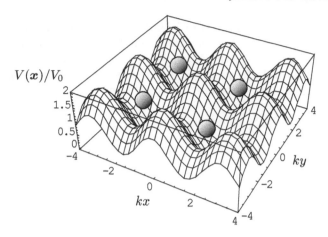

FIGURE 14.3
Profile of the confining potential (14.15) with $V_{0x} = V_{0y} \equiv V_0$ in the plane $z = 0$. It shows the scaled potential $V(\boldsymbol{x})/V_0$ as a function of kx and ky.

14.2.4   Optical Lattice

Suppose two linearly polarized electric fields $\boldsymbol{E}_{\pm} = (\boldsymbol{E}_0/2)\exp(\pm ikx)$, counterpropagating along the x-axis, are superposed. The resulting electric field is a sinusoidal standing wave $E_0 \cos(kx)$, and it follows from Eq. (14.13) that it produces a one-dimensional optical lattice potential $V(x) \propto \cos^2(kx)$ with the lattice constant $a = \pi/k = \lambda/2$, where $\lambda = 2\pi/k$ is the wave length of the laser beams. If, moreover, two sets of counterpropagating beams along the y- and z-axes are introduced, resulting in six beams in total, a three-dimensioanl optical lattice potential

$$V(\boldsymbol{x}) = V_{0x}\cos^2(kx) + V_{0y}\cos^2(ky) + V_{0z}\cos^2(kz) \tag{14.15}$$

is produced. Figure 14.3 shows the potential profile in the plane $z = 0$.

The ground state of alkali metal atoms is $nS_{1/2}$ ($F = I \pm 1/2$), and we consider two excited states $nP_{1/2}$ ($F = I \pm 1/2$) and $nP_{3/2}$ ($F = I \pm 1/2, 3/2$) corresponding to two resonance lines D_1 and D_2; see Fig. 14.1.

Although the ground state and excited state have such hyperfine states specified by F-spin quantum number, the contribution of the hyperfine structure energy to the light shift (14.13) is negligible. Furthermore the Rabi frequency of a transition between two ground states of different hyperfine F-spins is independent of F, after summing over the F-spin states of $nP_{1/2}$ and $nP_{3/2}$. Therefore the optical trapping potential is independent of the F-spin state. Note here that this applies only when all the laser beams are linearly polarized in the same direction.

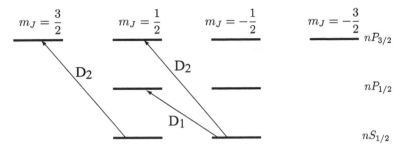

FIGURE 14.4
Fine structure of an alkali metal atom and excitation of the atom with the σ^+-polarized laser beam. There are two degenerate $m_J = \pm 1/2$ excitations over each of the ground states $nS_{1/2}$ and the excited states $nP_{1/2}$. There are $m_J = \pm 1/2$ and $m_J = \pm 3/2$ states in the excited states $nP_{3/2}$. It is possible to excite the ground state with $m_J = -1/2$ to states $nP_{1/2}$ and $nP_{3/2}$ both with $m_J = 1/2$, which corresponds to D_1 and D_2, respectively, while the state $m_J = 1/2$ may be excited to $nP_{3/2}$ only.

14.2.5 Spin-Dependent Optical Potential

It has been already mentioned above that two counterpropagating laser beams with the same linear polarization produce an optical potential independent of the hyperfine states. If, in contrast, the counterpropagating beams have different polarizations, the resulting potential is dependent on the hyperfine state of the atom. There have been proposed several two-qubit gates, in which state dependent lattice potential is employed.

Let us analyze the state dependent lattice, to begin with. Let $\hat{y}$ and $\hat{z}$ be unit vectors along the y- and z-axes, respectively. Suppose the polarizations of laser beams propagating along the x-axis and $-x$-axis are tilted by angles θ and $-\theta$ from the z-axis, respectively, which results in $\boldsymbol{E}_+ \propto e^{ikx}(\hat{z}\cos\theta + \hat{y}\sin\theta)$ and $\boldsymbol{E}_- \propto e^{-ikx}(\hat{z}\cos\theta - \hat{y}\sin\theta)$. Superposition of the two beams produces a standing wave potential

$$\boldsymbol{E}_+ + \boldsymbol{E}_- \propto \sigma^+ \cos(kx - \theta) - \sigma^- \cos(kx + \theta), \tag{14.16}$$

where $\sigma_\pm = \hat{y} \pm i\hat{z}$ denotes two circular polarizations. We consider a potential due to a σ^+-polarized laser beam first. The fine structure of an atom, which is related to D_1 and D_2 lines, has been shown in Fig. 14.4. The optical dipole potential is give by Eq. (14.12) when there is only one excited state involved. The trap potential is a sum of potentials due to excitations to $P_{1/2}$ and $P_{3/2}$ states when the ground state has $m_J = -1/2$, while it is solely made of the

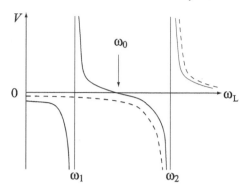

FIGURE 14.5
Light shift of the ground state $nS_{1/2}$. ω_1 and ω_2 are the resonance frequencies of the transitions to $nP_{1/2}$ and $nP_{3/2}$. The light shift with $m_J = -1/2$ is shown in a solid curve, and that with $m_J = 1/2$ in a broken curve. The light shift for the $m_J = -1/2$ vanishes at $\omega_L = \omega_0$.

potential corresponding to excitation to $P_{3/2}$ state when the ground state has $m_J = 1/2$ as shown in Fig. 14.4.

These two potentials are depicted in Fig. 14.5, where the natural width Γ_e of the excited states is put to zero. It is found from Fig. 14.5 that the potential for the state $S_{1/2}$ ($m_J = -1/2$) vanishes when $\omega_L = \omega_0$. A similar argument involving σ^- shows the potential for the state $S_{1/2}$ ($m_J = +1/2$) vanishes for $\omega_L = \omega_0$. In summary, it is possible to introduce optical potentials

$$V_\pm(x) \propto \cos^2(kx \pm \theta) \tag{14.17}$$

for the states $S_{1/2}$ with $m_J = \pm 1/2$. Therefore the optical potential for each hyperfine state is obtained by making use of the Clebsch-Gordan coefficients

$$|F = 2, m_F = 1\rangle = \frac{\sqrt{3}}{2}\left|\frac{3}{2}, \frac{1}{2}\right\rangle\left|\frac{1}{2}, \frac{1}{2}\right\rangle + \frac{1}{2}\left|\frac{3}{2}, \frac{3}{2}\right\rangle\left|\frac{1}{2}, -\frac{1}{2}\right\rangle$$

$$|F = 2, m_F = -1\rangle = \frac{\sqrt{3}}{2}\left|\frac{3}{2}, -\frac{1}{2}\right\rangle\left|\frac{1}{2}, -\frac{1}{2}\right\rangle + \frac{1}{2}\left|\frac{3}{2}, -\frac{3}{2}\right\rangle\left|\frac{1}{2}, \frac{1}{2}\right\rangle$$

$$|F = 1, m_F = 1\rangle = \frac{\sqrt{3}}{2}\left|\frac{3}{2}, \frac{3}{2}\right\rangle\left|\frac{1}{2}, -\frac{1}{2}\right\rangle - \frac{1}{2}\left|\frac{3}{2}, \frac{1}{2}\right\rangle\left|\frac{1}{2}, \frac{1}{2}\right\rangle$$

$$|F = 1, m_F = -1\rangle = -\frac{\sqrt{3}}{2}\left|\frac{3}{2}, -\frac{3}{2}\right\rangle\left|\frac{1}{2}, \frac{1}{2}\right\rangle + \frac{1}{2}\left|\frac{3}{2}, -\frac{1}{2}\right\rangle\left|\frac{1}{2}, -\frac{1}{2}\right\rangle$$

as

$$V_{(F=1, m_F=1)}(x) = V_{(F=2, m_F=-1)}(x) = \frac{3}{4}V_+(x) + \frac{1}{4}V_-(x), \tag{14.18}$$

$$V_{(F=1, m_F=-1)}(x) = V_{(F=2, m_F=1)}(x) = \frac{1}{4}V_+(x) + \frac{3}{4}V_-(x), \tag{14.19}$$

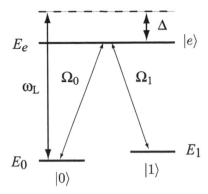

FIGURE 14.6
Hyperfine energy levels participating in the Raman transition. Ω_i is the Rabi frequency between the states $|i\rangle$ and $|e\rangle$, and Δ is the detuning frequency.

where m_F is the x-component of the F-spin.

14.3 One-Qubit Gates

Hyperfine states $|F = I + 1/2\rangle$ and $|F = I - 1/2\rangle$ of the ground state $nS_{1/2}$ of an alkali metal atom have degeneracies $2I + 2$ and $2I$, respectively. Qubit basis states $|0\rangle$ and $|1\rangle$ are chosen from these degenerate states. There are two strategies to introduce Rabi oscillation between the qubit states $|0\rangle$ and $|1\rangle$.

1. Transition between these states is introduced through the Rabi oscillation by applying a microwave pulse in resonance with the transition frequency. Selective addressing to each atom is possible by introducing inhomogeneous magnetic field. The Rabi oscillation frequency is determined by the coupling strength between the microwave and the atom [3].

2. Single qubit operations in a system of neutral atoms are also implemented by irradiating a laser beam to a particular atom so that the transition between two different hyperfine states in the ground state takes place with an excited state as an intermediate state (**Raman transition**); see Fig. 14.6.

We are concerned below with the derivation of the effective coupling between two energy levels following the second strategy listed above.

Levels with energies E_0, E_1 and E_e serve as the states $|0\rangle, |1\rangle$ and an auxiliary excited state, respectively. Let ω_L be the frequency of the laser beam and let

$$\hbar\Delta = \hbar\omega_L - (E_e - E_0) \tag{14.20}$$

be the detuning. By applying the laser beam, the state executes the Rabi oscillation between $|i\rangle$ $(i = 0, 1)$ and $|e\rangle$ with the frequency Ω_i. The most general quantum state described in Fig. 14.6 is of the form

$$|\psi\rangle = c_0|01\rangle + c_1|11\rangle + c_e|e0\rangle, \tag{14.21}$$

where the first "digit" in the ket decribes the atomic states $0, 1$ and e, while the second digit shows the number of photons. Note that photon number reduces by one when the atom is excited to $|e\rangle$ from $|0\rangle$ or $|1\rangle$ by absorbing a photon.

Equations of motion are

$$i\frac{dc_0}{dt} = -\frac{\Omega_0}{2}c_e$$

$$i\frac{dc_1}{dt} = \frac{E_1 - E_0}{\hbar}c_1 - \frac{\Omega_1}{2}c_e \tag{14.22}$$

$$i\frac{dc_e}{dt} = -\Delta c_e - \frac{\Omega_0}{2}c_0 - \frac{\Omega_1}{2}c_1.$$

We obtain, under the assumptions $|\Delta| \gg (E_1 - E_0)/\hbar, \Omega_i^2/|\Delta|$, the solution of the third equation

$$c_e = -\frac{\Omega_0}{2\Delta}c_0 - \frac{\Omega_1}{2\Delta}c_1. \tag{14.23}$$

By substituting this into the rest of Eq. (14.22), we find

$$i\frac{dc_0}{dt} = -\frac{\epsilon}{2}c_0 + \frac{\Omega_0\Omega_1}{4\Delta}c_1$$

$$i\frac{dc_1}{dt} = \frac{\epsilon}{2}c_1 + \frac{\Omega_0\Omega_1}{4\Delta}c_0, \tag{14.24}$$

where

$$\epsilon = E_1 - E_0 + \frac{\Omega_1^2}{4\Delta} - \frac{\Omega_0^2}{4\Delta} \tag{14.25}$$

and the origin of the energy has been redefined. The Hamiltonian leading to Eq. (14.24) is easily found to be

$$H = \frac{1}{2}\epsilon\sigma_z - \frac{\Omega_0\Omega_1}{4\Delta}\sigma_x, \tag{14.26}$$

where $\epsilon, \Omega_0, \Omega_1$ and Δ are all controllable parameters.

There are several possibilities for qubit basis states, and special attention has to be paid to choose appropriate states for qubits. Among the hyperfine states of the ground state, the states $|F = I-1/2, m_F\rangle$ and $|F = I+1/2, -m_F\rangle$ share the same magnetic moment if a small contribution ($\sim 10^{-3}$ relative to the electron contribution) from the nuclear spin is ignored. Let us consider the $I = 3/2$ state of ^{23}Na and ^{87}Rb, for example. Then possible pairs of states to form a qubit are, for example,

$$\begin{aligned} &\text{a)} \quad |F = 1, m_F = 1\rangle \text{ and } |F = 2, m_F = -1\rangle, \\ &\text{b)} \quad |F = 1, m_F = -1\rangle \text{ and } |F = 2, m_F = 1\rangle. \end{aligned} \qquad (14.27)$$

Dephasing between two levels due to fluctuations in the laser beams and environment magnetic field are made minimal, if a qubit is constructed out of these sets of two levels.

14.4 Quantum State Engineering of Neutral Atoms

14.4.1 Trapping of a Single Atom

An atom in MOT cannot be used as a qubit since it always radiates fluorescent light in MOT. Therefore we need to send the atom to an optical dipole trap and store it there to control the qubit state at our disposal.

The experiment starts with confining a small number of atoms in MOT. Atoms on the order of $10^3 \sim 10^{12}$ are trapped in an ordinary MOT. Subsequently, the trapping region is squeezed by increasing the field gradient dB_z/dz. The presessure of background vapor is also reduced so that only several atoms are left after this operation. Next, we apply a focused laser beam, whose width is on the order of the MOT size, to the center of the MOT by which transfer of atoms to an optical dipole trap is conducted with $\sim 100\%$ probability. Figure 14.7 shows the result of the confinement experiment reported in [4], in which a single atom has been successfully trapped for several seconds. The ordinate of the graph shows the photon counts/s, and the trap is switched between the MOT and the optical dipole trap every other second. The flourescent light photon count depends on the number of atoms N in the trap when the trap works as an MOT. Correspondingly the number of trapped atoms is read out from the reading of the photon counts. It is found from Fig. 14.7 that the numbers $N = 0, 1$ and 2 are observed in this experiment. The change in N is due to background vapor and cold collision [4].

14.4.2 Rabi Oscillation

There are several experiments in which a single Cs atom confined in an optical dipole trap is employed as a qubit. The quantum state of a single atom is controlled by making use of the Rabi oscillation. Two hyperfine states

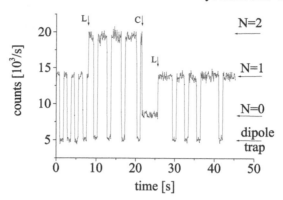

FIGURE 14.7
Loading atoms from MOT to the optical dipole trap. The ordinate shows
the photon counts obtained by switching between the optical dipole trap and
MOT every other second. The number of atoms in the trap is found from
the fluorescence photon counts when the trap is in the MOT mode. It is
observed from the graph that the atom number fluctuates due to loading of
a background vapor atom or a collision between a background atom and a
trapped atom. Reprinted figure with permission from D. Frese *etal.*, Physical
Review Letters, **85**, 3777 (2000). Copyright (2000) by the American Physical
Society.

$|F = 4, m_F = 4\rangle$ and $|F = 3, m_F = 3\rangle$ of the ground state $6S_{1/2}$ are chosen as
two levels of a qubit, where the propagation direction of the laser beam has
been taken as the quantization axis [5]. The experiment starts with an initial
state $|F = 4, m_F = 4\rangle$, which is obtained by manipulating laser beams. Then
a microwave pulse to drive the initial state to $|F = 3, m_F = 3\rangle$ state is applied
for a duration t_{pulse}. The probability of the state being in $|F = 3, m_F = 3\rangle$ is
subsequently measured. Figure 14.8 shows the result of the measurement as
a function of t_{pulse} [5], which indicates the clear Rabi oscillation observed in
this experiment.

The probability of the atom being in $|F = 3, m_F = 3\rangle$ is measured with
the following method. We first note that the dipole trap potential does not
trap atoms in $6P_{3/2}$. Therefore if a laser beam, the so-called push-out laser,
is applied to the atom to excite it from $|F = 4, m_F = 4\rangle$ state to $6P_{3/2}$, it is
expelled from the trap. Subsequently the trap is switched to MOT, and it is
oberved whether there are any atoms left in the trap.

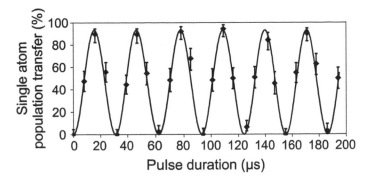

FIGURE 14.8

Rabi oscillation between the hyperfine states $|F = 4, m_F = 4\rangle$ and $|F = 3, m_F = 3\rangle$. The ordinate shows the single atom population of the state $|F = 3, m_F = 3\rangle$ as a function of pulse duration t_{pulse}. Atoms are initially in the state $|F = 4, m_F = 4\rangle$ and subsequently oscillate between $|F = 4, m_F = 4\rangle$ and $|F = 3, m_F = 3\rangle$ under irradiation of microwave pulse. Reprinted figure with permission from D. Schrader *et al.*, Physical Review Letters, **93**, 150501 (2004). Copyright (2004) by the American Physical Society.

14.4.3   Neutral Atom Quantum Regisiter

Atoms trapped in a one-dimensional optical lattice have been already selectively controlled [5]. Qubit states are encode as

$$|0\rangle = |F = 4, m_F = 4\rangle, \ |1\rangle = |F = 3, m_F = 3\rangle, \qquad (14.28)$$

for example. It has been demonstrated that a register, made of five atoms in the initial state $|00000\rangle$ in a one-dimensional trap along the z-axis of a quadrupole field, is successfully transformed to the final state $|01010\rangle$. There is a region in the vicinity of the center of the quadrupole field, where the field gradient is uniform along the trap axis, and therefore the microwave resonance frequency changes from atom to atom. Selective addressing to each atom is possible by making use of this field gradient. In the experiment mentioned above, only two qubits are driven from $|0\rangle$ to $|1\rangle$ by applying π-pulses in resonance with these two atoms. Readout was done by applying push-out laser and subsequently putting the atoms back to MOT.

14.5   Preparation of Entangled Neutral Atoms

Jaksch *et al.* proposed a method to prepare neutral-atom entangled states, Bell-like states to be more specific, by controlling atom-atom collision time [6]. Mandel *et al.* successfully created Bell-like states following this proposal [7]. Experiments employed ^{87}Rb atoms trapped in a three-dimensional optical lattice. The barrier of the lattice is high enough so that the atoms have negligibly small overlap. This state is called a **Mott insulator** state. The filling factor at each lattice site is close to 1, although there exist a small number of defects, i.e., empty sites and multiply occupied sites. Two atomic states forming a qubit are the hyperfine states $|F = 1, m_F = -1\rangle \equiv |0\rangle$ and $|F = 2, m_F = -2\rangle \equiv |1\rangle$.

Now let us follow Mandel *et al.* [7]. Single atoms at lattice points are initialized to $|0\rangle$, after which a microwave $\pi/2$-pulse around the y-axis ($Y = e^{-i(\pi/4)\sigma_y}$; see Eq. (12.29)) is applied to transform $|0\rangle$ to $(|0\rangle + |1\rangle)/\sqrt{2}$ (Step I). Note that the trapping potential is spin-dependent. Therefore it is possible to separate $|0\rangle$ atoms from $|1\rangle$ atoms by moving a potential for one of the spin states by a half of the original lattice constant (Step II).

Let us concentrate on the ith and $i + 1$st sites of the lattice and consider the above process. The initial state is $|0\rangle_i|0\rangle_{i+1}$ and the $\pi/2$-pulse transforms it to (Step I)

$$\frac{1}{2} \left(|0\rangle_i + |1\rangle_i\right)\left(|0\rangle_{i+1} + |1\rangle_{i+1}\right).$$

The shift of the potential (Step II) further transforms the state to

$$\frac{1}{2} \left(|0\rangle_i|0\rangle_{i+1} + |0\rangle_i|1\rangle_{i+2} + |1\rangle_{i+1}|0\rangle_{i+1} + |1\rangle_{i+1}|1\rangle_{i+2}\right). \tag{14.29}$$

As a result, there is a probability with which two atoms occupy the same lattice site and an on-site repulsion U is in action. Suppose this state is kept for a duration t_{hold}, and then the potential is shifted backward to its initial position. The resulting state has an extra relative phase only in the third term of the state (14.29) due to the U-interaction, and we are left with

$$\frac{1}{2} \left(|0\rangle_i|0\rangle_{i+1} + |0\rangle_i|1\rangle_{i+1} + e^{-i\theta}|1\rangle_i|0\rangle_{i+1} + |1\rangle_i|1\rangle_{i+1}\right), \tag{14.30}$$

where $\theta = U t_{\text{hold}}/\hbar$. If t_{hold} is adjusted so that $\theta = \pi$, we obtain

$$\frac{1}{2} \left(|0\rangle_i|0\rangle_{i+1} + |0\rangle_i|1\rangle_{i+1} - |1\rangle_{i+1}|0\rangle_{i+1} + |1\rangle_{i+1}|1\rangle_{i+1}\right). \tag{14.31}$$

This state is called a Bell-like state.

After these operations, a $\pi/2$-pulse

$$R\left(\frac{\pi}{2}, \frac{\pi}{2} + \alpha\right) = \frac{1}{\sqrt{2}} \begin{pmatrix} 1 & -e^{-i\alpha} \\ e^{i\alpha} & 1 \end{pmatrix}$$

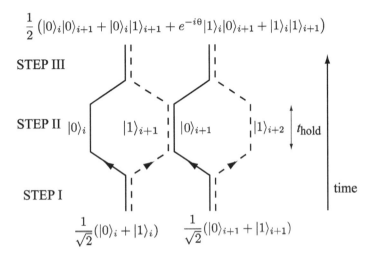

$$\frac{1}{2}\left(|0\rangle_i|0\rangle_{i+1} + |0\rangle_i|1\rangle_{i+1} + e^{-i\theta}|1\rangle_i|0\rangle_{i+1} + |1\rangle_i|1\rangle_{i+1}\right)$$

STEP III

STEP II $|0\rangle_i$ $\qquad$ $|1\rangle_{i+1}$ $\quad$ $|0\rangle_{i+1}$ $\qquad$ $|1\rangle_{i+2}$ $\quad$ t_{hold}

STEP I $\qquad\qquad\qquad\qquad\qquad\qquad\qquad\qquad$ time

$$\frac{1}{\sqrt{2}}(|0\rangle_i + |1\rangle_i) \qquad \frac{1}{\sqrt{2}}(|0\rangle_{i+1} + |1\rangle_{i+1})$$

FIGURE 14.9

Schematic diagram of the two-qubit operations by making use of interatomic interaction. All the atoms are in the superposition state $(|0\rangle + |1\rangle)/\sqrt{2}$ by applying a $\pi/2$-pulse on each qubit whose initial state is $|0\rangle$. Then a spin-dependent potential is manipulated so that the states $|0\rangle$ and $|1\rangle$ are shifted by half a lattice constant in the opposite directions, respectively. The states $|0\rangle_i$ and $|0\rangle_{i+1}$ of the atoms in the ith and the $i + 1$st lattice sites in Step I are transformed into the states $|0\rangle_i$ and $|0\rangle_{i+1}$ in Step II, respectively, while the states $|1\rangle_i$ and $|1\rangle_{i+1}$ in Step I are transformed into the states $|1\rangle_{i+1}$ and $|1\rangle_{i+2}$ in Step II, respectively. The states $|0\rangle_{i+1}$ and $|1\rangle_{i+1}$ in the same lattice sites interact for a duration t_{hold}, after which the potential is put back to that in Step I. Adapted from [7].

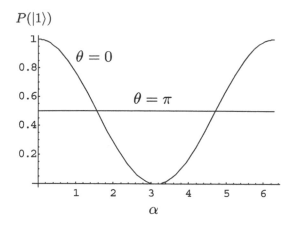

FIGURE 14.10
α-dependence of the probability $P(|1\rangle)$ for $\theta = 0$ and $\theta = \pi$.

is applied to transform

$$|0\rangle \rightarrow \tfrac{1}{\sqrt{2}}\left(|0\rangle + e^{i\alpha}|1\rangle\right)$$

$$|1\rangle \rightarrow \tfrac{1}{\sqrt{2}}\left(-e^{-i\alpha}|0\rangle + |1\rangle\right), \tag{14.32}$$

where R has been defined in Eq. (12.30). The state (14.30) transforms under this change as

$$\frac{1}{4}\Big[\left\{1 - \left(1 + e^{-i\theta}e^{-i\alpha}\right) + e^{-2i\alpha}\right\}|0\rangle_i|0\rangle_{i+1} + (1 - e^{i\phi} + 2i\sin\alpha)|0\rangle_i|1\rangle_{i+1}$$

$$+ \left(-1 + e^{-i\theta} + 2i\sin\alpha\right)|1\rangle_i|0\rangle_{i+1} + \left\{1 + \left(1 + e^{-i\theta}\right)e^{i\alpha} + e^{2i\alpha}\right\}|1\rangle_i|1\rangle_{i+1}\Big].$$

It follows from this wave function that the probability of the atom being in the state $|0\rangle$ or in the state $|1\rangle$ is independent of α for Bell-like states with the choice $\theta = \pi$. The probability of obtaining $|1\rangle$ upon a measurement is

$$P(|1\rangle) = \frac{1}{4}[2 + (1 + \cos\theta)\cos\alpha] \tag{14.33}$$

and it is clearly independent of α for $\theta = \pi$. Figure 14.10 shows $P(|1\rangle)$ as a function of α for $\theta = 0$ and $\theta = \pi$.

The above argument remains true for N atoms aligned in line. Suppose a product of wave functions $\prod_i(|0\rangle_i + |1\rangle_i)/2^{N/2}$. The Bell-like states are obtained by multiplying the phase $(-1)^n$ to each term of the expansion, where n is the number of pairs of the form $|1\rangle_i|0\rangle_{i+1}$ in this term.

Mandel *et al.* experimentally demonstrated the above theory and measured the probability $P(|1\rangle)$ as a function of the variable phase α [7] for $\theta = 0$ and

$\theta = \pi$. Their result reproduced Fig. 14.10 with a good precision. In particular, they have observed that $P(|1\rangle)$ is independent of α for a Bell-like state, for which $\theta = \pi$.

14.6 DiVincenzo Criteria

DiVIncenzo criteria for a neutral atom quantum computer are evaluated as follows.

1. A scalable physical system with well-characterized qubits:

   Atoms are loaded from a laser-cooled sample. Qubit basis vectors $|0\rangle$ and $|1\rangle$ are made of two atomic internal states. They are scalable, in principle, to a large number of qubits, possibly on the order of $\sim 10^6$, by trapping them in an optical lattice [8]. Single occupancy of a lattice point is made possible by making use of the transition to a Mott insulator state [9]. The lattice constant of the optical lattice is less than the wavelength of the laser beam, and special care my be taken for selective addressing to an atom.

2. The ability to initialize the state of the qubits to a simple fiducial state, such as $|00\ldots0\rangle$:

   Qubits may be initialized by optical pumping, which is already introduced in §13.2 for trapped ions. Loading of an optical lattice from a BEC via the superfluid-Mott insulator transition is also employed to obtain initialized qubits. It also serves to realize single occupancy of each lattice site.

3. Long decoherence times, much longer than the gate operation time:

   It is expected that neutral atoms have much longer coherence time compared to trapped ions thanks to their charge neutrality.

4. A "univeral" set of quantum gates:

   Single-qubit operations are implemented either by rf-pulses or by Raman transitions (§14.3). Although CNOT gates are demonstrated, it involves many pairs of qubits (§14.5). Implementation of a CNOT gate operation between a single pair of qubits is still a challenging task.

5. A qubit-specific measurement capability:

   Qubit state is measured by the electron-shelving method (or "quantum-jump" method) [10, 11, 12, 13], which is already introduced in §13.6.

Alternatively, a state-selective "push-out" laser beam may be applied to remove all atoms in state $|0\rangle$, and subsequently the CCD image of the residual atoms in state $|1\rangle$ is taken [5].

References

[1] D. Schrader, University of Bonn Dissertation (2004).

[2] E. L. Raab, et al., Phys. Rev. Lett. **59**, 2631 (1987).

[3] A. Derevianko and C. C. Cannon, Phys. Rev. A **70**, 062319 (2004).

[4] D. Frese, et al., Phys. Rev. Lett. **85**, 3777 (2000).

[5] D. Schrader, et al., Phys. Rev. Lett. **93**, 150501 (2004).

[6] D. Jaksch et al., Phys. Rev. Lett. **82**, 1975 (1999).

[7] O. Mandel et al., Nature **425**, 937 (2003).

[8] S. E. Hamann et al., Phys. Rev. Lett. **80**, 4149 (1998).

[9] M. Greiner, et al., Nature **415**, 39 (2002).

[10] N. Davidson et al., Phys. Rev. Lett. **74**, 1311 (1995).

[11] J. C. Bergquist et. al., Phys. Rev. Lett. **57**, 1699 (1986).

[12] TH. Sauter et al., Phys. Rev. Lett. **57**, 1696 (1986).

[13] W. Nagourney, J. Sandberg and H. Dehmelt, Phys. Rev. Lett. **56**, 2797 (1986).

15

Josephson Junction Qubits

15.1 Introduction

Josephson junction qubits are expected to be scalable within current lithography technology developed by the semiconductor industry. The resulting qubits are different from qubits in other realizations in that they are made up of a large number of particles (electrons in the present case) forming a mesoscopic condensate. Moreover, zero registance of superconducting metallic part implies the absence of dissipation. These facts, enhanced by a very low operating temperature, lead to the expectation that they are potentially robust against external noise and decoherence.

All the proposals of superconducting qubits make use of Josephson junctions, which are made of two superconductors connected through a thin oxide insulator in between. Quasiparticle excitations are largely suppressed, and tunneling between two superconductors is solely due to Cooper pair tunneling at very low temperature T, such that $\Delta \gg k_B T$, Δ being the gap of the superconductors. The qubit operates at such low temperatures.

Josephson junction qubits are roughly divided into two classes. One makes use of the charge degrees of freedom and the other, flux (or phase) degrees of freedom. In the former class, the number of excess Cooper pairs in an "island" between a small capacitance Josephson junction and a capacitor serially connected to the junction is a relevant quantum variable. Energies of states with different numbers of Cooper pairs are controlled by the gate voltage. It is then possible to have two states with almost degenerate energies by applying an appropriate gate voltage. These two states serve as a qubit in this proposal. The latter makes use of the number of magnetic flux quanta threading a loop made of several Josephson junctions as a quantum variable. Two flux states may be arranged to have almost degenerate energies by controlling the external magnetic field, and these two states work as a qubit.

General references for this chapter are [1] and [2].

We start our discussion with a brief introduction to nanoscale Josephson junctions and SQUIDs.

15.2   Nanoscale Josephson Junctions and SQUIDs

15.2.1   Josephson Junctions

When two superconductors are combined together through a thin insulator there flows an electric current whose direction and magnitude depend on the difference between the order parameter phases of the superconductors. Moreover, the current oscillates sinusoidally when a voltage across the junction is applied. These phenomena were first predicted by Josephson and called the **Josephson effect** [3]. We assume the reader is familiar with elements of superconductivity, such as the BCS theory [4].

Figure 15.1 shows (a) the schematic picture of a Josephson junction and (b) the electron microscope image of an actual Josephson junction. The superconductors in the image are made of Al. It is fabricated with the suspended shadow mask technique. Two beams with different deposition angles deposit aluminum on a substrate. Two slits on the mask yield four lines in total, two of which are arranged to have a tiny overlap. The first deposited metallic layer is exposed to oxygen gas to develop a thin layer of aluminum oxide, after which the second layer deposition is conducted. The tunnel junction charactersitics are easily controlled with this method by changing the oxidation time, overlapping area and so on. The superfluous metallic lines in both sides of the junction found in Fig. 15.1 (b) are inevitable with this method.

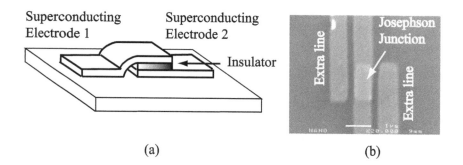

(a)   (b)

FIGURE 15.1

(a) Schematic diagram of a Josephson junction. (b) Actual nanoscale Josephson junction. The superconductors are made of aluminum, while the insulator is made of aluminum oxide. The white bar in the bottom is 1 μm long. Courtesy of Jukka Pekola, Heksinki University of Technology, Finland.

Let Ψ_1 and Ψ_2 be the condensate wave functions of electrodes 1 and 2, respectively. We assume both superconductors are made of the same material

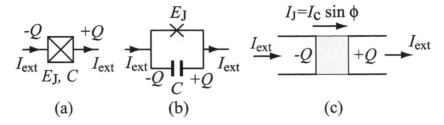

FIGURE 15.2
(a) Electrical symbol of a Josephson junction. (b) The cross ($\times$) shows an ideal junction. The capacitance C naturally arises due to the parallel plate geometry. (c) Details of the currents and charges to derive Eq. (15.1).

and write $\Psi_k = \Psi_0 e^{i\theta_k}$, $(k = 1, 2)$.

Let $\phi = \theta_2 - \theta_1$ be the difference in the phases of the order parameters. Let $\pm Q$ be the charges on the electrodes and introduce the number of Cooper pairs N by $Q = -2eN$, where the charge of an electron is denoted as $-e$. Then the Josephson current is given by the well-known formula $I_J = I_c \sin \phi$, where I_c is called the **critical current**. The junction cannot support a supercurrent greater than I_c. A Josephson junction has a similar geometrical structure as a capacitor and therefore has capacitance C. We often denote a Josephson junction with an equivalent circuit shown in Fig. 15.2.

Suppose an external current source I_{ext} is attached to a Josephson junction. The time derivative of the charge $Q = CV$ is derived from Fig. 15.2 (c) as

$$\frac{dQ}{dt} = -I_{\text{ext}} + I_c \sin \phi, \tag{15.1}$$

while that of the relative phase ϕ is given by

$$\frac{d\phi}{dt} = -\frac{2eV}{\hbar}. \tag{15.2}$$

It then follows from the Josephson equations (15.1) and (15.2) that

$$\frac{d^2\phi}{dt^2} = -\frac{2e}{\hbar}\frac{dV}{dt} = -\frac{2e}{\hbar}\frac{1}{C}\frac{dQ}{dt} = -\frac{2e}{\hbar}\frac{1}{C}(-I_{\text{ext}} + I_c \sin \phi). \tag{15.3}$$

We regard this equation as an equation of motion and obtain the Hamiltonian to quantize a Josephson junction through a relevant Lagrangian which yields the above equation as the Euler-Lagrange equation. It is immediately found that the Lagrangian

$$L = \frac{1}{2}\frac{C\hbar^2}{4e^2}\left(\frac{d\phi}{dt}\right)^2 + \frac{I_c\hbar}{2e}\cos\phi + I_{\text{ext}}\frac{\hbar}{2e}\phi \equiv K - U \tag{15.4}$$

leads to Eq. (15.3) as the associated Euler-Lagrange equation. Note that the eleectrostatic energy

$$K = \frac{1}{2}\frac{C\hbar^2}{4e^2}\left(\frac{d\phi}{dt}\right)^2 = \frac{1}{2}CV^2 \tag{15.5}$$

and

$$U = -\frac{I_c\hbar}{2e}\cos\phi - I_{\text{ext}}\frac{\hbar}{2e}\phi \tag{15.6}$$

are, respectively, the kinetic energy and the potential energy.

Our next task is to find the Hamiltonian. Canonically conjugate momentum to the 'coordinate' ϕ is

$$\pi = \frac{\partial L}{\partial(d\phi/dt)} = \frac{C\hbar^2}{4e^2}\frac{d\phi}{dt}, \tag{15.7}$$

from which we obtain the Hamiltonian

$$H = \pi\dot{\phi} - L = \frac{1}{2}\frac{E_C}{\hbar^2}\pi^2 - E_J\cos\phi - E_J\frac{I_{\text{ext}}}{I_c}\phi, \tag{15.8}$$

where

$$E_C = \frac{(2e)^2}{C}, \quad E_J = \frac{I_c\hbar}{2e} \tag{15.9}$$

have the dimension of energy. They are called the **Coulomb energy** and the **Josephson energy**, respectively. Note that $E_C/2$ is the energy required to add a Cooper pair to a charge neutral state.

Quantization of a Josephson junction, regarded as a dynamical system, proceeds by imposing the usual canonical commutation relation

$$[\pi, \phi] = \frac{\hbar}{i}. \tag{15.10}$$

It is important to clarify the physical meaning of the variable π. Observe that

$$\pi = \frac{C\hbar^2}{4e^2}\frac{d\phi}{dt} = -\frac{C\hbar^2}{4e^2}\frac{2e}{\hbar}V = -\frac{\hbar Q}{2e} = \hbar N, \tag{15.11}$$

where N is the number of Cooper pairs in the electrode. Therefore the variable π is the number of Cooper pairs, and the commutation relation (15.10) shows that there exists an uncertainty relation between the number and the phase of Cooper pairs.

Josephson equations are obtained as Heisenberg equations with the above Hamiltonian. In fact, we verify that

$$\frac{d\phi}{dt} = \frac{i}{\hbar}[H, \phi] = \frac{E_C}{\hbar^2}\pi = -\frac{2e}{\hbar}V \tag{15.12}$$

and

$$\frac{d\pi}{dt} = \frac{i}{\hbar}[H, \pi] = -E_J\sin\phi + \frac{\hbar}{2e}I_{\text{ext}}.$$

The second equation is put into the form

$$I = \frac{dQ}{dt} = -2e\frac{dN}{dt} = -\frac{2e}{\hbar}\frac{d\pi}{dt} = -I_{\text{ext}} + I_c \sin\phi. \qquad (15.13)$$

Josephson equations (15.1) and (15.2), if regarded as classical equations, may be obtained from Eqs. (15.12) and (15.13) through Ehrenfest's theorem.

It is instructive to look at these equations in view of the Schrödinger picture. Canonical commutation relation leads to

$$\pi = \frac{\hbar}{i}\frac{d}{d\phi}, \quad N = \frac{1}{i}\frac{d}{d\phi} \qquad (15.14)$$

when ϕ is diagonalized. Now the Hamiltonian is written as

$$\begin{aligned} H &= \frac{1}{2}E_C N^2 - E_J\left(\cos\phi + \frac{I_{\text{ext}}}{I_c}\phi\right) \\ &= -\frac{1}{2}E_C\frac{d^2}{d\phi^2} - E_J\cos\phi - E_J\frac{I_{\text{ext}}}{I_c}\phi. \end{aligned} \qquad (15.15)$$

15.2.2 SQUIDs

It is possible to adjust the parameters of a Josephson junction during its fabrication. However, once a junction is fabricated, it is impossible to change these parameters, set aside unwanted aging effects. It is certainly desirable to have more controls over the Hamiltonian for a Josephson junction to serve as a controllable qubit.

A **SQUID** (Superconducting QUantum Interference Device) is a device made of a superconducting loop which contains Josephson junctions and a controllable magnetic field which threads the loop. We will show below that the Josephson energy is tunable by changing the magnetic flux.

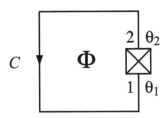

FIGURE 15.3
Circuit with a single Josephson junction. The flux threading the loop is Φ.

As a preliminary, we consider the circuit depicted in Fig. 15.3, where a superconducting loop contains a single Josephson junction. This circuit is called an **rf-SQUID** and may be employed as a prototypical flux qubit as will be shown later. Let $\boldsymbol{A}$ be a vector potential describing a magnetic field $\boldsymbol{B}$. Let $\Psi = |\Psi| e^{i\theta}$ be the macroscopic wave function and

$$\Phi = \int_S \boldsymbol{B} \cdot d\boldsymbol{S} = \oint_C \boldsymbol{A} \cdot d\boldsymbol{s} = \int_2^1 \boldsymbol{A} \cdot d\boldsymbol{s} \tag{15.16}$$

be the flux threading the loop with circumference C, which bounds an area S. Here we noted that the insulator is very thin, and hence the contribution to Φ from the junction is negligible. The supercurrent is

$$\boldsymbol{j}_s = -\frac{2e}{2(2m)} \left[\Psi^* \left(\frac{\hbar}{i} \nabla + 2e\boldsymbol{A} \right) \Psi + (\text{Hermitian conjugate}) \right]. \tag{15.17}$$

The supercurrent vanishes inside the superconductor and, by assuming $|\Psi|$ is constant, it holds

$$\boldsymbol{A} = -\frac{\hbar}{2e} \nabla \theta. \tag{15.18}$$

By substituting the above expression into Eq. (15.16), we find

$$\Phi = -\frac{\hbar}{2e} \int_2^1 \nabla \theta \cdot d\boldsymbol{s} = \frac{\Phi_0}{2\pi} (\phi + 2\pi n), \quad (n \in \mathbb{Z}), \tag{15.19}$$

where $\phi = \theta_2 - \theta_1$ is the phase discontinuity across the junction and

$$\Phi_0 = \frac{h}{2e} \simeq 2.0678 \times 10^{-15} \text{ Wb} \tag{15.20}$$

is the **flux quantum**.

Let us now consider a more widely used SQUID. Figure 15.4 (a) shows a schematic picture of a SQUID with two Josephson junctions, while (b) is an electron microscope image of an actual SQUID. The Josephson junction in Fig. 15.1 (b) is a part of this SQUID. Now the total flux threading the loop is expressed as

$$\begin{aligned}
\Phi &= -\frac{\Phi_0}{2\pi} \left(\int_2^3 \nabla \theta ds + \int_4^1 \nabla \theta ds \right) \\
&= \frac{\Phi_0}{2\pi} (\theta_2 - \theta_3 + \theta_4 - \theta_1) \\
&= \frac{\Phi_0}{2\pi} (\phi_1 - \phi_2),
\end{aligned} \tag{15.21}$$

where $\phi_1 = \theta_2 - \theta_1$ and $\phi_2 = \theta_3 - \theta_4$ are phase discontinuities across junctions 1 and 2, respectively. The Josephson energy of the SQUID is the sum of respective Josephson energies, given by

$$E = -\frac{E_J}{2} (\cos\phi_1 + \cos\phi_2) = -E_J \left(\cos\frac{\phi_1 + \phi_2}{2} \cos\frac{\phi_1 - \phi_2}{2} \right),$$

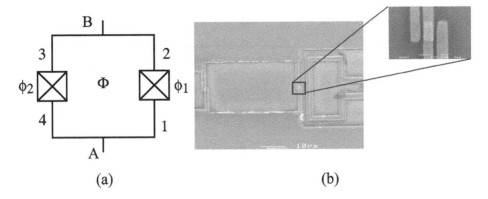

FIGURE 15.4

(a) Schematic diagram of a SQUID with two Josephson junctions. (b) Actual nanoscale SQUID. The enlarged structure is the Josephson junction in Fig. 15.1. The white line in the bottom is 10 μm long. Courtesy of Jukka Pekola, Heksinki University of Technology, Finland.

where we assumed that the two Josephson junctions are identical and each junction has the Josephson energy $E_J/2$. By introducing the parameters

$$\delta = \phi_1 - \phi_2 = 2\pi \frac{\Phi}{\Phi_0}, \quad \phi = \frac{\phi_1 + \phi_2}{2}, \tag{15.22}$$

the Josephson energy is rewritten as

$$E = -E_J \cos \frac{\delta}{2} \cos \phi. \tag{15.23}$$

The factor $\cos(\delta/2)$ in the RHS is due to the interference between Cooper pairs through two paths and hence is regarded as a manifestation of the Bohm-Aharonov effect. The current through junction 1, in the direction from A to B, is

$$I_1 = \frac{I_c}{2} \sin \phi_1,$$

while that through the junction 2, from A to B, is

$$I_2 = \frac{I_c}{2} \sin \phi_2,$$

where $I_c \equiv 2eE_J/\hbar$. Therefore the current through the SQUID from A to B in Fig. 15.4 (a) is

$$J = I_1 + I_2 = \frac{2e}{\hbar} \frac{\partial E}{\partial \phi}. \tag{15.24}$$

Note that a current flows along the SQUID loop even when $I_1 = -I_2$, for

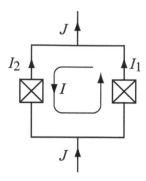

FIGURE 15.5

Currents through a SQUID. The total current flowing from bottom to top is $J = I_1 + I_2$. The circulating current flowing along the loop is $I = (I_1 - I_2)/2$.

which $J = 0$. This current is

$$I = \frac{I_1 - I_2}{2} = \frac{2e}{\hbar} \frac{\partial E}{\partial \delta} = \frac{I_c}{2}(\sin \phi_1 - \sin \phi_2) = \frac{I_c}{2} \sin \frac{\delta}{2} \cos \phi. \tag{15.25}$$

Now the total Josephson current is

$$J = \frac{I_c}{2}(\sin \phi_1 + \sin \phi_2) = I_c(\Phi) \sin \phi, \tag{15.26}$$

where Eq. (15.21) has been used to obtain

$$I_c(\Phi) = I_c \cos \left(\pi \frac{\Phi}{\Phi_0} \right). \tag{15.27}$$

The above expression shows that the effective Josephson energy is a function of the flux threading the loop and is given by

$$E_J(\Phi) = E_J \cos \left(\pi \frac{\Phi}{\Phi_0} \right). \tag{15.28}$$

Therefore a SQUID behaves as if it were a single Josephson junction with a controllable E_J. This is consistent with our previous observation (15.23).

We assume that the two junctions are identical and hence have the same capacitance C. The electrostatic energy of the Josephson junctions is then given by

$$K = \frac{1}{2}CV_1^2 + \frac{1}{2}CV_2^2 = \frac{1}{2} \left(\frac{\hbar}{2e} \right)^2 C(\dot{\phi}_1^2 + \dot{\phi}_2^2), \tag{15.29}$$

where use has been made of the relation $V_k = -(\hbar/2e)\dot{\phi}_k$. Here we assume that the superconducting loop has a small inductance so that $\Phi = \Phi_{\text{ext}} + LI \simeq \Phi_{\text{ext}}$. By noting the definitions

$$\phi_1 = \phi + \pi \frac{\Phi_{\text{ext}}}{\Phi_0}, \quad \phi_2 = \phi - \pi \frac{\Phi_{\text{ext}}}{\Phi_0}$$

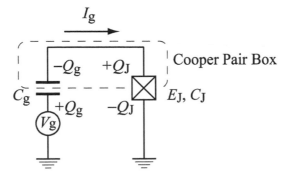

FIGURE 15.6

Schematic diagram of a Josephson charge qubit. $\pm Q_J$ denotes charges induced at the electrodes of the Josephson junction with the capacitance C_J and the Josephson energy E_J, while $\pm Q_g$ denotes charges induced at the electrodes of the gate capacitor C_g. The circuit is biased by the voltage V_g. The part enclosed by a broken line comprises the Cooper pair island whose excess charge (in units of $2e$) corresponds to the qubit degree of freedom. The current flowing through the circuit is denoted by I_g.

we obtain a simple expression for the electrostatic energy

$$K = \frac{1}{2}\left(\frac{\hbar}{2e}\right)^2 (2C)\dot{\phi}^2, \tag{15.30}$$

where we have dropped the term $\propto \dot{\Phi}_{\text{ext}}^2$ since it is a c-number and does not affect the quantum dynamics.

15.3 Charge Qubit

15.3.1 Simple Cooper Pair Box

Figure 15.6 shows the schematic diagram of a Josephson charge qubit in its simplest form. A Josephson junction with the Josephson energy E_J and the capacitance C_J is connected to gate voltage V_g through a gate capacitance C_g. The **Cooper pair box** (**CPB**) is made of an electrode of the Josephson junction and an electrode of the gate capacitor and the superconducting lead connecting them. The CPB works as a qubit as we show below.

We first find the Lagrangian of the system. The electrostatic energy plays the role of the kinetic energy, while the Josephson energy is identified as the potential energy. Thus we obtain

$$L = \frac{1}{2}C_J V_J^2 + \frac{1}{2}C_g(V_g - V_J)^2 + E_J \cos\phi. \tag{15.31}$$

Recall that the voltage V_J across the junction is related to the time derivative of the phase discontinuity as

$$V_J = -\frac{\hbar}{2e}\dot{\phi}.$$

Then the kinetic energy (the electrostatic energy) is

$$K = \frac{1}{2}\left(\frac{\hbar}{2e}\right)^2 C_J \dot{\phi}^2 + \frac{1}{2}C_g\left(V_g + \frac{\hbar}{2e}\dot{\phi}\right)^2$$
$$= \frac{1}{2}(C_J + C_g)\left(\frac{\hbar}{2e}\right)^2 \dot{\phi}^2 + \frac{\hbar}{2e}C_g V_g \dot{\phi}, \tag{15.32}$$

where the constant term $C_g V_g^2/2$ has been dropped. The momentum canonically conjugate to ϕ is

$$\pi = \frac{\partial K}{\partial \dot{\phi}} = \left(\frac{\hbar}{2e}\right)^2 (C_J + C_g)\dot{\phi} + \frac{\hbar}{2e}C_g V_g. \tag{15.33}$$

We will consider the dynamics of Cooper pairs only since we assume that $k_B T \ll \Delta$ and accordingly the quasiparticle excitations are totally negligible. At such a low temperature, the charge Q in the box is identified with the number of excess Cooper pairs relative to a charge-neutral background. Let us introduce a variable N defined as $\pi = \hbar N$ and work out its physical meaning. The charge stored in the junction is $Q_J = C_J V_J$, while that in the gate capacitor is $Q_g = C_g(V_g - V_J)$. Therefore the total charge in the island is

$$Q = Q_J - Q_g = (C_J + C_g)V_J - C_g V_g$$
$$= -\frac{\hbar}{2e}(C_J + C_g)\dot{\phi} - C_g V_g$$
$$= -\frac{2e}{\hbar}\pi = -2eN.$$

It has been shown that N is the number of Cooper pairs in the island, which should be compared with Eq. (15.11).

Now the Hamiltonian is derived as

$$H = \pi\dot{\phi} - L$$
$$= \frac{1}{2}\left(\frac{2e}{\hbar}\right)^2 \frac{1}{C_J + C_g}\left(\pi - \frac{\hbar}{2e}C_g V_g\right)^2 - E_J \cos\phi$$
$$= \frac{1}{2}\frac{(2e)^2}{C_J + C_g}(N - N_g)^2 - E_J \cos\phi, \tag{15.34}$$

where
$$N_g = \frac{C_g V_g}{2e}. \tag{15.35}$$

In summary, the Hamiltonian of this system is written as
$$H = \frac{1}{2} E_C (N - N_g)^2 - E_J \cos \phi, \tag{15.36}$$

where
$$E_C = \frac{(2e)^2}{C_g + C_J}. \tag{15.37}$$

Let us consider the case in which the electrostatic energy dominates over the Josephson energy,
$$E_C \gg E_J. \tag{15.38}$$

This means physically that tunneling between states with different N is suppressed and N is a good quantum number in practice. Note that this assumption breaks down when $N \sim N_g$, where the tunneling energy dominates over the Coulomb energy. Naturally we employ the complete set of the eigenstates of the operator N, which we denote as $|N\rangle$, to write down the Hamiltonian. Ordinary ket notation $|k\rangle$ will be reserved to denote the kth eigenstate of H. The Josephson coupling energy leads to tunneling current. It therefore changes the number of excess Cooper pairs in the CPB. It follows from the canonical commutation relation $[\phi, N] = i$ that $\phi = i\partial/\partial N$ and
$$e^{\pm i\phi}|N\rangle = e^{\mp \partial/\partial N}|N\rangle = |N \mp 1\rangle. \tag{15.39}$$

From these considerations, we obtain the Hamiltonian (15.34) in N-represetation,
$$H = \sum_{N \in \mathbb{Z}} \left[\frac{E_C}{2}(N - N_g)^2 |N\rangle\langle N| - \frac{E_J}{2}\left(|N\rangle\langle N+1| + |N+1\rangle\langle N|\right) \right]. \tag{15.40}$$

Figure 15.7 shows the spectrum of this Hamiltonian.

It is important to notice here that N_g in Eq. (15.40) is a controllable parameter by adjusting the gate voltage V_g. In particular, N_g may be set in the vicinity of $N + 1/2$, $(N \in \mathbb{N})$ so that the states $|N\rangle$ and $|N+1\rangle$ have almost degenerate energies. All other states have higher energies and can be ignored in the following discussions. Now the Hamiltonian is represented in the subspace of two states $|N\rangle = (1,0)^t$ and $|N+1\rangle = (0,1)^t$ as

$$\begin{aligned}
H &= \frac{E_C}{2}\left[N_g^2|N\rangle\langle N| + (1-N_g)^2|N+1\rangle\langle N+1|\right] \\
&\quad - \frac{E_J}{2}\left[|N\rangle\langle N+1| + |N+1\rangle\langle N|\right] \\
&= \frac{E_C}{2}\begin{pmatrix} N_g^2 & 0 \\ 0 & (1 - 2N_g + N_g^2) \end{pmatrix} - \frac{E_J}{2}\begin{pmatrix} 0 & 1 \\ 1 & 0 \end{pmatrix} \\
&= -\frac{1}{2}B_z\sigma_z - \frac{1}{2}B_x\sigma_x,
\end{aligned} \tag{15.41}$$

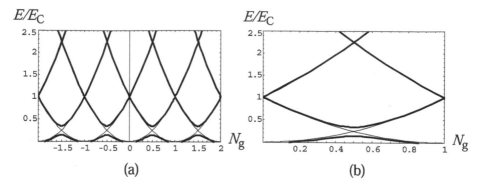

FIGURE 15.7
Energy spectrum of the Hamiltonian (15.40). (a) Thick curves show the spectrum with $E_J/E_C = 0.2$ and thin curves, with $E_J/E_C = 0$. (b) Details of the spectrum for $0 \le N_g \le 1$, which is relevant for charge qubit construction.

where

$$B_z = \frac{E_C}{2}(1 - 2N_g), \ B_x = E_J, \tag{15.42}$$

and we dropped an irrelevant constant term $(E_C/4)(1 - 2N_g + 2N_g^2)I$, I being the 2×2 unit matrix, which merely contributes to the overall phase. From now on, we will be concerned with the subspace in which $N = 0$, and denote $|0\rangle$ and $|1\rangle$ by $|N = 0\rangle$ and $|N = 1\rangle$, respectively. The eigenvalues are

$$E_0 = -\frac{1}{2}\sqrt{B_x^2 + B_z^2}, \ E_1 = \frac{1}{2}\sqrt{B_x^2 + B_z^2}, \tag{15.43}$$

and the corresponding eigenvectors are

$$|0\rangle = \cos\left(\frac{\alpha}{2}\right)|N = 0\rangle + \sin\left(\frac{\alpha}{2}\right)|N = 1\rangle$$
$$|1\rangle = -\sin\left(\frac{\alpha}{2}\right)|N = 0\rangle + \cos\left(\frac{\alpha}{2}\right)|N = 1\rangle, \tag{15.44}$$

respectively, where

$$\alpha = \tan^{-1}\left[\frac{B_x}{B_z(N_g)}\right]. \tag{15.45}$$

Note the similarity between this Hamiltonian and the controllable part of a single-spin NMR Hamiltonian (12.14), except that this Hamiltonian contains $\{\sigma_x, \sigma_z\}$, instead of $\{\sigma_x, \sigma_y\}$. From the commutation relation $[\sigma_z, \sigma_x] = i\sigma_y$, the absense of σ_y does not necessarily imply SU(2) rotations around the y-axis are impossible to implement with Eq. (15.41). However, a serious inconveience with Eq. (15.41) is that E_J is *not* controllable. We need to introduce a SQUID to make E_J tunable as we see next.

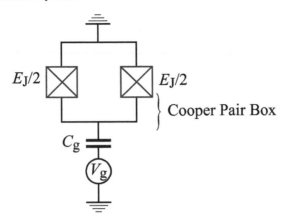

FIGURE 15.8
Charge qubit with a tunable Josephson energy.

15.3.2 Split Cooper Pair Box

It has been shown in §15.2.2 that a superconducting loop with two identical Josephson junctions with identical Josephson energy $E_J/2$ has the total Josephson energy

$$-\frac{E_J}{2}(\cos\phi_1 + \cos\phi_2) = -E_J \cos\frac{\phi_1 + \phi_2}{2} \cos\frac{\phi_1 - \phi_2}{2}, \qquad (15.46)$$

where ϕ_1 and ϕ_2 are the phase differences across the junctions 1 and 2, respectively. Let L be the self-inductance of the loop and I be the current circulating the loop. Then the difference $\phi_1 - \phi_2$ is related to the magnetic flux $\Phi = \Phi_{\text{ext}} + LI$ threading the loop as

$$\frac{\phi_1 - \phi_2}{2} = \pi\frac{\Phi}{\Phi_0} \simeq \pi\frac{\Phi_{\text{ext}}}{\Phi_0}, \qquad (15.47)$$

where we assumed that the self-inductance L is very small for a charge qubit under consideration. Now the Josephson energy is expressed as

$$-E_J \cos\left(\pi\frac{\Phi_{\text{ext}}}{\Phi_0}\right)\cos\phi, \qquad (15.48)$$

where $\phi = (\phi_1 - \phi_2)/2$. It has been shown that the effective Josephson energy is given by

$$E_J(\Phi) = E_J \cos\left(\pi\frac{\Phi_{\text{ext}}}{\Phi_0}\right). \qquad (15.49)$$

Figure 15.8 shows a schematic picture of a charge qubit with a tunable Josephson energy, called a **split Cooper pair box**. The total Hamiltonian

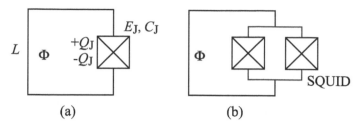

FIGURE 15.9
Simplest flux qubits. (a) An rf-SQUID and (b) a flux qubit with a tunable Josephson energy E_J.

is again given by

$$H = -\frac{1}{2}B_z(N_g)\sigma_z - \frac{1}{2}B_x(\Phi_{\text{ext}})\sigma_x, \tag{15.50}$$

where both coefficients

$$B_z(N_g) = \frac{E_C}{2}(1 - 2N_g), \quad B_x(\Phi_{\text{ext}}) = E_J \cos\left(\pi\frac{\Phi_{\text{ext}}}{\Phi_0}\right) \tag{15.51}$$

are controllable.

The eigenvalues and the corresponding eigenvectors of the Hamiltonian (15.50) are given by Eqs. (15.43) and (15.44) with B_x in Eq. (15.45) replaced by $B_x(\Phi_{\text{ext}})$. Observe that $|0\rangle \to |N = 0\rangle$ and $|1\rangle \to |N = 1\rangle$ as $\Phi_{\text{ext}} \to \Phi_0/2$ with $B_z(N_g) \neq 0$, while $|0\rangle \to (|N = 0\rangle + |N = 1\rangle)/\sqrt{2}$ and $|1\rangle \to (-|N = 0\rangle + |N = 1\rangle)/\sqrt{2}$ as $N_g \to 1/2$ with $B_x(\Phi_{\text{ext}}) \neq 0$.

15.4 Flux Qubit

In contrast with a charge qubit, the flux threading a SQUID is the relevant degree of freedom in a flux qubit. Here we need to employ a circuit with

$$E_J \gg E_C. \tag{15.52}$$

15.4.1 Simplest Flux Qubit

The simplest flux qubit, the **rf-SQUID**, is shown in Fig. 15.9. This is nothing but a SQUID with no input/output current. It is made of a superconducting loop with a Josephson junction where the loop supports a persistent current I and a flux Φ threading the loop. The flux is related to the external magnetic flux Φ_{ext} and the current I along the loop as

$$\Phi = \Phi_{\text{ext}} + LI, \tag{15.53}$$

where L is the self-inductance of the circuit. It was shown in §15.2.2 that the phase difference ϕ across the junction is related to the flux Φ as

$$\Phi = \frac{\Phi_0}{2\pi}\phi, \tag{15.54}$$

where we have taken $n = 0$ in (15.19).

The Hamiltonian of the circuit is made of the electrostatic energy, the inductance energy and the Josephson energy as

$$
\begin{aligned}
H &= \frac{1}{2}CV^2 + \frac{1}{2}LI^2 - E_J\cos\phi \\
&= \frac{E_C}{2}N^2 + \frac{1}{2L}(\Phi - \Phi_{\text{ext}})^2 - E_J\cos\phi \\
&= -\frac{E_C}{2}\frac{d^2}{d\phi^2} + \frac{1}{2}E_L(\phi - \phi_{\text{ext}})^2 - E_J\cos\phi,
\end{aligned}
\tag{15.55}
$$

where

$$E_C = \frac{(2e)^2}{C}, \quad E_L = \frac{1}{L}\left(\frac{\Phi_0}{2\pi}\right)^2, \quad \phi_{\text{ext}} = \frac{2\pi\Phi_{\text{ext}}}{\Phi_0}. \tag{15.56}$$

The potential energy of the Hamiltonian is therefore

$$U(\phi) = -E_J\cos\phi + \frac{1}{2}E_L(\phi - \phi_{\text{ext}})^2. \tag{15.57}$$

Let us define $\varepsilon \equiv E_J/E_L - 1$ and $f \equiv \phi_{\text{ext}} - \pi$ and assume that

$$|f| \ll 1, \quad |\varepsilon| \ll 1. \tag{15.58}$$

Moreover we define $\tilde{\phi} \equiv \phi - \pi$. The potential energy, scaled by E_L, is approximated in terms of these parameters as

$$\frac{U(\tilde{\phi})}{E_L} = -f\tilde{\phi} - \frac{1}{2}\varepsilon\tilde{\phi}^2 + \frac{1}{24}(1+\varepsilon)\tilde{\phi}^4, \tag{15.59}$$

where a constant term has been dropped. Figure 15.10 (a) shows the potential minima are degenerate when $f = 0$, while Fig. 15.10 (b) shows this degeneracy is lifted when $f \neq 0$. The minima are at $\tilde{\phi} \simeq \pm\sqrt{6\varepsilon}$ when $f = 0$. The minima for $f \neq 0$ is obtained by iteration assuming $0 \leq f \ll \varepsilon$. By writing $\tilde{\phi} = \pm\sqrt{6\varepsilon} + \delta$ and substituting this to $\partial U(\tilde{\phi})/\partial\tilde{\phi} = -f - \varepsilon\tilde{\phi} + (1+\varepsilon)\tilde{\phi}^3/6 = 0$, we obtain the minima

$$\tilde{\phi}_\pm = \pm\sqrt{6\varepsilon} + \frac{f}{2\varepsilon}. \tag{15.60}$$

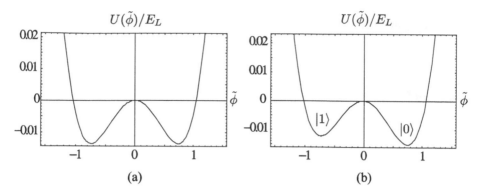

FIGURE 15.10
Potential energy $U(\tilde{\phi})/E_L$ for a flux qubit. (a) $f = 0$ and $\varepsilon = 0.1$ and (b) $f = 0.002$ and $\varepsilon = 0.1$.

The curvature at the minima

$$\left.\frac{\partial^2 U(\tilde{\phi})}{\partial \tilde{\phi}^2}\right|_{\tilde{\phi}_\pm} = \left.-\varepsilon + \frac{1}{2}(1+\varepsilon)\tilde{\phi}^2\right|_{\tilde{\phi}_\pm} \simeq 2\varepsilon \qquad (15.61)$$

is independent of f within the harmonic approximation. The difference between the minimum values of $U(\tilde{\phi})$ is easily obtained as

$$\Delta U = U(\tilde{\phi}_-) - U(\tilde{\phi}_+) = 2\sqrt{6\varepsilon} f E_L \qquad (15.62)$$

to the lowest order.

At sufficiently low temperatures, we need to consider the lowest energy eigenstate of each potential well. Taking the tunneling between these two states into account, the effective Hamiltonian is given by

$$H = -\frac{B_z}{2}\sigma_z - \frac{B_x}{2}\sigma_x. \qquad (15.63)$$

Here

$$B_z(\phi_{\text{ext}}) = \Delta U(\phi_{\text{ext}}) = 2\sqrt{6\varepsilon}(\phi_{\text{ext}} - \pi)E_L \qquad (15.64)$$

is the energy difference between the states $|0\rangle$ and $|1\rangle$ and is controllable by changing the external magnetic flux Φ_{ext}. The parameter B_x is determined by the tunneling probability between two wells and dependes on the barrier height, which in turn depends on the Josephson energy E_J through ε. Therefore B_x is also controllable if the Josephson junction is replaced by a SQUID whose E_J depends on the external magnetic flux threading the SQUID, see Fig. 15.9 (b). We identify $|0\rangle$ and $|1\rangle$ with $|\tilde{\phi} = \tilde{\phi}_+\rangle$ and $|\tilde{\phi} = \tilde{\phi}_-\rangle$, respectively.

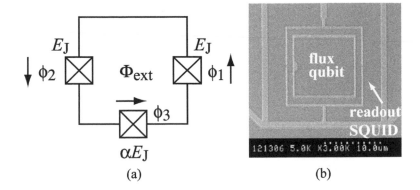

FIGURE 15.11
Three-junction SQUID. (a) Schematic diagram of a three-junction SQUID.
Phase change ϕ_k across the junction k is measured in the direction of the
arrow. (b) Electron microscope image of a three-junction SQUID. There is an
extra readout SQUID with two Josephson junctions surrounding the three-
junction SQUID. Reprinted figure with permission from J. Johansson *et al.*,
Physical Review Letters **96**, 127006 (2006). Copyright (2006) by the American
Physical Society.

15.4.2 Three-Junction Flux Qubit

It was required in the simplest design (Fig. 15.9) of a flux qubit that E_J/E_L
is on the order of 1 so that $|\varepsilon| \ll 1$. This condition requires that the self-
inductance of the SQUID be sufficiently large to make E_L small and at the
same time the Josephson energy E_J be large enough. It is required to fabricate
a Josephson junction with a large area and a large SQUID loop to satisfy these
conditions. As a result, the coupling between a flux qubit and its environment
is enhanced, leading to fast decoherence. This is considered to be the reason
why macroscopic quantum coherent oscillation has not been observed earlier.

Mooij and his collaborators proposed a three-junction SQUID and a four-
junction SQUID to overcome this difficulty [7]. A three-junction SQUID is
depicted in Fig. 15.11. These SQUIDs may be fabricated to have small self-
inductance, so that decoherence is suppressed.

A simple generalization of the argument on SQUID leads to the the following
expression for the total flux threading the loop with three junctions;

$$\Phi = \oint \boldsymbol{A} \cdot d\boldsymbol{s} = \frac{\Phi_0}{2\pi} \oint \nabla\theta \cdot d\boldsymbol{s}$$

$$= \frac{\Phi_0}{2\pi}(\phi_1 + \phi_2 + \phi_3), \tag{15.65}$$

where $\Phi = \Phi_{\text{ext}} + LI$ as before, and ϕ_k denotes the phase change across the kth
junction. The self-inductance L is made small so that the condition $E_L \gg E_J$

is satisified. Then we find $\Phi \simeq \Phi_{\text{ext}}$ and accordingly

$$\phi_1 + \phi_2 + \phi_3 = 2\pi \frac{\Phi_{\text{ext}}}{\Phi_0} \equiv \pi + f. \tag{15.66}$$

This equation is solved for f to yield

$$f = \frac{2\pi}{\Phi_0}\left(\Phi_{\text{ext}} - \frac{\Phi_0}{2}\right). \tag{15.67}$$

Note that $|f| \ll 1$ when Φ_{ext} is tuned in the vicinity of $\Phi_0/2$, which we will always assume from now on.

We further assume junctions 1 and 2 are identical, and hence $E_{J1} = E_{J2} \equiv E_J$, while junction 3 is a SQUID and $E_{J3} = \alpha E_J$ with $1/2 < \alpha < 1$.

The potential energy U associated with the three-junction loop is

$$\begin{aligned} U(\phi_+, \phi_-) &= -E_J(\cos\phi_1 + \cos\phi_2) - \alpha E_J \cos\phi_3 \\ &= -E_J\left[2\cos\phi_+\cos\phi_- - \alpha\cos(2\phi_+ - f)\right] \\ &\simeq -E_J\left[2\cos\phi_+\cos\phi_- - \alpha\cos 2\phi_+ - \alpha f \sin 2\phi_+\right], \end{aligned} \tag{15.68}$$

where $|f| \ll 1$ is assumed and

$$\phi_\pm \equiv \frac{\phi_1 \pm \phi_2}{2}. \tag{15.69}$$

The parameter f is controllable by manipulating Φ_{ext}. Minimization of U with respect to ϕ_- yields the condition $\phi_- = 0$, viz. $\phi_1 = \phi_2$.

When $f = 0$, minimization of U with respect to ϕ_+ yields $\sin\phi_{+0} = \alpha\sin 2\phi_{+0}$, or

$$\cos\phi_{+0} = \frac{1}{2\alpha}. \tag{15.70}$$

This equation has a solution if and only if $|\alpha| \geq 1/2$. This solution corresponds to the condition

$$I_p = \frac{2e}{\hbar}\frac{\partial U}{\partial\phi_1} = \frac{2e}{\hbar}\frac{\partial U}{\partial\phi_2} = \frac{2e}{\hbar}\frac{\partial U}{\partial\phi_3}, \tag{15.71}$$

which guarantees that the persistent currents through the junctions are all identical. There are two solutions $\pm\phi_{+0}$ corresponding to two currents I_p and $-I_p$; see Fig. 15.12.

Let us consider the case $f \neq 0$ next. If follows from Eq. (15.68) that minima of $U(\phi_+, \phi_-)$ are attained at $\phi_- = 0$ and $\phi_+ = \phi_{+0} + \delta\phi_+$, where $\delta\phi_+$ is on the order of f. By noting that ϕ_{+0} gives the minimum of $U(\phi_+, 0)$ when $f = 0$, we find that the first order contribution of $\delta\phi_+$ to $U(\phi_{+0} + \delta\phi_+, 0)$ vanishes. Therefore we may write Eq. (15.68) as

$$\begin{aligned} U(\pm\phi_{+0}, 0) &= -E_J\left[2\cos\phi_{+0} - \alpha\cos 2\phi_{+0}\right] \\ &\quad \pm I_p\left(\Phi_{\text{ext}} - \frac{1}{2}\Phi_0\right). \end{aligned} \tag{15.72}$$

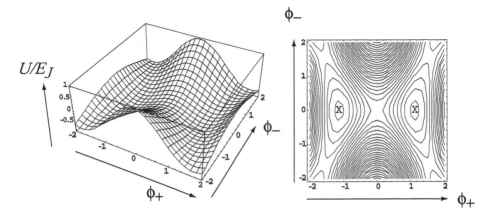

FIGURE 15.12
Potential profile U/E_J of a three-junction SQUID, in which $\alpha = 0.6$ and $f = 0$. Points marked with x are minima of the potential.

The "effective magnetic field" along the z-axis is now given by

$$B_z = 2I_p \left(\Phi_{\text{ext}} - \frac{1}{2}\Phi_0 \right), \tag{15.73}$$

where

$$I_p = \frac{2\pi}{\Phi_0} E_J \sin\phi_{+0} = I_c \sin\phi_{+0}, \tag{15.74}$$

while B_x is fixed by the tunneling probability with which the state tunnels between two minima.

In summary, we have obtained a single-qubit Hamiltonian

$$H = -\frac{1}{2} \left(B_z\sigma_z + B_x\sigma_x \right). \tag{15.75}$$

15.5 Quantronium

Saclay Quantronics group proposed a qubit having a very long decoherence time and called it a **quantronium** [8]. Here a SQUID with approximately equal electrostatic energy and Josephson energy ($E_C \simeq E_J$) is employed. In this sense, it is a system intermediate between a charge qubit and a flux qubit.

It follows from Eq. (15.50) that a split Cooper pair box charge qubit has a Hamiltonian

$$H = -\frac{1}{2} \left[B_z(N_g)\sigma_z + B_x(\Phi_{\text{ext}})\sigma_x \right], \tag{15.76}$$

which has eigenvalues

$$E_0(N_g, \Phi_{\text{ext}}) = -\frac{1}{2}\sqrt{B_z(N_g)^2 + B_x(\Phi_{\text{ext}})^2},$$

$$E_1(N_g, \Phi_{\text{ext}}) = \frac{1}{2}\sqrt{B_z(N_g)^2 + B_x(\Phi_{\text{ext}})^2}.$$

(15.77)

Let us consider fluctuation of the eigenvalues due to noise. Let δN and $\delta \Phi$ be fluctuations in N_g and Φ_{ext}, respectively. Then it is found that

$$E_{0,1}(N_g+\delta N, \Phi_{\text{ext}}+\delta\Phi) = E_{0,1}(N_g, \Phi_{\text{ext}})+\frac{\partial E_{0,1}}{\partial N_g}\delta N+\frac{\partial E_{0,1}}{\partial \Phi_{\text{ext}}}\delta\Phi+O[\delta N^2, \delta\Phi^2].$$

The influence of the fluctuations in N_g and Φ_{ext} on the eigenvalues is optimized when the conditions $\partial E_{0,1}/\partial N_g = \partial E_{0,1}/\partial \Phi_{\text{ext}} = 0$ or equivalently $\partial B_z^2/\partial N_g = \partial B_x^2/\partial \Phi_{\text{ext}} = 0$ are satisfied. Written explicitly, these conditions yield, by using Eq. (15.51),

$$B_z = 0 \Rightarrow N_g = \frac{1}{2}, \quad \frac{\partial B_x}{\partial \Phi_{\text{ext}}} = 0 \Rightarrow \Phi_{\text{ext}} = 0. \qquad (15.78)$$

By substituting these parameters into Eq. (15.45), we find $\alpha = \pi/2$, and the corresponding eigenvectors are

$$|0\rangle = \frac{1}{\sqrt{2}}(|N = 0\rangle + |N = 1\rangle), \ |1\rangle = \frac{1}{\sqrt{2}}(-|N = 0\rangle + |N = 1\rangle). \quad (15.79)$$

A qubit with a long decoherece time is obtained if the gate voltage and external flux are adjusted as above. This qubit is called the **quantronium**, and the point (15.78) is called the optimal working point.

We evaluate the SQUID loop current I in the states $|0\rangle$ and $|1\rangle$. The current has already been derived in Eq. (15.25). We note that ϕ is an operator in this equation to obtain

$$\langle 0|I|0\rangle = I_c \sin\left(\pi\frac{\Phi_{\text{ext}}}{\Phi_0}\right), \ \langle 1|I|1\rangle = -I_c \sin\left(\pi\frac{\Phi_{\text{ext}}}{\Phi_0}\right). \qquad (15.80)$$

Gate operation of a quantronium is conducted at the optimal working point $(N_g = 1/2, \Phi_{\text{ext}} = \Phi_0 = 0)$. However, this working point is shifted to $N_g = 1/2$ and $\Phi_0 \neq 0$ during measurement, in which case $|0\rangle$ and $|1\rangle$ support currents flowing in opposition directions.

15.6 Current-Biased Qubit (Phase Qubit)

Rabi oscillation has been observed recently in a **current-biased Joesephson junction** [9]. This system has been actively investigated for more than two decades since Leggett suggested that macroscopic quantum tunneling (MQT) might be observable in this system [5, 6]. The system in which Rabi oscillation has been observed employed a relatively large Josephson junction of dimension $\sim 10\ \mu\mathrm{m}$, for which much effort had been devoted to suppress noise from the environment. At the same time, the large size of the junction implies fabrication technique is well estabished within the current technology, and it is expected that this system could be scalable to a multiqubit system. A current-biased qubit is also known as a **phase qubit**.

The Josephson energy dominates over the electrostatic energy so that $E_J \gg E_C$ in such a large Josephson junction. The Hamiltonian associated with the phase ϕ is

$$H = -\frac{E_C}{2}\frac{d^2}{d\phi^2} + U(\phi), \tag{15.81}$$

where the potential energy in the presence of a bias current I_{ext} is obtained from Eq. (15.15) as

$$U(\phi) = -E_J\left(\cos\phi + \frac{I_{\mathrm{ext}}}{I_c}\phi\right). \tag{15.82}$$

The extremal condition for $U(\phi)$ is

$$\sin\phi = \frac{I_{\mathrm{ext}}}{I_c}. \tag{15.83}$$

The potential energy in the presence of I_{ext} is plotted in Fig. 15.13, which justifies the nickname "washboard potential." We have introduced in the figure a parameter $x = \phi - \pi/2$.

Consider the limit $I_{\mathrm{ext}} \to I_c$ and put $\phi = \pi/2 + x$ ($|x| \ll 1$) with application to qubits in our mind. Then the potential is expanded up to the third order in x as

$$U(x) \simeq E_J\left[\left(1 - \frac{I_{\mathrm{ext}}}{I_c}\right)x - \frac{x^3}{6} - \frac{I_{\mathrm{ext}}}{2I_c}\pi\right]. \tag{15.84}$$

The extrema are found to be at

$$x_\pm = \pm\sqrt{2\left(1 - \frac{I_{\mathrm{ext}}}{I_c}\right)}. \tag{15.85}$$

The minimum is attained at $x = x_-$, at which the curvature is

$$E_J\sqrt{\left(1 - \frac{I_{\mathrm{ext}}}{I_c}\right)}. \tag{15.86}$$

Therefore the harmonic oscillator frequency (**plasma frequency**) of a small oscillation in the neighborfood of x_- is

$$\bar{\omega}_p = \omega_p\left[2\left(1 - \frac{I_{\mathrm{ext}}}{I_c}\right)\right]^{1/4}, \tag{15.87}$$

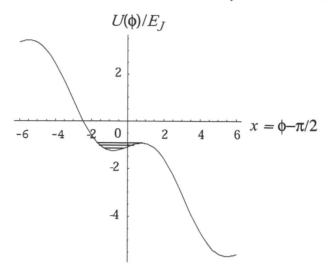

FIGURE 15.13

Potential profile of a current-biased Josephson junction as a function of $x = \phi - \pi/2$. We have taken $I_{\text{ext}}/I_c = 0.8$. Three quasibound states at the central potential well are indicated.

where $\omega_p = \sqrt{E_C E_J}/\hbar$.

The above analysis shows that the potential barrier height, namely the potential difference between the minimum and the maximum, is

$$\frac{\Delta U}{E_J} = U(x_+) - U(x_-)$$

$$= \frac{2}{3}\left[2\left(1 - \frac{I_{\text{ext}}}{I_c}\right)\right]^{3/2}. \tag{15.88}$$

It is found from Eqs. (15.87) and (15.88) that $\bar{\omega}_p$ approaches 0 more slowly compared to ΔU as $I_{\text{ext}} \to I_c$, and accordingly a small number of bound states remains in the potential wall as shown in Fig. 15.13. We construct a qubit out of the ground state and the first excited state among these quasibound states.

The rf-SQUIDs in Fig. 15.9 are also used as Josephson phase qubits, whose potential is similar to the washboard potential depicted in Fig. 15.13 [10]. Let L be the self-inductance of the circuit. The potential energy of this circuit is given by (cf. Eq. (15.57)),

$$U(\phi) = E_L\left[\frac{1}{2}(\phi - \phi_{\text{ext}})^2 - \beta_L \cos\phi\right], \tag{15.89}$$

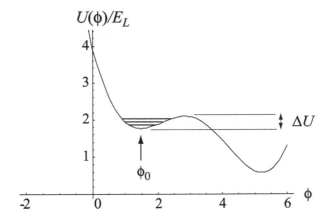

FIGURE 15.14
Potential energy of an rf-SQUID qubit. Parameters are chosen as $\phi_{\text{ext}} = 1.1\pi$ and $\beta_L = 2$. Three quasibound states trapped in the left potential well at $\phi = \phi_0$ are also depicted. The energy difference between the bottom of the local minimum and the local maximum is denoted as ΔU.

where

$$\beta_L = \frac{E_J}{E_L}, \quad \phi = 2\pi\frac{\Phi}{\Phi_0}, \quad \phi_{\text{ext}} = 2\pi\frac{\Phi_{\text{ext}}}{\Phi_0}.$$

Figure 15.14 shows the potential (15.89) with $\phi_{\text{ext}} > \pi$ and $1 < \beta_L$. Note that the potential has two minima for the choice of the parameters, $\phi_{\text{ext}} = 1.1\pi$ and $\beta_L = 2$. Let us analyze the shallow minimum ϕ_0 of the potential. A point ϕ_0 is a minimum of $U(\phi)$ when it satisfies

$$\left.\frac{\partial U}{\partial \phi}\right|_{\phi=\phi_0} = \phi_0 - \phi_{\text{ext}} + \beta_L \sin\phi_0 = 0$$

and

$$\left.\frac{\partial^2 U}{\partial \phi^2}\right|_{\phi=\phi_0} = 1 + \beta_L \cos\phi_0 > 0.$$

It ceases to be a minimum when $\left.\partial^2 U/\partial\phi^2\right|_{\phi=\phi_0}$ vanishes by changing ϕ_{ext}. (Note that ϕ_0 is a function of ϕ_{ext}.) The inflection point of $U(\phi)$ satisfies $\left.\partial U/\partial\phi\right|_{\phi=\phi_{0c}} = \left.\partial^2 U/\partial\phi^2\right|_{\phi=\phi_{0c}} = 0$. These two conditions fix parameter ϕ_{0c} and the external flux $\phi_{\text{ext}}^{(0)}$ at which the inflection point appears. Straightforward calculation yields

$$\cos\phi_{0c} = -\frac{1}{\beta_L}, \quad \phi_{\text{ext}}^{(0)} = \phi_{0c} - \beta_L \sin\phi_{0c}. \tag{15.90}$$

A minimum appears as we change ϕ_{ext} above $\phi_{\text{ext}}^{(0)}$. We obtain, by expanding $U(\phi)$ around ϕ_{0c},

$$\frac{U(\phi)}{E_L} = \frac{1}{2}(\phi_{0c} - \phi_{\text{ext}})^2 + 1 + (\phi_{\text{ext}} - \phi_{\text{ext}}^{(0)})(\phi - \phi_{0c})$$
$$-\frac{1}{6}\beta_L \sin\phi_{0c}(\phi - \phi_{0c})^3. \tag{15.91}$$

Then the flux ϕ giving the shallow minimum satisfies

$$\phi_0 - \phi_{0c} = -\sqrt{\frac{2(\phi_{\text{ext}} - \phi_{\text{ext}}^{(0)})}{\beta_L \sin\phi_{0c}}},$$

when the external flux is ϕ_{ext}. The curvature at this point is

$$\sqrt{2(\phi_{\text{ext}} - \phi_{\text{ext}}^{(0)})\beta_L \sin\phi_{0c}}.$$

The difference between the shallow minimum and the absolute maximum energies is

$$\Delta U = \frac{4}{3}\sqrt{\frac{2(\phi_{\text{ext}} - \phi_{\text{ext}}^{(0)})^3}{\beta_L \sin\phi_{0c}}}. \tag{15.92}$$

Therefore an argument similar to the case of a washboard potential is applicable so that the number of bound states trapped in the minimum reduces as $\phi_{\text{ext}} \to \phi_{\text{ext}}^{(0)}$ as shown in Fig. 15.14. The qubit is made of the ground state $|0\rangle$ and the first excited state $|1\rangle$.

Gate operation of this qubit is realized by using the Rabi oscillation by applying microwave pulses to the bias current. [10]

15.7   Readout

There are several readout schemes proposed to date. We briefly summarize these schemes in this section.

15.7.1   Charge Qubit

Charge qubit measurement with the circuit depicted in Fig. 15.15 has been first employed by Nakamura *et al.* [11] to demonstrate the Rabi oscillation in CPB. A tunnel junction is directly connected to a CPB, and a small current I associated with quasiparticles flowing out through the tunnel junction has been measured.

To observe the Rabi oscillation, the gate voltage V_g is adjusted so that N_g is sufficiently smaller than 0.5, a typical value being 0.25. Then the ground

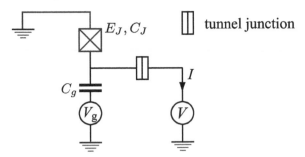

FIGURE 15.15
Direct observation of the charge qubit state by measurement of a current through a tunnel junction.

state is approximately the same as the state $|N = 0\rangle$, and the system is in this state with a large probability. Next a gate pulse with the pulse width Δt is applied so that N_g is now equal to 0.5 ($B_z = 0$) during this period. Finally the gate voltage is put back to its initial value. Suppose the risetime and falltime of pulses are much shorter than $\hbar/E_J$. Let us analyze the state evolution associated with this pulse, following Fig. 15.16. (i) The system is in the state $|N = 0\rangle = (|0\rangle - |1\rangle)/\sqrt{2}$ immediately after the pulse is turned on, and the Bloch vector starts to rotate around the x-axis for the period Δt due to the E_J-term (B_x), which implies that the sytem oscillates between $|N = 0\rangle$ and $|N = 1\rangle$ (**Rabi oscillation**). (ii) When the pulse is turned off, the state is a superposition of $|N = 0\rangle$ and $|N = 1\rangle$ whose coefficients are determined by Δt. The state $|N = 1\rangle$ has considerably higher energy than $|N = 0\rangle$ without the pulse, viz $N_g = 0.25$. Suppose the tunneling bias voltage V is adjusted so that the quasiparticle energy outside the Josephson junction sits between those of $|N = 0\rangle$ and $|N = 1\rangle$ states. Then the state $|N = 1\rangle$ decays into two quasiparticles and the electric current I, associated with these quasiparticles, is detected, while the $|N = 0\rangle$ state, in contrast, is stable against this decay. This is how the qubit states $|N = 0\rangle$ and $|N = 1\rangle$ are discriminated by measuring the tunneling current. Another advantage of this method is that if one waits for some duration of time required for the $|N = 1\rangle$ state to dacay into two quasiparticles after the pulse is turned off, the qubit is definitely in the initial state $|N = 0\rangle$. Therefore, by repeating the above process many times with a fixed Δt, the probability of the qubit in the state $|N = 1\rangle$ is found by tunneling current measurements.

Figure 15.17 shows the Rabi oscillation observed in this way. The phase decoherence time T_2 has been also measured by using the same readout method [12].

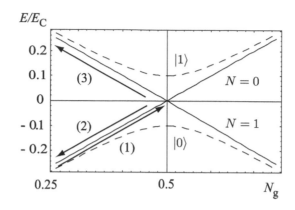

FIGURE 15.16

Change of a charge qubit state under readout pulses. (i) Gate voltage V_g is adjusted so that $N_g = 0.25$ initially. V_g is then moved to the degeneracy point $V_g = 0.5$ of the electrostatic energy in a time shorter than $\hbar/E_J$. As a result, the state changes along the solid line (1) and subsequently executes the Rabi oscillaton between the states $|N = 0\rangle$ and $|N = 1\rangle$. (ii) After the Rabi oscillation for a period Δt, the gate voltage is put back to its initial value $N_g = 0.25$ in a time shorter than $\hbar/E_J$. The state $|N = 0\rangle$ changes along the solid line (2) while $|N = 1\rangle$ changes along (3). Adapted by permission from Macmillan Publishers Ltd: Nature **398**, 786-788, copyright 1999.

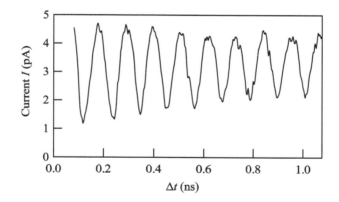

FIGURE 15.17

Rabi oscillations of the Cooper pair box state measured by pulse-induced current. Δt is the duration of the pulse, during which the system oscillates between $|N = 0\rangle$ and $|N = 1\rangle$. The current I is measured immediately after the pulse is switched off. Current measurement reveals the probability of the system being in $|N = 1\rangle$ when the pulse is turned off. Courtesy of NEC Nano Electronics Research Laboratories, Japan.

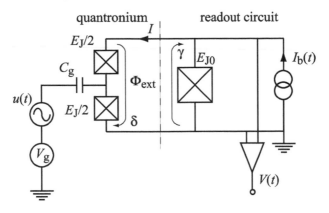

FIGURE 15.18

Measurement circuit of quantronium state. The left part prepares the system in a desired state of a quantronium in the central part. The rightmost part is a readout circuit. From D. Vion *et al.*, SCIENCE **296**: 886-889 (3 May 2002). Reprinted with permissions from AAAS.

15.7.2 Readout of Quantronium

The readout circuit for a quantronium is schematically shown in Fig. 15.18 [8]. Two identical Josephson junctions with Josephson energy $E_J/2$ and a capacitor with capacitance C_g comprise the split Cooper pair box, while the readout circuit is made of the Josephson junction with E_{J0} and current source $I_b(t)$.

Let Φ_{ext} be the external flux threading the circuit. Then the phase jumps ϕ_1 and ϕ_2 across the junctions satisify

$$\phi_1 - \phi_2 = \delta, \ \frac{\phi_1 + \phi_2}{2} = \phi, \ \delta - \gamma = 2\pi \frac{\Phi_{\text{ext}}}{\Phi_0}. \tag{15.93}$$

Current flowing through the split Cooper pair box is

$$I = \frac{I_c}{2} \sin \frac{\delta}{2} \cos \phi = I_0(e^{-\partial/\partial N} + e^{\partial/\partial N}) = I_0 \sigma_x, \ I_0 = \frac{I_c}{4} \sin \frac{\delta}{2}, \tag{15.94}$$

where

$$I_c = \frac{2\pi}{\Phi_0} \frac{E_J}{2}, \tag{15.95}$$

γ and δ are c-numbers and ϕ is treated as a q-number acting on the qubit state space. The current I satisfies

$$I_b(t) - I = I_{c0} \sin \gamma, \tag{15.96}$$

with

$$I_{c0} = \frac{2\pi}{\Phi_0} E_{J0}. \tag{15.97}$$

The Josephson energy of a split Cooper pair box is

$$-\frac{E_J}{2}(\cos\phi_1 + \cos\phi_2) = -E_J \cos\frac{\delta}{2}\cos\phi = -\frac{E_J}{2}\cos\frac{\delta}{2}\sigma_x. \qquad (15.98)$$

The contribution I is negligible in case $I_{c0} \gg I_c$, and the phase γ is approximately given by

$$\gamma = \arcsin\left[\frac{I_b(t)}{I_{c0}}\right]. \qquad (15.99)$$

Note that γ vanishes when $I_b(t) = 0$, namely when the qubit is under operation.

Let us recall our analysis, made in §15.5, that the optimal working point condition is $B_z = 0$, leading to $N_g = 1/2$, while the condition

$$\frac{\partial B_x}{\partial \delta}\frac{\partial \delta}{\partial \Phi_{\text{ext}}} = 0$$

is satisfied at $\delta = 0$ and hence $\Phi_{\text{ext}} = 0$. The eigenstates, when the condition $B_z = 0$ is satisfied, are

$$|0\rangle = \frac{1}{\sqrt{2}}(|N = 0\rangle + |N = 1\rangle), \ |1\rangle = \frac{1}{\sqrt{2}}(-|N = 0\rangle + |N = 1\rangle) \qquad (15.100)$$

as shown in Eq. (15.79). Experimentally observed current I through the qubit in Fig. 15.18 in the states $|0\rangle$ and $|1\rangle$ is

$$\langle 0|I|0\rangle = I_0, \ \langle 1|I|1\rangle = -I_0, \qquad (15.101)$$

respectively. It follows from Eq. (15.94) that the current vanishes when $\delta = 0$.

There is an experiment [8, 14] in which T_1 is directly measured at the optimal point from the relaxation of the state $|1\rangle$ to $|0\rangle$, where $|1\rangle$ is prepared by applying a radio frequency π-pulse $u(t)$ on the initial state $|0\rangle$. The qubit is in the state $c_0|0\rangle + c_1|1\rangle$ during the relaxation process. It is difficult to discriminate two states $|0\rangle$ and $|1\rangle$ since $\delta = 0$ implies the vanishing expectation values of the current I for both cases. The coefficients $|c_0|^2$ and $|c_1|^2$ are measured as follows. Suppose a pulse current $I_b(t)$ is switched on with the delay time t_d after a π-pulse is switched off. Then we find $\gamma \neq 0$ and accordingly $\delta \neq 0$ while the pulse is applied. Note that the conidition $N_g = 1/2$ is maintained even when $\delta \neq 0$. Therefore the Hamiltonian takes the same form as Eq. (15.98) and hence the eigenstates are still given by Eq. (15.100). The expectation values of the current I in the states $|0\rangle$ and $|1\rangle$ are non-vanishing because $\delta \neq 0$. The current through the junction E_{J0} depends on the Cooper pair box state so that it is given by $I_b(t) + I_0$ when the CPB state is $|1\rangle$, while it is $I_b(t) - I_0$ when $|0\rangle$. If the pulse current I_b is taken slightly smaller than the critical current I_{c0}, the current $I_b + I$ exceeds I_{c0} and the junction undergoes phase transition to a normal state resulting in the voltage $V(t)$ across the junction. Since the transition to the normal state is probabilistic, there is

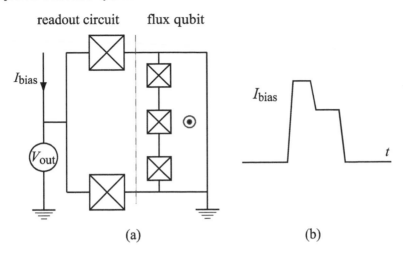

FIGURE 15.19
(a) A flux qubit with readout SQUIDs. (b) Bias current I_{bias} for readout. Transition to normal state manifests itself as finite V_{out}.

a small probability with which the voltage $V(t) \neq 0$ is observed even in the CPB state $|0\rangle$. Let p_0 and p_1 be the probabilities with which the junction undergoes a transition to the normal state when the CPB is in the states $|0\rangle$ and $|1\rangle$, respectively. A theoretical estimate shows that $p_1 - p_0 = 0.95$ while the observed value is ~ 0.6. However this discrepancy does not matter in the estimation ~ 1.8 μs of T_1. The phase decoherence time T_2 has also been measured by making use of the Ramsey fringe experiment [8] and spin-echo-type technique. The value they obtained is $T_2 \sim 0.50$ μs, corresponding to approximately 8,000 free precessions.

15.7.3 Switching Current Readout of Flux Qubits

Readout of a flux qubit is conducted in a similar manner as that of a quantronium [16, 17]. Figure 15.19 (a) shows a typical circuit for flux qubit readout, in which a readout dc SQUID with two junctions is connected to a three-junction flux qubit [17]. The circuit may support a bias current I_{bias}, and a voltmeter measures the voltage V_{out} while the bias current is applied.

Readout is carried out with the bias current I_{bias}. The current I_{bias} comprises a short pulse and a succedent trailing plateau as shown in Fig. 15.19 (b). The flux qubit is driven to normal state by a short pulse depending on its qubit state, and the trailing plateau is applied to prevent the qubit from reentering the superconducting state.

The Hamiltonian of a flux qubit has been given in Eq. (15.75) as

$$H = -\frac{1}{2}\left(B_z\sigma_z + B_x\sigma_x\right), \quad B_z = 2I_p\left(\Phi_{\text{ext}} - \frac{1}{2}\Phi_0\right), \tag{15.102}$$

whose eigenvalues are

$$E_{0,1} = \mp\sqrt{B_z^2 + B_x^2}. \tag{15.103}$$

There exists an optimal point, similar to that of a quantronium, for a flux qubit. Suppose Φ_{ext} fluctuates by $\delta\Phi$. The energy of the qubit then changes by $(\partial E_{0,1}/\partial\Phi_{\text{ext}})\delta\Phi$. Therefore the optimal point satisfies $\partial E_{0,1}/\partial\Phi_{\text{ext}} = 0$, which is solved to yield $\Phi_{\text{ext}} = \Phi_0/2$, viz $B_z = 0$.

Both $|0\rangle$ and $|1\rangle$ states are superpositions of clockwise current and anti-clockwise current with equal weights at the optimal point. Therefore the expectation value of the persistent current is extremely small near the optimal point, and it is difficult to measure the qubit state $c_0|0\rangle + c_1|1\rangle$ as it is. When I_{bias} is applied, however, the circulating current along the SQUID changes. Since the flux qubit is coupled with the SQUID through the mutual inductance, the flux threading the flux qubit also changes. The eigenstates of the Hamiltonian are given by Eq. (15.75), provided that the change in the flux is adiabatic. The coefficients c_0 and c_1 do not change if the time required for readout pulse sequence is sufficiently shorter than the energy relaxation time from $|1\rangle$ to $|0\rangle$.

Furthermore, if the qubit state is driven far away from the optimal point, the states $|0\rangle$ and $|1\rangle$ at this point essentially correspond to clockwise and counterclockwise persisitent currents, respectively. Finite persistent current through the qubit induces finite current through the dc SQUID due to mutual inductance. Therefore the bias current at which the SQUID makes transition to the finite voltage state (normal state) depends on whether the state to be measured is $|0\rangle$ or $|1\rangle$. The bias current works as a switching current. The transition of the SQUID to the normal state is probabilistic. The coefficients $|c_0|^2$ and $|c_1|^2$ are measured with a good precision if I_{bias} is tuned so that the difference in the transition probabilities associated with $|0\rangle$ and $|1\rangle$ are maximized.

Rabi oscillation, energy relaxation rate $\Gamma_1 = 1/T_1$ and dephasing rate $\Gamma_2 = 1/T_2$ from the spin echo measurement are observed in this way [17].

15.8 Coupled Qubits

Needless to say, we should be able to couple qubits for practical quantum information processing and quantum computation. Note that entanglement can be produced only through interaction between qubits. In this section, we analyze several types of coupling strategies: capacitively coupled charge

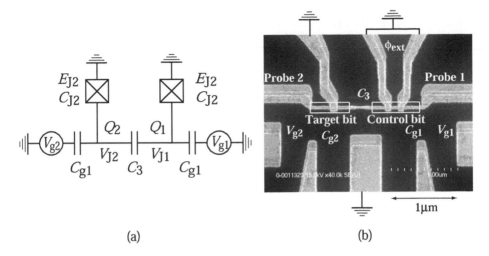

(a)   (b)

FIGURE 15.20

(a) Schematic picture of capacitively coupled two charge qubits. Readout circuits and gate pulse inputs are omitted. (b) Electron microscope image of actual capacitively coupled charge qubits. Reprinted by permission from Macmillan Publishers Ltd: Nature **425**, 941–944, copyright 2003.

qubits, inductively coupled charge qubits, tunable flux qubit coupling and flux qubit coupling with an LC resonator.

15.8.1 Capacitively Coupled Charge Qubits

Two charge qubits may be coupled through a capacitor as shown in Fig. 15.20 [18]. The electrostatic energy of the coupling capacitor is $(1/2)C_3V_3^2$, where V_3 is the voltage across the capacitor, $V_3 = (\hbar/2e)(\dot{\phi}_1 - \dot{\phi}_2)$. Now the total electrostatic energy is

$$K = \frac{1}{2}\left(\frac{\hbar}{2e}\right)^2 \sum_{i,j=1,2} C_{ij}\dot{\phi}_i\dot{\phi}_j + \frac{\hbar}{2e}\sum_{i=1,2} C_{gi}V_{gi}\dot{\phi}_i, \qquad (15.104)$$

with

$$C_{ii} = C_{Ji} + C_{gi} + C_3, \quad C_{12} = C_{21} = -C_3.$$

The variable π_i canonically conjugate to ϕ_i is

$$\pi_i = \left(\frac{\hbar}{2e}\right)^2 \sum_j C_{ij}\dot{\phi}_j + \frac{\hbar}{2e}C_{gi}V_{gi}. \qquad (15.105)$$

Now $\dot{\phi}_i$ is eliminated from the electrostatic energy in favor of π_i through the relation

$$\dot{\phi}_i = \left(\frac{2e}{\hbar}\right)^2 \sum_j (C^{-1})_{ij} \left(\pi_j - \frac{\hbar}{2e} C_{gj} V_{gj}\right). \tag{15.106}$$

By substituting this into Eq. (15.104), we obtain the electrostatic part of the Hamiltonian expressed in terms of the momenta,

$$\begin{aligned}
\sum_i \pi_i \dot{\phi}_i - K &= \frac{1}{2} \left(\frac{\hbar}{2e}\right)^2 \sum_{i,j=1,2} C_{ij} \dot{\phi}_i \dot{\phi}_j \\
&= \frac{1}{2} \left(\frac{2e}{\hbar}\right)^2 \sum_{i,j=1,2} C_{ij} (C^{-1})_{ii'} (C^{-1})_{jj'} \\
&\quad \times \left(\pi_{i'} - \frac{\hbar}{2e} C_{gi'} V_{gi'}\right) \left(\pi_{j'} - \frac{\hbar}{2e} C_{gj'} V_{gj'}\right) \\
&= \frac{1}{2} \left(\frac{2e}{\hbar}\right)^2 \sum_{i,j=1,2} (C^{-1})_{ij} \\
&\quad \times \left(\pi_i - \frac{\hbar}{2e} C_{gi} V_{gi}\right) \left(\pi_j - \frac{\hbar}{2e} C_{gj} V_{gj}\right). \tag{15.107}
\end{aligned}$$

The total Hamiltonian is obtained from Eq. (15.107) and the replacement $\pi_i = \hbar N_i$ as

$$\begin{aligned}
H &= \frac{1}{2} E_{C1}(N_1 - N_{g1})^2 + \frac{1}{2} E_{C2}(N_2 - N_{g2})^2 \\
&\quad + E_{C12}(N_1 - N_{g1})(N_2 - N_{g2}) - E_{J1} \cos\phi_1 - E_{J2} \cos\phi_2, \\
&= -\frac{1}{4} \left[(1 - 2N_{g1})E_{C1} + (1 - 2N_{g2})E_{C12}\right] (\sigma_z \otimes I) \\
&\quad - \frac{1}{4} \left[(1 - 2N_{g2}) E_{C2} + (1 - 2N_{g1})E_{C12}\right] (I \otimes \sigma_z) \\
&\quad + \frac{1}{4} E_{C12}\sigma_z \otimes \sigma_z - \frac{1}{2} E_{J1}(\sigma_x \otimes I) - \frac{1}{2} E_{J2}(I \otimes \sigma_x), \tag{15.108}
\end{aligned}$$

where

$$E_{C1} = (2e)^2 \left(C^{-1}\right)_{11}, \quad E_{C2} = (2e)^2 \left(C^{-1}\right)_{22}, \quad E_{C12} = (2e)^2 \left(C^{-1}\right)_{12}$$

$$N_{g1} = \frac{1}{2e} C_{g1} V_{g1}, \quad N_{g2} = \frac{1}{2e} C_{g2} V_{g2}.$$

The parameter E_{Ji} is the Josephson energy of the ith qubit.

The NEC group has implemented the CNOT gate by employing a pulse technique in two coupled charge qubits [18]. The gate voltages are initially adjusted as $N_{g1}(0) \simeq N_{g2}(0) \simeq 0.25$ so that the initial state, in the absence of pulses, is the lowest energy state $|N_1 = 0, N_2 = 0\rangle$. The state of qubit 1 depends on the state of qubit 2 in the presence of the coupling E_{C12}.

Suppose qubit 2 is in the state $|N_2 = 0\rangle$. States $|N_1 = 0\rangle$ and $|N_1 = 1\rangle$ of qubit 1 have degenerate electrostatic energy when $N_{g1} = N_{g1}^0 \equiv 1/2 - (E_{C12}/E_{C1})N_{g2}(0)$ for a given $N_{g2}(0)$. On the other hand, the electrostatic energy of the state $|N_1 = 0\rangle$ and $|N_1 = 1\rangle$ are degenerate at $N_{g1} = N_{g1}^1 \equiv 1/2 + (E_{C12}/E_{C1})(1 - N_{g2}(0))$ when qubit 2 is in the state $|N_2 = 1\rangle$. Let qubit 2 be in the state $|N_2 = 0\rangle$. Suppose qubit 1 is initially in the state $|N_1 = 0\rangle$ by adjusting the gate voltage so that $N_{g1}(0) = 0.25$. The gate voltage is subsequently changed by $\Delta V_{g1}^0 \equiv (C_{g1}/2e)(N_{g1}^0 - N_{g1}(0))$ so that the electrostatic energies of qubit 1 states $|N_1 = 0\rangle$ and $|N_1 = 1\rangle$ are degenerate. As a result, qubit 1 executes oscillations between the states $|N_1 = 0\rangle$ and $|N_1 = 1\rangle$ through the term $-E_{J1}\cos\phi_1$. The pulse width of the applied voltage may be controlled so that the resulting rotating angle around the x-axis is π. As a result, we realize the gate operation

$$|N_1 = 0\rangle|N_2 = 0\rangle \to |N_1 = 1\rangle|N_2 = 0\rangle. \tag{15.109}$$

Suppose qubit 2 is in the state $|N_2 = 1\rangle$ next. Then application of a voltage ΔV_{g1}^0 does not lead to degenerate electrostatic energy in qubit 1. In fact, the energy difference between the qubit 1 states $|N_1 = 0\rangle$ and $|N_1 = 1\rangle$ is E_{C12}. Therefore the above π-pulse introduces a rotation of the qubit 1 Bloch vector around an axis tilted from the z-axis by an angle $\alpha = \tan^{-1}(E_{J1}/E_{C12})$. Therefore this tilting of the Bloch vector is negligible provided that the condition $E_{C12} \gg E_{J1}$ is satisfied and the state remains close to $|N_1 = 0\rangle$ up to the global phase. Conversely, the qubit 1 state is transformed from $|N_1 = 0\rangle$ to $|N_1 = 1\rangle$ if a π-pulse with the gate voltage change $\Delta V_{g1}^1 \equiv (C_{g1}/2e)(N_{g1}^1 - N_{g1}(0))$ is applied when qubit 2 is in the state $|N_2 = 1\rangle$.

It is also true that the electrostatic energy of qubit 2 depends on whether qubit 1 is in the state $|N_1 = 0\rangle$ or $|N_1 = 1\rangle$. Therefore the gate voltage change to make the energies of $|N = 0\rangle$ and $|N = 1\rangle$ degenerate depends on N_1 in such a way as

$$\Delta V_{g2}^i = \left(\frac{C_{g2}}{2e}\right)\left(N_{g2}^i - N_{g2}(0)\right), \tag{15.110}$$

where $i = N_1 \in \{0, 1\}$ and

$$N_{g2}^0 = \frac{1}{2} - \frac{E_{C12}}{E_{C2}}N_{g1}(0), \quad N_{g2}^1 = \frac{1}{2} + \frac{E_{C12}}{E_{C2}}(1 - N_{g1}(0)).$$

The NEC group has experimentally generated entangled states

$$\alpha|N_1 = 0, N_2 = 1\rangle + \beta|N_1 = 1, N_2 = 0\rangle$$

and

$$\alpha|N_1 = 0, N_2 = 0\rangle + \beta|N_1 = 1, N_2 = 1\rangle$$

by applying pulses depicted in Fig. 15.21 [18].

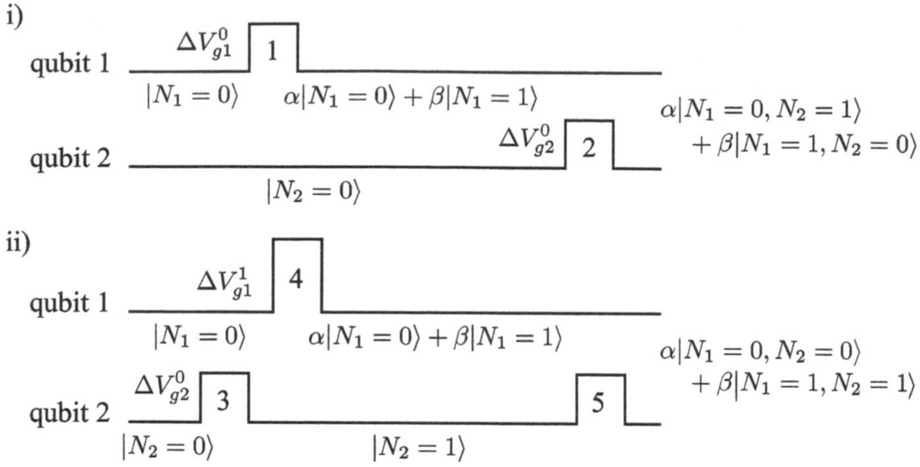

FIGURE 15.21

Pulse sequences to generate the states i) $\alpha|N_1 = 0, N_2 = 1\rangle + \beta|N_1 = 1, N_2 = 0\rangle$ and ii) $\alpha|N_1 = 0, N_2 = 0\rangle + \beta|N_1 = 1, N_2 = 1\rangle$. Pulses 2, 3 and 5 are π-pulses. Pulses 1 and 4 transform $|N_i = 0\rangle$ of the ith qubit to $\alpha|N_i = 0\rangle + \beta|N_i = 1\rangle$. Adapted by permission from Macmillan Publishers Ltd: Nature **425**, 941-944, copyright 2003.

15.8.2   Inductive Coupling of Charge Qubits

Figure 15.22 shows a scalable design of coupling n charge qubits [19]. The electrostatic energy of the ith qubit is

$$K_i = \frac{1}{2}C_J V_{Ji}^2 + \frac{1}{2}C_g(V_{Ji} - V_L - V_{gi})^2, \qquad (15.111)$$

where

$$V_{Ji} = -\frac{\hbar}{2e}\dot{\phi}_i, \ V_L = -\dot{\Phi} = -\frac{\hbar}{2e}\dot{\phi},$$

and it is assumed that all the gate capacitors have the same capacitance C_g and all the junctions have the same capacitance C_J for simplicity. The parameter ϕ has been introduced as a scaled magnetic flux threading the common inductance L. Now K_i is rewritten in terms of ϕ and ϕ_i as

$$K_i = \frac{1}{2}\left(\frac{\hbar}{2e}\right)^2\left[C_J\dot{\phi}_i^2 + C_g\left(\dot{\phi}_i - \dot{\phi} + \frac{2e}{\hbar}V_{gi}\right)^2\right]. \qquad (15.112)$$

Let us make the change of variables

$$\phi_i \rightarrow \phi_i + \frac{C_g}{C_g + C_J}\phi$$

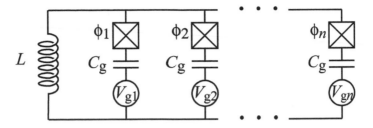

FIGURE 15.22
Inductively coupled n charge qubits. It is assumed that all the gate capacitances are C_g for simplicity.

to get rid of a coupling term $\propto \dot{\phi}_i \dot{\phi}$ in Eq. (15.112). Then K_i takes a decoupled form

$$
K_i = \frac{1}{2}\left(\frac{\hbar}{2e}\right)^2 \left[C_J \left(\dot{\phi}_i + \frac{C_g}{C_g + C_J}\dot{\phi} \right)^2 + C_g \left(\dot{\phi}_i - \frac{C_J}{C_g + C_J}\dot{\phi} + \frac{2e}{\hbar}V_{gi} \right)^2 \right]
$$

$$
= \frac{1}{2}\left(\frac{\hbar}{2e}\right)^2 \left[(C_J + C_g)\dot{\phi}_i^2 + \frac{C_J C_g}{C_g + C_J}\dot{\phi}^2 + 2\frac{2e}{\hbar}C_g V_{gi}\left(\dot{\phi}_i - \frac{C_J}{C_g + C_J}\dot{\phi} \right) \right],
$$
(15.113)

where we have dropped a physically irrelevant term proportioanl to V_{gi}^2. The variables canonically conjugate to ϕ_i and ϕ are

$$
\pi_i = \left(\frac{\hbar}{2e}\right)^2 (C_J + C_g)\dot{\phi}_i + \frac{\hbar}{2e}C_g V_{gi}
$$
(15.114)

$$
\pi = \left(\frac{\hbar}{2e}\right)^2 n C_t \dot{\phi} - \frac{\hbar}{2e}C_t \sum_i V_{gi},
$$
(15.115)

respectively, where n is the number of charge qubits and

$$
\frac{1}{C_t} \equiv \frac{1}{C_g} + \frac{1}{C_J}.
$$

It is convenient, as before, to introduce operators N_i and N by $\pi_i = \hbar N_i$ and $\pi = \hbar N$, where

$$
N_i = \frac{\hbar}{(2e)^2}(C_J + C_g)\dot{\phi}_i + \frac{C_g V_{gi}}{2e},
$$
(15.116)

$$
N = \frac{\hbar}{(2e)^2}n C_t \dot{\phi} - \frac{C_t}{2e}\sum_i V_{gi}.
$$
(15.117)

The Hamiltonian is then written as

$$
H = \pi\dot{\phi} + \sum_i \pi_i \dot{\phi}_i - K + U
$$

$$= \frac{1}{2}(C_J + C_g)\left(\frac{\hbar}{2e}\right)^2 \dot{\phi}_i^2 + \frac{1}{2}nC_t \left(\frac{\hbar}{2e}\right)^2 \dot{\phi}^2 + U$$

$$= \frac{1}{2}\frac{(2e)^2}{C_J + C_g}\sum_i (N_i - N_{gi})^2 + \frac{1}{2}\frac{(2e)^2}{nC_t}N^2 + U, \qquad (15.118)$$

where U is the potential energy

$$U = \frac{1}{2L}\left(\frac{\hbar}{2e}\right)^2 \phi^2 - \sum_i E_{Ji} \cos\left(\phi_i + \frac{C_g}{C_g + C_J}\phi\right), \qquad (15.119)$$

and we have made the change of a variable

$$N + \frac{C_t}{2e}\sum_i V_{gi} \to N.$$

Let us consider the dynamics of the variable ϕ. The relevant parts of the Hamiltonian are made into the form of a harmonic oscillator by making replacement $N \to -i\partial/\partial\phi$ as

$$H_\phi = -\frac{1}{2}\frac{(2e)^2}{nC_t}\frac{\partial^2}{\partial\phi^2} + \frac{1}{2L}\left(\frac{\hbar}{2e}\right)^2 \phi^2. \qquad (15.120)$$

Oscillation associated with this mode is called the **plasma oscillation**, whose frequency is

$$\omega_p = \sqrt{\frac{\frac{1}{L}\left(\frac{\hbar}{2e}\right)^2}{nC_t\left(\frac{\hbar}{2e}\right)^2}} = \frac{1}{\sqrt{nC_t L}}. \qquad (15.121)$$

We consider a low temperature regime where $kT \ll \hbar\omega_p$ is satisfied. Then we may assume that the harmonic oscillator is in the ground state with a good accuracy.

Let us consider the coupling between ϕ and ϕ_i next. The potential energy (the Josephson energy) of the ith qubit in Eq. (15.119) is

$$-E_{Ji} \cos\left(\phi_i + \frac{C_g}{C_g + C_J}\phi\right), \qquad (15.122)$$

where E_{Ji} is controllable. If a condition

$$\frac{1}{L}\left(\frac{\hbar}{2e}\right)^2 = \frac{1}{L}\left(\frac{\Phi_0}{2\pi}\right)^2 \gg \frac{(2e)^2}{nC_t} \qquad (15.123)$$

is satified in (15.120), the potential energy dominates over the kinetic energy and we obtain $\langle\phi^2\rangle \ll 1$ and hence

$$\frac{C_g}{C_g + C_J}\sqrt{\langle\phi^2\rangle} \ll 1. \qquad (15.124)$$

The coupling between the inductance and the ith qubit is

$$-E_{Ji} \cos\left(\phi_i + \frac{C_g}{C_g + C_J}\phi\right)$$
$$= -E_{Ji}\left[\cos\phi_i - \left(\frac{C_g}{C_g + C_J}\phi\right)\sin\phi_i\right]$$
$$\rightarrow \lambda_i \phi \sin\phi_i, \tag{15.125}$$

where

$$\lambda_i = \frac{C_g}{C_g + C_J}E_{Ji},$$

and a term $\propto \cos\phi_i$ has been ignored since it does not contribute to ϕ-ϕ_i interaction. The coupling (15.125) between the plasma oscillation and the ith qubit induces the coupling between qubits.

For charge qubits, in particular, it holds that $E_C \gg E_{Ji}$, where

$$E_C = \frac{(2e)^2}{C_J + C_g}.$$

We take $C_J \gg C_g$ to reduce decoherence due to gate voltage noise, in which case we have $C_t \simeq C_g$ and

$$\frac{(2e)^2}{nC_t} \simeq \frac{(2e)^2}{nC_g} \gg E_C.$$

The number of qubits n cannot be too large for the condition

$$\frac{C_J}{C_g} \gg n \tag{15.126}$$

to be satisifed. These analyses require that the condition

$$\hbar\omega_p \gg E_C \gg E_{Ji} \gg \lambda_i \tag{15.127}$$

be satisfied.

It follows from Eq. (15.127) that the coupling between two qubits is described by the second order perturbation with a good precision, in which the first excited state of the harmonic oscillator is the only intermediate state. In fact, let $c^\dagger$ and c be the creation and annihilation operators of the harmonic oscillator, respectively, and rewrite ϕ in terms of them as

$$\phi = \sqrt{\frac{\hbar\omega_p}{2E_L}}(c + c^\dagger), \tag{15.128}$$

where

$$E_L = \frac{1}{L}\left(\frac{\Phi_0}{2\pi}\right)^2$$

is the characteristic energy of the inductance L. Second order perturbation then produces the interaction term

$$-\left(\sum_i \lambda_i \sin \phi_i\right)^2 \frac{\langle 0|\phi|1\rangle\langle 1|\phi|0\rangle}{\hbar\omega_p} = -\frac{1}{2E_L}\left(\sum_i \lambda_i \sin \phi_i\right)^2. \qquad (15.129)$$

By the replacement

$$\sin \phi_i = \frac{e^{i\phi} - e^{-i\phi}}{2i} = \frac{e^{-\partial/\partial N} - e^{\partial/\partial N}}{2i} \to \sigma_{iy}$$

with

$$\sigma_{iy} = I \otimes I \otimes \ldots \otimes \underbrace{\sigma_y}_{i\text{th position}} \otimes \ldots \otimes I \otimes I,$$

we obtain the coupling between the ith and the jth qubits,

$$H_{ij}^{\text{int}} = -\frac{1}{E_L}\lambda_i\lambda_j\sigma_{iy} \otimes \sigma_{jy}. \qquad (15.130)$$

It is important to note that the interaction may be selectively turned on between the ith qubit and the jth qubit by setting all $\lambda_k = 0$ by tuning $E_{Jk} = 0$ except for $k = i, j$.

15.8.3   Tunable Coupling between Flux Qubits

Tunable coupling between two flux qubits has been introduced by the NEC group by making use of three coupled flux qubits [20, 21] as shown in Fig. 15.23. Each of their flux qubits has four Josephoson junctions, whose Hamiltonian takes exactly the same form as the three-junction flux qubit Hamiltonian

$$H = -\frac{1}{2}\left(B_x\sigma_x + B_z\sigma_z\right). \qquad (15.131)$$

Here

$$B_z = 2I_p\left(\Phi_{\text{ext}} - \frac{1}{2}\Phi_0\right) \qquad (15.132)$$

is the energy difference between the states $|r\rangle$ and $|l\rangle$, where persistent current with magnitude I_p circulates the loop clockwise in $|r\rangle$ and counterclockwise in $|l\rangle$. We fix the external flux $\Phi_{\text{ext}} \geq \Phi_0/2$ for definiteness in the following and assign the notation $|r\rangle = |\sigma_z = 1\rangle$ and $|l\rangle = |\sigma_z = -1\rangle$. The parameter B_x is fixed by the tunneling energy between two potential wells as before.

The Hamiltonian (15.131) has energy eigenvalues

$$E_0 = -\frac{\omega}{2}, \; E_1 = \frac{\omega}{2}, \; \omega \equiv \sqrt{B_x^2 + B_z^2} \qquad (15.133)$$

Qubit 1   Qubit 3   Qubit 2

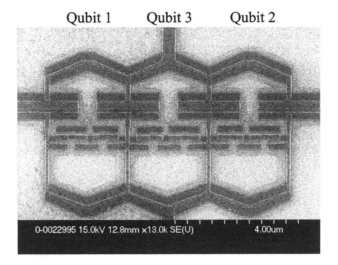

FIGURE 15.23
Tunable coupling of flux qubits 1 and 2 through an auxiliary flux qubit 3.
Courtesy of NEC Nano Electronics Research Laboratories, Japan.

with the corresponding eigenvectors (cf. Eq. (15.44))

$$|0\rangle = \cos\frac{\alpha}{2}|r\rangle + \sin\frac{\alpha}{2}|l\rangle$$
$$|1\rangle = -\sin\frac{\alpha}{2}|r\rangle + \cos\frac{\alpha}{2}|l\rangle \tag{15.134}$$
$$\text{where} \quad \cos\alpha = \frac{B_z}{\omega}, \ \sin\alpha = \frac{B_x}{\omega}.$$

Qubit 1 and qubit 2 in Fig. 15.23 are set in the optimal points $B_z = 0$ to achieve long coherence time. Coupling between qubits 1 and 2 is tunable by controlling qubit 3 between them. The controllability of the coupling can be shown by calculating the effective coupling between qubits 1 and 2 mediated by qubit 3, with the conditions $B_{1z} = B_{2z} = 0$ and $\omega_3 \gg \omega_1 = B_{1x}, \omega_2 = B_{2x}$ satisfied. Three qubits in Fig. 15.23 are coupled with each other through mutual inductances. The interaction energy is given by

$$\sum_{j,j=1,i>j}^{3} L_{ij}I_iI_j, \tag{15.135}$$

with L_{ij} being the mutual inductance of qubits i and j, and therefore the Hamiltonian of three coupled flux qubits is

$$H = -\frac{1}{2}\sum_{i=1}^{3}(B_{iz}\sigma_{iz} + B_{ix}\sigma_{ix}) + \sum_{i>j}J_{ij}\sigma_{iz} \otimes \sigma_{jz}, \tag{15.136}$$

where $J_{ij} = L_{ij} I_{pi} I_{pj}$, I_{pi} being the amplitude of the persistent current flowing around the ith qubit, and $\sigma_{2x} = I \otimes \sigma_x \otimes I$, for example. Note that σ_{iz} specifies the sense of the current flowing around the ith qubit.

The Hamiltonian (15.136) is written with the basis $\{|r\rangle, |l\rangle\}$. It turns out to be convenient to switch the basis from $\{|r\rangle, |l\rangle\}$ to $\{|0\rangle, |1\rangle\}$. We find from the identity

$$|r\rangle\langle r| - |l\rangle\langle l| = \cos\alpha(|0\rangle\langle 0| - |1\rangle\langle 1|) - \sin\alpha(|1\rangle\langle 0| + |0\rangle\langle 1|)$$

that the Hamiltonian (15.136) is mapped to an operator with respect to the basis $\{|0\rangle, |1\rangle\}$ as

$$H = -\sum_{i=1}^{3} \omega_i(\sigma_{iz} - I) + J_{12}\sigma_{1x}\sigma_{2x}$$

$$- (J_{13}\sigma_{1x} + J_{23}\sigma_{2x})\left(\frac{B_{3z}}{\omega_3}\sigma_{3z} - \frac{B_{3x}}{\omega_3}\sigma_{3x}\right) \qquad (15.137)$$

under the basis change, where the energy of the state $|0\rangle$ is set to zero and we noted that qubits 1 and 2 are working at the optimal points at which $\alpha = \pi/2$.

Let us write a general state as $\sum_{i,j,k \in \{0,1\}} C_{ijk} |i\rangle_1 |j\rangle_2 |k\rangle_3$ and consider the Schrödinger equation for C_{ijk}. We find

$$i\frac{\partial C_{000}}{\partial t} = J_{12}C_{110} + \frac{B_{3x}}{\omega_3}(J_{13}C_{101} + J_{23}C_{011}), \qquad (15.138)$$

where contributions from $\sigma_{1x}\sigma_{3z}$ and $\sigma_{2x}\sigma_{3z}$ are ignored. This is justified by the facts that (1) we may replace $\sigma_{1x}\sigma_{3z}$ and $\sigma_{2x}\sigma_{3z}$ by σ_{1x} and σ_{2x}, respectively, in the equation of motion for C_{ij0} since qubit 3 is in the state $|0\rangle$, which is the eigenstate σ_{3z} with eigenvalue 1, and (2) the coupling $\sigma_{1x}\sigma_{3z}$ and $\sigma_{2x}\sigma_{3z}$ are negligible in the equaiton of motion for C_{ij1} since $\omega_3 \gg J_{ij}$. Next, we obtain the Schrödinger equations

$$i\frac{\partial C_{101}}{\partial t} = (\omega_1 + \omega_3)C_{101} + J_{12}C_{011} + \frac{B_{3x}}{\omega_3}(J_{13}C_{000} + J_{23}C_{110})$$

$$(15.139)$$

$$i\frac{\partial C_{011}}{\partial t} = (\omega_2 + \omega_3)C_{011} + J_{12}C_{101} + \frac{B_{3x}}{\omega_3}(J_{13}C_{110} + J_{23}C_{000}).$$

If the conditions

$$\omega_3 \gg \frac{1}{C_{101}}\frac{\partial C_{101}}{\partial t}, \frac{1}{C_{011}}\frac{\partial C_{011}}{\partial t}, \omega_1, \omega_2$$

are taken into account in the above equations, the solutions are obtained as

$$C_{101} = -\frac{1}{\omega_3}\left[J_{12}C_{011} + \frac{B_{3x}}{\omega_3}(J_{13}C_{000} + J_{23}C_{110})\right]$$

$$(15.140)$$

$$C_{011} = -\frac{1}{\omega_3}\left[J_{12}C_{101} + \frac{B_{3x}}{\omega_3}(J_{13}C_{110} + J_{23}C_{000})\right].$$

Substituting Eq. (15.140) into Eq. (15.138), the effective coupling J_{12}^{eff} is found from the coefficient of C_{110}. We obtain

$$J_{12}^{\text{eff}} = J_{12} - \frac{2}{\omega_3} \left(\frac{B_{3x}}{\omega_3} \right)^2 J_{13} J_{23}. \tag{15.141}$$

It can be shown from the study of the effective interactions in the equations of motion for C_{100}, C_{010} and C_{110} that J_{12}^{eff} takes the same form (15.141) for any C_{ij0}. The coupling J_{12}^{eff} is tunable by controlling ω_3 through $\Phi_{3\text{ext}}$. It is even possible to adjust $\Phi_{3\text{ext}}$ so that J_{12}^{eff} vanishes when the coupling is not required.

Two-qubit operation is introduced by microwave radiation. Let ω be the microwave frequency and let $\delta\Phi_{3\text{ext}} e^{i\omega t}$ be the change in the magnetic flux through qubit 3 due to the microwave. Then it induces the interaction

$$\left(\frac{\partial J_{12}^{\text{eff}}}{\partial \omega_3} \right) \left(\frac{\partial \omega_3}{\partial \Phi_{3\text{ext}}} \right) \delta\Phi_{3\text{ext}} e^{i\omega t} \tag{15.142}$$

between qubits 1 and 2. Therefore, it becomes possible to have operations $|00\rangle \leftrightarrow |11\rangle$ and $|10\rangle \leftrightarrow |01\rangle$ by setting $\omega = |B_{1x} \pm B_{2x}|$.

15.8.4 Coupling Flux Qubits with an LC Resonator

Two flux qubits interacting with a common LC resonator may be coupled and entangled by first entangling one of the qubits with the LC resonator and then entangling the resonator with the other qubit. The LC resonator may couple far remote flux qubits and is able to entangle them. Therefore it works as a qubit bus interacting with all the flux qubits. The LC resonator mode should be compared with the harmonic oscillator modes in a trapped ion quantum computer, which couple with all the ionic qubits.

We first introduce entanglement of a flux qubit with an LC resonator and then outline the interaction of two flux qubits via the LC resonator.

Coupling between a flux qubit and an LC resonator has been demonstrated with a circuit shown in Fig. 15.24 [22]. There is a three-junction flux qubit in the middle of Fig. 15.24 (b), which is surrounded by an LC circuit. The flux qubit and the LC circuit are coupled through the mutual inductance. The qubit is also surrounded by a readout SQUID. The microwave to manipulate the qubit is supplied by the MW line shown in Fig. 15.24.

The LC circuit comprises two capacitors with capacitance C and two inductors with inductance L, all in series. The energy (Hamiltonian) of the circuit is

$$H_{\text{plasmon}} = LI^2 + \frac{1}{C}Q^2, \quad I = \dot{Q}. \tag{15.143}$$

The resonance frequency (plasmon frequency) of this circuit is $\omega_p = 1/\sqrt{LC}$.

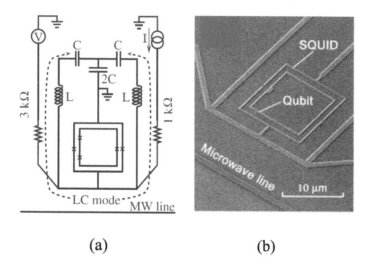

(a)   (b)

FIGURE 15.24

Coupled flux qubit and an LC resonator. (a) Schematic diagram of a flux qubit
and an LC resonator, which are coupled through the mutual inductance. The
flux qubit is surrounded by a readout SQUID. The flux qubit is coherently
controlled by the microwave line. The LC mode is plotted in a dotted curve.
Reprinted figure with permission from J. Johansson *et al.*, Physical Review
Letters **96**, 127006 (2006), Copyright (2006) by the American Physical Society.
(b) Electron microgram of a flux qubit, a microwave (MW) line and a readout
SQUID. Reprinted figure with permission from S. Saito *et al.*, Physical Review
Letters **96**, 107001 (2006), Copyright (2006) by the American Physical Society.

Let c ($c^\dagger$) be the annihilation (creation) operator of the plasmon. Then the Hamiltonian is written as

$$H_{\text{plasmon}} = \hbar \omega_p c^\dagger c. \tag{15.144}$$

The charge and the current operators are

$$Q = \frac{1}{2}\sqrt{\hbar \omega_p C}(c^\dagger + c), \quad I = \dot{Q} = \frac{i}{2}\sqrt{\frac{\hbar \omega_p}{L}}(c^\dagger - c), \tag{15.145}$$

respectively.

The flux qubit Hamiltonian, on the other hand, is given as Eq. (15.75);

$$H_{\text{qubit}} = -\frac{1}{2}(B_x \sigma_x + B_z \sigma_z), \quad B_z = 2I_p\left(\Phi_{\text{ext}} - \frac{\Phi_0}{2}\right), \tag{15.146}$$

using Eq. (15.145). The eigenstates $|0\rangle$ and $|1\rangle$ are given by Eq. (15.134), whose energies are $\mp\omega/2$ with $\omega = \sqrt{B_x^2 + B_z^2}$. They support a persistent current $\pm I_p$. Let M be the mutual inductance between the flux qubit and the LC circuit. Then the interation Hamiltonian between them is expressed as a current-current interaction, taking the form

$$H_i = i\hbar\lambda(c^\dagger - c), \quad \hbar\lambda = \frac{MI_p}{2}\sqrt{\frac{\hbar \omega_p}{L}}. \tag{15.147}$$

The oscillation between a state with ground state qubit $|0\rangle$ with plasmon number $n = 1$ ($|0\rangle|n = 1\rangle$) and a state with excited state qubit $|1\rangle$ with plasmon number $n = 0$ ($|1\rangle|n = 0\rangle$) is called the **vacuum Rabi oscillation**. The vacuum Rabi oscillation experiment in the coupled flux qubit-LC resonator system has been conducted to demonstrate the coupling between them [22]. The experiment was carried out at low temperature satisfying $k_B T \ll \hbar\omega, \hbar\omega_p$, under which plasmon excitation with $n \geq 2$ may be ignored.

The energy gap $\hbar\omega$ changes while the plasmon frequency remains unchanged as Φ_{ext} changes. Therefore it is possible to adjust Φ_{ext} so that the states $|0\rangle|n = 1\rangle$ and $|1\rangle|n = 0\rangle$ have the degenerate energies. This is done by properly choosing ω and ω_p so that $-\omega/2 + \omega_p = +\omega/2$.

Actual experimental steps are depicted in Fig. 15.25. The state $|0\rangle|n = 0\rangle$ denoted as 1 is prepared to begin with. Then a microwave π-pulse is applied to convert the state to the state $|1\rangle|n = 0\rangle$ marked as 2. Then a shift pulse with the width t_p, corresponding to Φ_{ext}, is applied to drive the system to the degenerate point of $|0\rangle|n = 1\rangle$ and $|1\rangle|n = 0\rangle$ denoted as 3 so that the Rabi oscillation takes place during the period t_p. The rise time τ_r of this pulse is chosen adiabatic compared to the characterstic times of the qubit and the LC circuit ($\tau_r \gg 2\pi/\omega, 2\pi/\omega_p$) while, at the same time, nonadiabatic compared to the coupling strength between the qubit and the LC circuit ($\tau_r \ll 2\pi/\lambda$). As a result, the Rabi oscillation with the initial condition $|1\rangle|n = 0\rangle$ will be observed.

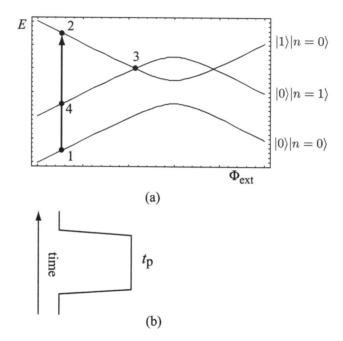

(a)

(b)

FIGURE 15.25

(a) Schematic diagram of pulse operations. The initial state 1 makes transition to state 2 by a microwave π-pulse in resonance. Subsequently a pulsed magnetic field (b) is applied to send state 2 to state 3, where $|0\rangle|n=1\rangle$ and $|1\rangle|n=0\rangle$ have degenerate energy, leading to the Rabi oscillation between the two states. Readout is made by turning off the pulse (b) and putting the state back from 3 to 2 and 4. Reprinted figure with permission from J. Johansson *et al.*, Physical Review Letters **96**, 127006 (2006), Copyright (2006) by the American Physical Society.

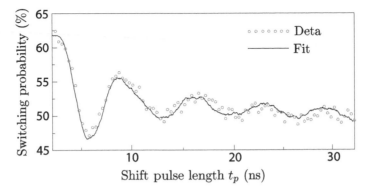

FIGURE 15.26

Switching probability, which reveals vacuum Rabi oscillation, as a function of the pulse length t_p [ns]. Reprinted figure with permission from J. Johansson *et al.*, Physical Review Letters **96**, 127006 (2006), Copyright (2006) by the American Physical Society.

The Hamiltonian at the degeneracy point 3 is H_i, and it follows from Eq. (15.134) that the relevant matrix element is

$$(\langle 0|\langle n = 1|)\; H_i\; (|1\rangle|n = 0\rangle) = i2\hbar\lambda \sin\frac{\alpha}{2}\cos\frac{\alpha}{2}. \tag{15.148}$$

The eigenstates of the Hamiltonian are $(|1\rangle|n = 0\rangle \pm i|0\rangle|n = 1\rangle)/\sqrt{2}$, and the corresponding eigen energies are $\pm\hbar\lambda\sin\alpha$. Therefore the Rabi oscillation frequency is given by $\Omega_R = 2\lambda\sin\alpha$. Finally the shift pulse is turned off so that Φ_{ext} assumes its initial value at $t = 0$. The resulting states are $|0\rangle|n = 1\rangle$ and $|1\rangle|n = 0\rangle$ denoted as 4 and 2, respectively, in Fig. 15.25. Then a readout process is applied to the state. Figure 15.26 shows the measurement result.

The CNOT gate in trapped ions has been implemented by introducing effective coupling between distant ions through mediating phonon mode. It has been pointed out recently that a similar mechanism might work for flux qubits if the phonon mode in trapped ions is replaced by the plasmon mode in the LC resonator. The coupling between the qubit and the resonator is made possible through the mutual inductance.

It is shown in a recent proposal by the NTT group [23] that a flux qubit-flux qubit coupling may be introduced by coupling them selectively to an LC resonator whose resonace frequency is controllable by adjusting the bias current through the Josephson junction in the LC circuit. An explicit protocol to entangle arbitrary two flux qubits among many qubits is proposed, in which a resonating circuit surrounding these qubits is made use of. The theoretical proposal of this scenerio has been already reported, and experiments demonstrating this qubit-qubit coupling are planned [23].

Recently, entanglement of two current-biased qubits via a resonant cavity is demonstrated at NIST [24]. They reported that two qubits separated by ~ 1.1 mm have been successfully entangled.

15.9 DiVincenzo Criteria

1. A scalable physical system with well-characterized qubits:

   Superconducting qubits are largely divided into two classes. One makes use of the charge degrees of freedom as a qubit variable while the other uses the flux degrees of freedom to construct a qubit. They are expected to be scalable up to a larger register within the current fabrication technology, but currently the largest register has merely a few qubits.

2. The ability to initialize the state of the qubits to a simple fiducial state, such as $|00\ldots0\rangle$:

   Superconducting qubits work at very low temperature and they are in the ground state with a good probability as one waits for a long enough time. Alternatively, one measures the qubit state and flips it to $|0\rangle$ if it is found in $|1\rangle$.

3. Long decoherence times, much longer than the gate operation time:

   It is expected that superconducting qubits are strong against decoherence due to the absence of dissipation. Quantronium works at the optimal point so that the coherence time reaches as long as 1 μs. A typical gate operation time for a charge qubit is ~ 0.1 ns, which means $\sim 10^4$ gate operations are possible. This number is yet small for practical applications.

4. "Univeral" set of quantum gates:

   Single-qubit gates are already confirmed in superconducting qubits. They use either Rabi oscillation or nonadiabatic switching. Coupling between two qubits is introduced capacitively, inductively or by introducing an intermediate qubit. Conditional gate operations, such as a CNOT gate, are demonstrated [18]. Tunable coupling between two qubits is recently demonstrated [21]. Coupling separated superconducting qubits via a resonant cavity has been demonstrated recently [24].

5. A qubit-specific measurement capability:

   There are several readout schemes proposed to date. They all make use of charge or flux degrees of freedom. Not all readout schemes have good efficiency (visibility).

References

[1] Y. Makhlin, G. Schön and A. Shnirman, Rev. Mod. Phys. **73**, 357 (2001).

[2] G. Wendin and V. S. Shumeiko, eprint cond-mat/0508729 (2005).

[3] B. D. Josephson; Phys. Lett. **1**, 251 (1962), Rev. Mod. Phys. **46**, 251 (1974).

[4] M. Tinkham, *Introduction to Superconductivity* (2nd ed.), Mcgraw-Hill College (1995).

[5] A. O. Caldeira and A. J. Leggett, Phys. Rev. Lett. **46**, 221 (1981).

[6] A. O. Caldeira and A. J. Leggett, Ann. Phys. (N. Y.) **149**, 374 (1983).

[7] J. E. Mooij, *et al.*, Science **285**, 1036 (1999).

[8] D. Vion *et al.*, Science **296**, 886 (2002).

[9] J. M. Martinis, S. Nam and J. Aumentado, Phys. Rev. Lett. **89**, 117901 (2002).

[10] R. W. Simmonds *et al.*, Phys. Rev. Lett. **93**, 077003 (2004).

[11] Y. Nakamura, Yu. A. Pashkin and J. S. Tsai, Nature **398**, 786 (1999).

[12] Y. Nakamura, *et al.*, Phys. Rev. Lett. **88**, 047901 (2002).

[13] D. Vion in *Quantum Entanglementand Information Processing: Les Houches2003*, ed. D. Estève, J.-M. Raimond and J. Dalibard, Elsevier, Amsterdam (2004).

[14] A. Cottet, Ph. D. Thesis (2002). Available at http://www.lps.u-psud.fr/Utilisateurs/cottet/ACottetThesis.pdf

[15] T. Duty *et al.*, Phys. Rev. B **69**, 140503 R (2004).

[16] C H. van der Wal *et al.*, Science **290**, 773 (2000).

[17] L. Chiorescu *et al.*, Science **299**, 1869 (2003).

[18] T. Yamamoto *et al.*, Nature **425**, 941 (2003).

[19] Y. Makhlin, G. Schön and A. Shnirman, Nature **398**, 305 (1999).

[20] A. O. Niskanen, Y. Nakamura and J.-S. Tsai, Phys. Rev. B **73**, 094506 (2006).

[21] A. O. Niskanen, *et al.*, Science **316**, 723 (2007).

[22] J. Johansson *et al.*, Phys. Rev. Lett. **96**, 127006 (2006).

[23] N. Nakano *et al.*, Appl. Phys. Lett. **91**, 032501 (2007).

[24] M. A. Sillanpää, J. I. Park and R. W. Simmonds, Nature **449**, 438 (2007).

16

Quantum Computing with Quantum Dots

16.1 Introduction

It has become possible, with the progress of the semiconductor fabrication technology, to confine an electron in an artificial nano-scale potential well embedded in a semiconductor. This structure is called a **quantum dot** (QD).

There are two types of qubits making use of an electron trapped in a QD. One uses two adjacent QDs, called a **double quantum dot** (DQD), in which the qubit states corresponds to in which QD the electron resides. For example, a state in which the electron resides in the right QD may be called $|0\rangle$, while the electron occupies the left QD in $|1\rangle$. In the other qubit realization, an electron always occupies a QD and the two spin states $|\uparrow\rangle$ and $|\downarrow\rangle$ are qubit basis vectors. The former is called a **charge quantum dot** and the latter a **spin quantum dot**. We consider these two types of qubits separately.

16.2 Mesoscopic Semiconductors

A quantum dot is an artificially fabricated semiconductor structure of submicron size ($\sim 10^{-5}$ cm), which contains $10^3 \sim 10^9$ atoms and approximately the same number of electrons. Among many types of quantum dots, layered semiconductors which support two-dimensional electron gas in the inversion layer are mainly employed for quantum computing devices, in which case the number of electrons in a QD may be reduced down to $\simeq 10^0 \sim 10^2$.

16.2.1 Two-Dimensional Electron Gas in Inversion Layer

Let us start our exposition with the semiconductor inversion layer of Si MOS-FET (Metal-Oxide-Semiconductor Field-Effect Transistor), on which extensive research including device applications has been conducted. Figure 16.1 shows a schematic picture of a typical n-channel MOSFET. Here "n-channel" denotes that the carrier in the inversion layer is an electron. It is called p-channel in case the carrier is a hole. It is observed from the figure that

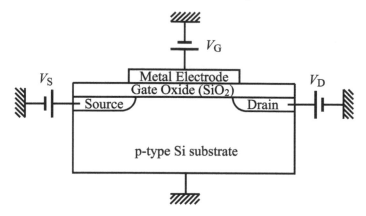

FIGURE 16.1
Schematic picture of an n-channel MOSFET. The source and the drain are
made of n-type Si while the substrate is made of p-type Si. The inversion layer
is formed in the region between the gate oxide (SiO_2) and the gate electrode.
The gate voltage controls the density of the two-dimensional electron gas in
the inversion layer.

MOSFET has a similar structure as a capacitor. An insulator made of SiO_2
is deposited on a Si semiconductor of p-type, over which a metal electrode
(gate) is attached. Electric charge is induced at the Si/SiO_2 interface when
a gate voltage $V_G > 0$ is greater than a threshold voltage. Then electrons
flow from the source to the drain along the interface. Figure 16.2 shows the
band structure in the vicinity of the Si/SiO_2 interface. The band spectrum
of the p-type Si substrate is bent in the vicinity of the interface so that the
bottom of the conduction band at the interface has lower energy than the top
of the valence band for large z. As a result, electron bound states along the
z-axis in Fig. 16.2 develop, leading to a two-dimensional electron subband in
the inversion layer. The electron number in the inversion layer is controllable
by changing the gate voltage V_G.

16.2.2   Coulomb Blockade

U. Meirav *et al.* measured conductance of narrow channels of GaAs/AlGaAs
[1]. See [2] for a review on mesoscopic conductance. Figure 16.3 shows the
schematic picture of the device employed in their experiments. An insulating
AlGaAs layer is deposited on a heavily doped GaAs substrate. Over the
AlGaAs layer is a pure GaAs layer. Note the similarity of this structure with
that of MOSFET: the metal layer in MOSFET corresponds to the doped
GaAs layer and the p-type Si layer in MOSFET to the pure GaAs layer.
An inversion layer is formed in the interface between the pure GaAs layer

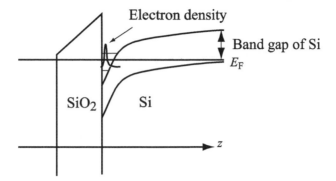

FIGURE 16.2
Band structure of the Si/SiO$_2$ interface. A two-dimensional electron gas is formed in the inversion layer, when the gate voltage V_G exceeds the threshold voltage. The density of the electron gas can be controlled by the gate voltage. The z-axis is perpendicular to the Si/SiO$_2$ interface.

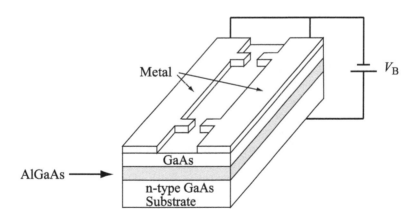

FIGURE 16.3
Schematic drawing of a device structure employed in [1]. A two-dimensional electron gas forms at the top GaAs/AlGaAs interface. Electrons underneath the electrods are repelled by the negative bias voltage V_B.

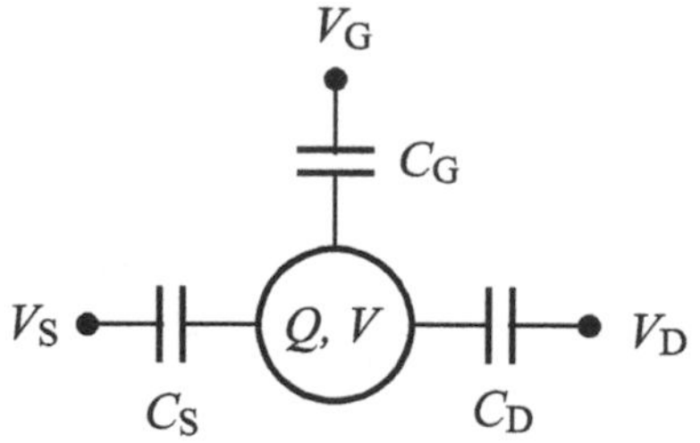

$$V_G$$

FIGURE 16.4
Schematic diagram of a single quantum dot in a Coulomb blockade state. No current flows from S to D in this state.

and the AlGaAs layer. The charge density of the inversion layer may be controllable by manipulating the gate voltage V_G between the doped GaAs and the pure GaAs layers, in a similar manner as in MOSFET. It should be noted, however, that the electron density of this system is much smaller than that of a Si MOSFET. As a result, it is easier to control the electron number in the dot in this system compared to that in Si MOSFET, and accordingly this system is suitable for quantum computing. There are two metal strips in Fig. 16.3, each of which has two small metal pieces to form narrow channels, and the overall structure is considered to have two gates. A narrow channel is formed if negative voltage is applied to these metal strips so that no inversion layer is formed underneath these strips.

This structure with two gates, employed to study the conductance fluctuations in narrow channels, may be regarded as a quantum dot formed between the two gates. They have measured the conductance of a narrow channel which contains a single quantum dot while the gate voltage, and hence the electron number, is varied. They observed periodic oscillations in conductance. This phenomenon is explained by a contact interaction model in which the total energy is a sum of single-electron energies and the charging energy, which is approximated by the classical electrostatic energy. Figure 16.4 shows a schematic picture of the circuit employed in experiments by Meirav *et al.* [1]. Let Q and V be the charge and voltage, respectively, of the QD and V_G be the gate voltage. The source (drain) voltage of the left electrode (right electrode) is denoted as V_S (V_D). The gate, source and drain are connected to the QD with capacitances C_G, C_S and C_D. It is found from Fig. 16.4 that

$$Q = C_G(V - V_G) + C_S(V - V_S) + C_D(V - V_D), \qquad (16.1)$$

from which we obtain

$$CV = Q + C_G V_G + C_S V_S + C_D V_D, \qquad (16.2)$$

where $C = C_G + C_S + C_D$. The electrostatic energy $U(Q)$ is found from the condition $dU/dQ = V$ as

$$U(Q) = \frac{(Q + C_G V_G + C_S V_S + C_D V_D)^2}{2C}. \tag{16.3}$$

Let N be the number of electrons in a QD. Then $Q = -e(N - N_0)$, where N_0 is the number of electrons on the dot when all the external voltages are set to zero. This compensates the positive background charge provided by the donor in the heterostructure. Contribution of the electrostatic energy to chemical potential is evaluated as $U(N) - U(N - 1)$. Therefore the chemical potential with vanishing bias voltages $V_S \simeq V_D \simeq 0$ is found as

$$\begin{aligned} E_N &= \epsilon_N + U(N) - U(N - 1) \\ &= \epsilon_N + \left(N - N_0 - \frac{1}{2} \right) E_C - \frac{E_C}{e}(C_G V_G + C_S V_S + C_D V_D) \end{aligned} \tag{16.4}$$

by taking the contribution from single-particle energy ϵ_N into account. Here $E_C = e^2/C$ and ϵ_N stands for the single-particle level when the electron levels up to the Nth level are occupied. When the condition $V_S \simeq V_D \simeq 0$ is satisfied, the chemical potential difference between $N+1$-electron and N-electron states is obtained from Eq. (16.4) as

$$E_{N+1} - E_N = \frac{e^2}{C} + \Delta\epsilon, \quad \Delta\epsilon = \epsilon_{N+1} - \epsilon_N. \tag{16.5}$$

The difference $\Delta\epsilon$ is nothing but the energy spacing between two discrete quantum levels when the Nth electron and $N + 1$st electron occupy different energy levels. For QDs employed in [1], the inequality $E_C \gg \Delta\epsilon$ is satisfied. We will drop $\Delta\epsilon$, for simplicity, in this subsection.

Now the levels in the dot are discrete with the level splitting $\simeq E_C$. Let us consider the energy levels depicted in Fig. 16.5, in which case conduction by way of the dot is suppressed. Let μ_S and μ_D be the chemical potentials of electrons in the left metal (source) and the right metal (drain), respectively. Assuming μ_S and μ_D take values in the vicinity of E_N, we find that the relation

$$E_N < \mu_S, \mu_D < E_{N+1} \tag{16.6}$$

is satisfied in Fig. 16.5. Under the condition (16.6), a state with N electrons in the dot is stabilized, and processes in which the dot obtains an extra electron from the source or the drain to have $N + 1$ electrons, or in which an electron escapes from the dot to have $N - 1$ trapped electrons in the dot, are forbidden. Therefore electron conduction through a QD is prohibited. This phenomenon is called the **Coulomb blockade**.

The condition under which an N electron state permits electron conduction through a dot is (a) $\mu_S > E_N > \mu_D$ or (b) $\mu_D > E_N > \mu_S$. Let us analyze these conditions in the case $V_S \simeq V_D$, following experiments done by Meirav *et al.* [1]. We assume, for simplicity, that the drain is grounded so that $V_D =$

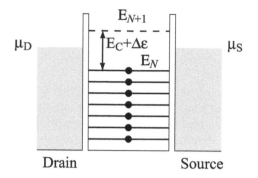

FIGURE 16.5

Energy level of electrons in a QD with conditions $E_N < \mu_S, \mu_D < E_{N+1}$. Energy $E_C + \Delta\epsilon$ is required to add an electron to a QD with N electrons to make it a QD with $N+1$ electrons, where E_C is the charging energy and $\Delta\epsilon$ is the single electron energy spacing. When the source chemical potential μ_S and the drain chemical potentials μ_D take values in the energy gap $E_C + \Delta\epsilon$, electrons are in the Coulomb blockage state, in which electron number is fixed at N. Therefore current does not flow through the dot. Note that this simplified argument applies only when the spin degrees of freedom are ignored.

0. Electron energy is measured with respect to the drain chemical potential $\mu_D = E_F$, where E_F is the bulk electron Fermi energy. Since $V_{SD} \equiv V_S \simeq 0$, it is required that the condition $E_N \simeq E_F$ must be satisfied for the conditions (a) and (b) to be fulfilled. Thus the gate voltage $V_G(N)$ which satisfies the condition $E_N = E_F$ is fixed and the current flows when $V_G \simeq V_G(N)$. Such $V_G(N)$ is obtained for each $N \geq 0$, and consequently the periodic behavior of conductance in V_G observed in the experiments [1] is accounted for by Coulomb blockade.

Let us consider the case with finite V_{SD} next. No electric current is allowed to flow when the Coulomb blockade condition (16.6) is satisfied and hence the state with N electons in the dot is stabilized. The current-supporting region is outside this Coulomb blockade domain and the boundaries between them are four lines specified by the following conditions

(a) $\mu_S = E_N$, (b) $\mu_D = E_N$, (c) $\mu_S = E_{N+1}$, (d) $\mu_D = E_{N+1}$.   (16.7)

The boundaries for a given N define a parallelogram called the **blockade diamond** as depicted in Fig. 16.6. States with N electrons in the dot are stabilized inside this parallelogram in the $V_G V_S$-plane. It is found from Eq. (16.4) that the slope of lines given by the conditions (a) and (c) is $C_G/(C - C_S)$ and that given by (b) and (d) is $-C_G/C_S$ since the relation $\mu_S = E_F - eV_S$ is satified in (a) and (c) while $\mu_D = E_F$ in (b) and (d). These blockade diamonds line up in the $V_G V_S$-plane. The electron number N attached to

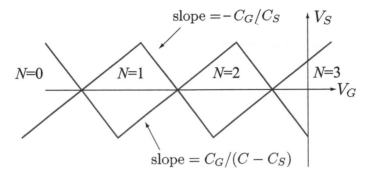

FIGURE 16.6
Coulomb blockade diamond. The parallelogram made of lines with a slope $-C_G/C_D$ and lines with a slope $C_G/(C - C_S)$ is a Coulomb blockade region in the V_G-V_S plane. The number of electrons is fixed in the Coulomb blockade region. The diagram shows the region where $N = 0, 1, 2$ and 3. The region corresponding to $N = 0$ is exceptional in that the diamond does not close, and hence it is discriminated from the other diamonds.

each parallelogram decreases as V_G is lowered; see Fig. 16.6. The state with $N = 0$ appears in the left end of V_G axis, which is easily identifiable since the corresponding "parallelogram" does not close. The failure of the closure is attributed to the fact that the conditions (a) and (b) are not necessary in this region. States with small N, such as $N = 1$ or 2, are prepared by making use of this fact.

It is found from the above observation that the Coulomb blockade phenomena, with which a given electron number state stabilizes the QD, plays a very important role when a QD is employed as a qubit.

16.3 Electron Charge Qubit

Two types of qubits that make use of electronic states in quantum dots are proposed to date. One uses two neighboring quantum QDs, which we call the **double quantum dots** or DQD for short in the following, and the two different states $|0\rangle$ and $|1\rangle$ of a qubit correspond to which of the two QDs is occupied. This qubit is called a **charge qubit**. The other qubit, called a **spin qubit**, makes use of two spin states, spin up and spin down, of an electron trapped in a single quantum dot.

300 nm

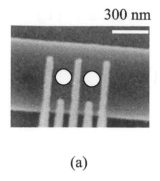

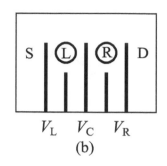

(a)   (b)

FIGURE 16.7
(a) Scanning electron micrograph of a double quantum dot. Reprinted from
[3] with permission from World Scientific Publishing Co. Pte. Ltd. (b)
Schematic diagram of the DQD.

16.3.1   Electron Charge Qubit

Figure 16.7 shows a scanning electron micrograph of an electron charge qubit
fabricated at NTT Basic Research Laboratory [4]. A qubit is made of two
quantum dots (DQD), five control gate electrodes to control the charge, the
source (S) and the **drain** (D). A negative voltage is applied to the gate
electrodes to control the electric potential of each dot separately and the
tunneling barrier between the dots. Let us denote a state of a DQD with
N_L electrons in the left QD and N_R electrons in the right QD by (N_L, N_R).
Suppose the gate electrode voltages are adjusted so that two states $(N_L +
1, N_R)$ and $(N_L, N_R + 1)$ have almost degenerate energies. Furthermore this
DQD is isolated from S and D, with a good precision, by making use of
Coulomb blockade. Then an arbitrary state of the DQD is a superposition
of $|L\rangle$, which corresponds to $(N_L + 1, N_R)$, and $|R\rangle$, which corresponds to
$(N_L, N_R + 1)$, as

$$|\psi\rangle = \cos\frac{\theta}{2}|L\rangle + e^{i\phi}\sin\frac{\theta}{2}|R\rangle. \qquad (16.8)$$

Figure 16.8 shows the schematic energy diagram of the DQD depicted in
Fig. 16.7. The energy levels of the left (the right) QD in the presence of $N+1$
electrons is denoted as E_L (E_R), while t denotes the overlap integral between
the QDs. The electrochemical potential of the left S (the right D) is denoted
as μ_S (μ_D). The Hamiltonian describing this DQD is

$$H = -\frac{1}{2}\varepsilon\sigma_z - \frac{1}{2}\Delta\sigma_x, \qquad (16.9)$$

where $\varepsilon = E_R - E_L$ and $\Delta = 2t$.

It should be kept in mind that there are not only the ground state but
also excited states in an actual physical system. Therefore, for a DQD to

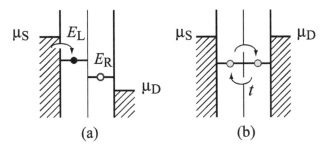

FIGURE 16.8
(a) Energy diagram of DQD when $\mu_D < E_R < E_L < \mu_S$. (b) The same for the case $E_L = E_R < \mu_S = \mu_D$, where t is the overlap integral.

be described by Eqs. (16.8) and (16.9), temperature must be low enough and moreover $|\varepsilon|$ and Δ must be small enough compared to the energy level separation due to Coulomb blockade.

16.3.2　Rabi Oscillation

Rabi oscillation in an electron charge qubit has been first observed by the NTT group [4]. Initialization and measurements were made in the state shown in Fig. 16.8 (a). This state satisfies the conditions $\mu_S > E_L > E_R > \mu_D$ and $|\varepsilon| \gg \Delta$. The gate electrode voltages are controlled first to make the system in the state shown in Fig. 16.8 (a). An electron is supplied from the source S, and the relevant energy level of the left QD is occupied so that the state $(N_L + 1, N_R)$ is realized. This electron does not hop to the right QD due to the assumption $|\varepsilon| \gg \Delta$. Next the DQD is steered to a state with $\mu_S = \mu_D > E_L = E_R$ so that $\varepsilon = 0$ as shown in Fig.16.8 (b). Then the DQD is detached from S and D and allowed to execute Rabi oscillation for the duration t_P, after which the system is further driven back to the state in Fig. 16.8 (a). The coupling between the QDs is turned off in this state and only the electron residing in the right dot flows into D and the system is put back to the initial state. The signal obtained in a single measurement is not strong enough to be observed. Hayashi *et al.* measured the current I flowing out from D by repeating the above cycle many times [4]. Figure 16.9 shows experimentally observed Rabi oscillation.

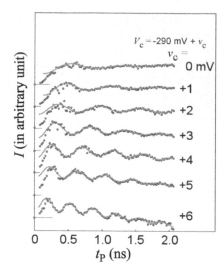

FIGURE 16.9

Coherent charge oscillation. The barrier height between the left and the right dots is controlled by changing the paramter V_C, which changes the overlap intergral t and hence Δ, leading the change in the Rabi oscillation frequency. Courtesy of NTT Basic Research Laboratory.

16.4 Electron Spin Qubit

16.4.1 Electron Spin Qubit

Let us consider an electron confined in a QD. We define the logical state $|0\rangle$ as the spin-↑ state and the logical $|1\rangle$ as the spin-↓ state. Technology making use of electron spins thus defined is called **spintronics** in the literature. The electron number N for a spintronics is 1 in an ideal situation. A QD with $N = 1$ is easily realized not only in a QD but also in a DQD as mentioned in the previous section. Let us consider a DQD and let S_L (S_R) be the spin operator of the left (right) QD. Two-qubit coupling is given by

$$H_{int} = J \sum_{k=x,y,z} S_k \otimes S_k, \tag{16.10}$$

where the exchange energy J is controllable by changing the overlap intergral between the electron wave functions. This is done by changing the barrier height between the dots by controlling the voltage applied to the electrodes separating the dots.

Although coherent motion of a large number of spins is well established in

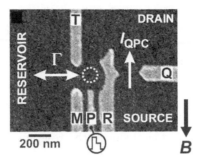

FIGURE 16.10

Scanning electron micrograph of a quantum dot used in the single-shot readout measurement. Electrodes M, P, R and T are attached to the dot. Bias voltage V_P is applied through P to control the dot potential. The current I_{QPC} from the source to the drain is very sensitive to the charge of the dot, and hence the number of electrons trapped by the dot can be found by measuring I_{QPC}. Reprinted by permission from Macmillan Publishers Ltd: Nature **430**, 431–435, copyright 2004.

ESR, for example, individual control of an electron spin, which is necessary for qubit operations, is rather difficult to implement experimentally, in spite of several theoretical proposals. Qubit operations are easy, however, if we restrict ourselves within single qubit operations. We outline the experiment carried out by Elzerman *et al.* [5] in the following.

16.4.2 Single-Qubit Operations

Elzerman *et al.* [5] conducted single-shot readout measurement of a spin qubit state. Figure 16.10 shows the scanning electron micrograph image of the GaAs/AlGaAs heterostructure employed in their experiments. A QD is formed in the area surrounded by gates T, M and R, where T and M control the electron flow from/to the reservoir. It is possible to realize a state with electron number $N = 0$ or $N = 1$ of a dot by controlling these gate voltages, as well as the voltage of the gate P. **Quantum point contact (QPC) current** I_{QPC} flows from the source to the drain through the narrow channel formed by the gates R and Q. This current is very sensitive to the dot charge, and therefore measurement of I_{QPC} enables us to tell whether $N = 0$ or 1 by single shot measurement.

Spin of an electron in the dot may be measured by employing such equipment along with a "spin-to-charge" conversion method. First a large magnetic field of ~ 10 T is applied to lift the electron spin degeneracy. The Zeeman splitting of electron energies in the dot due to this large magnetic field is

~ 2 K. The temperature must be lowered so that $k_B T$ is much smaller than the Zeeman splitting energy as well as the difference in the charging energy E_C of the dot with $N = 0$ and $N = 1$.

Measurement of the electron spin is made by controlling the P gate voltage in three steps as described in Fig. 16.11 (a).

1. The voltage V_P is set to 0 in the first step. The other gate voltages are adjusted so that the ↑-spin and the ↓-spin have higher energies relative to the Fermi energy of electrons in the reservoir.

2. A positive voltage $V_P > 0$ is applied in the second step. Then both the ↑-spin and the ↓-spin have lower energies than the Fermi energy of electrons in the reservoir. If the system is kept in this state for some duration of time, an electron is injected to the dot from the reservoir. either in the ↑-spin state or the ↓-state. This time duration is controllable by the gate voltages applied to the gates M and T and is adjusted so that a proper amount of injection is made. Note that the $N = 2$ state has much higher energy due to Coulomb blockade, and only states with $N = 1$ are realized in reality.

3. Readout of the spin of an electron injected in the dot is done by adjusting V_P in the third step so that the Fermi energy of the reservoir sits between the ↑-spin energy and the ↓-spin energy. If the electron in the dot is in the ↑-spin state, the electron does not escape to the reservior. On the other hand, if it is in the ↓-spin state, it escapes to the reservoir. Subsequently, a ↑ electron is newly injected to the dot from the reservoir.

This behavior can be also observed by measuring the change in current I_{QPC} as follows. Figure 16.11 shows the schematic QPC current response under the above steps. I_{QPC} increases during the transition from Step 1 to Step 2 as a positive voltage is applied to the gate P and accordingly the barrier formed by gates R and Q is lowered. However I_{QPC} slightly decreases when an electron is injected into the dot. The state of the dot does not change provided that the electron is in the ↑-state when readout is made in Step 3. Accordingly the current I_{QPC} does not change after it drops off as V_P is reduced as depicted by the solid line in Fig. 16.11. In contrast, when the electron is in the ↓-state, I_{QPC} increases temporarily as the electron escapes to the reservoir, and subsequently I_{QPC} comes back to its initial value due to an ↑-electron supplied from the reservoir as depicted by the broken line in Fig. 16.11. It is thus possible to tell whether the electron spin is ↑ or ↓ by single-shot measurement.

The relaxation time T_1, defined as the average time for a spin to flip from ↓ to ↑ in the second step, is measured using this strategy. Time required for an electron to escape to the reservoir or to be supplied from the reservoir is negligiblly small compared to T_1. (Note that these times are exaggerated in Fig. 16.11.) Therefore T_1 is obtained by measuring the probability with which

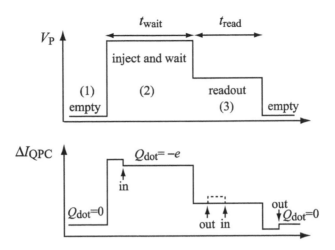

FIGURE 16.11

(a) Pulses and (b) response to the pulses to measure the state of an individual electron spin in a quantum dot. (a) The voltage V_P to control the potential of the dot changes in three steps. (1) $V_P = 0$ so that the dot is empty (vanishing electron numbers), (2) V_P at the "injection and wait" period is adjusted so that the energy levels of the spin-up state and the spin-down state in the dot are lower than the Fermi energy of electrons in the reservoir. As a result, a spin-up or a spin-down electron is injected to the dot. There is a finite probability with which a spin-down electron flips to a spin-up electron. The probability depends on the waiting time t_{wait}. The spin relaxation time T_1 is measured by changing t_{wait}. (3) Readout is conducted by adjusting V_P so that the energy of the spin-up state (spin-down state) is lower (higher) than the electron Fermi energy of the reservoir. Then only a spin-up electron remains in the dot. (b) ΔI_{QPC} is the change of I_{QPC} as V_P changes as in (a). The current I_{QPC} increases by applying a positive V_P in Step (2). It slightly decreases when an electron is injected to the dot as denoted with "in" in (b). Readout of an electron spin trapped by the dot is done by observing if the electron escapes the dot in (3). When the electron spin is down, the current I_{QPC} increases shortly after V_P is lowered since the down-spin electron escapes from the dot. The current I_{QPC} decreases since a spin-up electron from the reservoir fills the dot. Adapted by permission from Macmillan Publishers Ltd: Nature **430**, 431–435, copyright 2004.

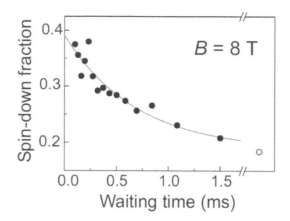

FIGURE 16.12
Measurement of the spin-relaxation time T_1. Relaxation from the spin-down state to the spin-up state takes place in waiting time t_{wait}. Electron injection to the dot takes place slightly after V_P is set to 2). This delay is so short, however, that it may be negligible in practice [6]. Courtesy of Jeroen Elzerman, ETH, Switzerland.

the spin is observed in the ↓-state as the pulse width in Step 2 is changed. The relaxation time T_1 measured in this way is $\sim 0.85 \pm 0.11$ ms at a magnetic field of 8 T.

16.4.3   Coherence Time

Suppose there is a single electron in a quantum dot. It is reported that the longitudinal relaxation time T_1 of the electron is as long as ~ 10 ms [5]. The coherence time T_2^*, which is defined as the phase relaxation time experimentally observed, is on the order of ~ 10 ns. An electron in a quantum dot is under fluctuating magnetic field produced by Ga and As nuclear spins, which couples with the electron through the hyperfine interaction. Each dot has $\sim 10^6$ nuclear spins of Ga and As, which thermally fluctuate. The effective magnetic field on the electron is known to fluctuate with the time scale on the order of ~ 10 μs.

Let us turn to the measurement of the dephasing time T_2. Single-shot signal strength is, however, not strong enough to directly read out T_2. Signals are accumulated during the interval longer than the typical fluctuation time of the nuclei. Therefore the observed phase decoherence time is obtained by averaging a snap-shot T_2 over ensemble under fluctuating magnetic field. The decoherence time thus obtained is denoted as T_2^*. It is found from the observed value $T_2^* = 10$ ns that the amplitude of the nuclear magnetic field

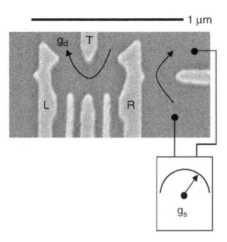

FIGURE 16.13

Scanning electron micrograph of a DQD. The electron numbers in the DQD are measured from the current g_d through the DQD and the current g_s through the quantum point contact. The gates L and R fix the electronic charge in the left and the right dots and the gate T controls the coupling strength between the left and the right dots [7]. Reprinted from Physica E, **35**, J. R Petta *et al.*, "Preparing, manipulating, and measuring quantum states on a chip", 251–256, Copyright (2006), with permission from Elsevier. Also from J. R. Petta *et al.*, SCIENCE, **309**, 2180–2184 (30 September 2005). Reprinted with permission from AAAS.

fluctuation is ~ 2 mT.

Measurement of T_2, with the nuclear spin fluctuation removed, has been made for a coupled double quantum dot [7]. Let (m, n) be a state of a DQD, employed for measurements, in which there are m electrons in the left QD and n electrons in the right QD. Figure 16.14 shows the charge stability diagram of the DQD as a function of bias voltages V_L and V_R. Let us consider the boundary region $(1, 1)$ and $(0, 2)$ on this diagram. The parameter $\varepsilon = E_R - E_L$ is chosen, instead of V_L and V_R, in such a way that the state $(0, 2)$ is stabilized when $\varepsilon > 0$ while $(1, 1)$ is stabilized when $\varepsilon < 0$.

Let us consider electron spins in these states. Electron spins in the state $(0, 2)$ are either parallel ($|0\rangle |\uparrow\uparrow\rangle$, for example) or antiparallel ($|0\rangle |\uparrow\downarrow\rangle$). One of the electrons in the state $|\uparrow\uparrow\rangle$ occupies the first excited state whose energy is higher than that of the ground state by $\Delta\epsilon$. The state $|\uparrow\uparrow\rangle$, i.e., a triplet state, has higher energy than a singlet state when $\Delta\epsilon$ is sufficiently large. We may ignore the contribution from triplet states in this case. This implies that an electron with up-spin in the left dot cannot hop to the right dot if the right dot is already occupied by a spin-up electron. This phenomenon

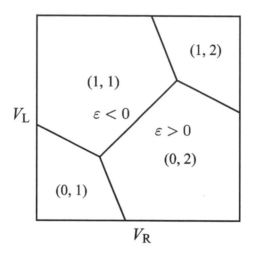

FIGURE 16.14

Charge stability diagram showing the electron numbers (m, n) in a DQD as a function of the gate voltages V_R and V_L. Reprinted from Physica E, **35**, J. R Petta *et al.*, "Preparing, manipulating, and measuring quantum states on a chip", 251–256, Copyright (2006), with permission from Elsevier. Also from J. R. Petta *et al.*, SCIENCE, **309**, 2180–2184 (30 September 2005). Reprinted with permission from AAAS.

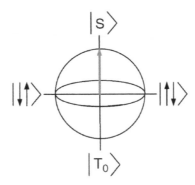

FIGURE 16.15
Bloch sphere, which represents a DQD state in $(1, 1)$. The north pole and the south pole of the Bloch sphere corresponds to the singlet state $|S\rangle$ and the triplet state $|T_0\rangle$, respectively, while $|\uparrow\downarrow\rangle$ and $|\downarrow\uparrow\rangle$ are in the xy-plane. The arrow shows the initial state $|S\rangle$. Reprinted from Physica E **35**, J. R. Petta et al., "Preparing, manipulating, and measuring quantum states on a chip", 251–256, Copyright (2006), with permission from Elsevier.

is called the **spin blockade** and is employed as a readout method of qubit states in this exeperiment. In the state $(1, 1)$ on the other hand, the energies of the three states of a triplet electron pair and that of the singlet pair are almost degenerate if the exchange energy J is negligible. If a magnetic field is applied in this state, the $|\uparrow\uparrow\rangle$ and $|\downarrow\downarrow\rangle$ states have different energies, while that of the residual states, a singlet state $|S\rangle = (|\uparrow\downarrow\rangle - |\downarrow\uparrow\rangle)/\sqrt{2}$ and a triplet state $|T_0\rangle = (|\uparrow\downarrow\rangle + |\downarrow\uparrow\rangle)/\sqrt{2}$, remain degenerate. We choose $|S\rangle$ and $|T_0\rangle$ as the qubit states $|0\rangle$ and $|1\rangle$, respectively. Any qubit state is expressed as a point on a Bloch sphere as shown in Fig. 16.15. In fact, the degeneracy between $|S\rangle$ and $|T_0\rangle$ is lifted if the exchange interaction J between the dots is taken into account. As a result, the energy of the singlet state S is lower than that of the triplet state T_0 by J. The parameter J is controllable by changing ε (≤ 0). It takes maximum value $J(0)$ when $\varepsilon = 0$ and $J(\varepsilon) \to 0$ in the limit $|\varepsilon| \gg J(0)$. Let B_{nuc}^L (B_{nuc}^R) be the component of the hyperfine field in the left (right) dot. Then the Hamiltonian of this qubit is

$$H = \begin{pmatrix} 0 & g^* \mu_B \Delta B_{\text{nuc}} \\ g^* \mu_B \Delta B_{\text{nuc}} & J(\varepsilon) \end{pmatrix}, \tag{16.11}$$

where $g^* \simeq -0.44$ is the electron g factor in GaAs, μ_B is the Bohr magneton, $\Delta B_{\text{nuc}} \equiv B_{\text{nuc}}^L - B_{\text{nuc}}^R$ and $J(0) \gg |\Delta B_{\text{nuc}}|$.

The intrinsic phase coherence time T_2 of this qubit has been measured by eliminating the effect of the nuclear magnetic field fluctuation. The parameter ε has been scanned in the spin-echo experiment as shown in Fig. 16.16.

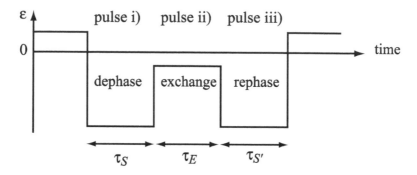

FIGURE 16.16

Spin-echo pulse sequence. Pulse i) changes the voltage ε from a positive value to $\varepsilon \ll -J(0)$. Then the initial singlet state of $(0,2)$ is transformed into a singlet state of $(1,1)$, which corresponds to the north pole of the Bloch sphere. The spin which represents the qubit state rotates around the x-axis during the pulse duration τ_S under the influence of the hyperfine field ΔB_{nuc}. As a result, the spin state has a finite y-component $\langle \sigma_y \rangle$. Next, pulse ii) is applied so that ε is slightly less than 0. Then the spin rotates around the z-axis due to the exchange coupling $J(0)$ between two qubits. The pulse width τ_E is taken so that the resulting rotation around the z-axis is π. Then the expectation value of the y-component of the spin is flipped to $-\langle \sigma_y \rangle$. Finally pulse iii) is applied to put the voltage ε back to the value under the pulse i). The magnetic field ΔB_{nuc} rotates the spin so that the expectation value $-\langle \sigma_y \rangle$ approaches zero. The time required for the above spin-echo pulse sequence is much shorter than the characteristic time of fluctuations in ΔB_{nuc}. The effect of ΔB_{nuc} may be removed if we take $\tau_S = \tau_{S'}$, where $\tau_{S'}$ is the width of the pulse iii).

1) Singlet electron pair $(0, 2)$ is prepared in the right dot by taking $\varepsilon > 0$.

2) The parameter ε is changed to $\varepsilon < 0, |\varepsilon| \gg J(0)$. The change is slower than the time scale $\simeq \hbar/J(0)$ for an electron to tunnel to the left dot and faster than the singlet-triplet mixing time $\sim \hbar/(g^*\mu_B \Delta B_{\mathrm{nuc}})$. Singlet state $|S\rangle$ is obtained as a result. The initial state $|S\rangle$ then mixes with $|T_0\rangle$ through $g^*\mu_B \Delta B_{\mathrm{nuc}}$ term in H since the system stays in the parameter region in which $J(\varepsilon) \simeq 0$. Let τ_S be the duration of this mixing step.

3) Now ε is put back to the value $\varepsilon \sim 0$. Then $J(\varepsilon) \gg \Delta B_{\mathrm{nuc}}$ is satisfied and a rotation around the z-axis is in action. The duration τ_E of this process is taken such that $J(\varepsilon)\tau_E/\hbar = \pi$ so that a π-pulse around the z-axis results in. Note that $\tau_E \ll \tau_S$.

4) The system is put to the same state as in 2) for the same duration τ'_S.

5) Readout is conducted by putting ε at the same positive value as in 1). Suppose the state contains the triplet state $|T_0\rangle$. Then the corresponding component does not come back to $(0, 2)$ due to spin blockade, which is active only in the triplet state. The probability P_S of the state being in the singlet state $|S\rangle$ after operation 4) is measured in this way.

Figure 16.17 shows the observed result of the probability P_S, where P_S is plotted against $\tau_S - \tau_{S'}$ for a given value of $\tau_S + \tau_{S'}$. It is expected that the effect of ΔB_{nuc} is eliminated when $\tau_S = \tau_{S'}$. This is because the time required for operations 1) $\sim$ 5) above is much shorter than the characteristic time of fluctuation in ΔB_{nuc}. The probability P_S therefore takes the maximum value at this position. The phase relaxation time T_2 is obtained to be 1.2 μs by plotting the maximum of P_S as a function of $\tau_S + \tau_{S'}$ to measure the decay rate, and subsequently the effect of nuclear spins are subtracted.

The Rabi oscillation between $|\uparrow\downarrow\rangle$ and $|\downarrow\uparrow\rangle$ is also observed in this system by manipulating the pulse sequence for ε. This guarantees that the SWAP gate and the $\sqrt{\mathrm{SWAP}}$ gate may be implemented by controlling the Rabi oscillation time by regarding the DQD as a two-qubit system, in which up-spin and down-spin states of each dot are basis vectors of a qubit.

An attempt has been made to control an individual electron spin of each dot by using ESR (electron spin resonance) to make a DQD a two-qubit system [9]. We note that the observed $T_2 \simeq 1.2$ μs is roughly 7,000 times longer than the $\sqrt{\mathrm{SWAP}}$ gate operation time.

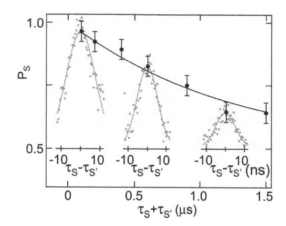

FIGURE 16.17

Spin echo recovery probability P_S plotted against $\tau_S - \tau_{S'}$ for given $\tau_S + \tau_{S'}$. Observe that the probability takes the maximum value at $\tau_S - \tau_{S'} = 0$. The coherence time T_2 is obtained by plotting the maximum value as a function of $\tau_S + \tau_{S'}$. From J. R. Petta *et al.*, SCIENCE **309**: 2180-2184 (30 September 2005) and J. R. Petta *et al.*, Physica E **35**, 251-256 (December 2006). Reprinted with permissions from AAAS and Elsevier.

16.5   DiVincenzo Criteria

DiVincenzo criteria for a quantum dot quantum computer are evaluated separately for charge qubits and spin qubits.

16.5.1   Charge Qubits

Let us evaluate the DiVincenzo criteria for charge qubits.

1. A scalable physical system with well-characterized qubits:

   A qubit is made of a double quantum dot fabricated in a GaAs/Al/GaAs heterostructure. It is potentially scalable using currently available semiconductor lithography technology. A two-qubit system has been fabricated, and coupling between two qubits has been demonstrated so far [10]. The two basis states of the qubit correspond to states in which an electron sits in the right (left) quantum dot.

2. The ability to initialize the state of the qubits to a simple fiducial state, such as $|00\ldots0\rangle$:

   The qubit can be initialized by electron injection [4]; see Fig. 16.8.

3. Long decoherence times, much longer than the gate operation time:

Decoherence time for a charge qubit is short due to charge fluctuation and phonon emission. The former restricts the phase relaxation time T_2 and the latter the energy relaxation time T_1 [4].

4. A "univeral" set of quantum gates:

One-qubit gate operations have already been demonstrated [4]. Although two-qubit coupling has been demonstrated [10], two-qubit gates have not been implemented yet.

5. A qubit-specific measurement capability:

Readout is possible by making use of tunneling current [4]. Single-shot measurement with the radio-frequency single-electron transistor (rf-SET) is also proposed, although this proposal has not been demonstrated experimentally [11].

16.5.2   Spin Qubits

1. A scalable physical system with well-characterized qubits:

A qubit in a spin qubit is an electron spin trapped in a quantum dot. It is also expected to be scalable within the current lithography technology. A double quantum dot may be regarded as a two-spin qubit system. Individual spin control in a double quantum dot has not been demonstrated yet.

2. The ability to initialize the state of the qubits to a simple fiducial state, such as $|00\dots0\rangle$:

Initialization is done with electron injection similarly to the charge qubit case [5].

3. Long decoherence times, much longer than the gate operation time:

Energy relaxation time T_1 is very long, on the order of 1 ms, under a strong magnetic field [5]. Phase relaxation time T_2 is measured to be $\sim 1\ \mu$s, if the contribution from the background hyperfine interaction is subtracted [8].

4. A "univeral" set of quantum gates:

ESR may be used to control a single spin qubit [12], although selective addressing may be difficult. As for two-qubit operations, $\sqrt{\text{SWAP}}$ gate, making use of the exchange coupling, has been demonstrated [7]. Implementation of general two-qubit gates requires one-qubit gates, which is difficult due to lack of individual addressing.

5. A qubit-specific measurement capability:

Tunneling current measurement to readout the qubit state has been experimentally demonstrated. Single-shot measurement making use of a quantum point contact (QPC) electrometer has been also demonstrated [5].

References

[1] U. Meirav, M. A. Kastner, and S. J. Wind, Phys. Rev. Lett. **65**, 771 (1990).

[2] M. A. Kastner, Rev. Mod. Phys. **64**, 849 (1992).

[3] T. Fujisawa in *Physical Realizations of Quantum Computing: Are the DiVincenzo Criteria Fulfilled in 2004?*, ed. M. Nakahara *et al.*, World Scientific, Singapore (2006).

[4] T. Hayashi, *et al.*, Phys. Rev. Lett. **91**, 226804 (2003).

[5] J. M. Elzerman *et al.*, Nature **430**, 431 (2004).

[6] J. M. Elzerman, Technische Universiteit Delft Thesis (2004).

[7] J. R. Petta, *et al.*, Science **309**, 2180 (2005).

[8] J. R. Petta, *et al.*, Physica E **35**, 251 (2006).

[9] F. H. L. Koppens, *et al.*, Nature **442** (2006) 766.

[10] G. Shinkai *et al.*, Appl. Phys. Lett. **90** 103116 (2007).

[11] A. Aassime *et al.*, Phys. Rev. Lett. **86**, 3376 (2001).

[12] H.-A. Engel and D. Loss, Phys. Rev. Lett. **86**. 4648 (2001).

A

Solutions to Selected Exercises

Chapter 1

1.1 For these vectors to be linearly dependent, we must have $|v_1\rangle = c|v_2\rangle$. The solution is $c = 3$, for which we obtain $x = 6$ and $y = 9/2$. Therefore, they are linearly independent if $x \neq 6$ or $y \neq 9/2$.

1.2 We may show that the determinant of a matrix $(|v_1\rangle, |v_2\rangle, |v_3\rangle)$ does not vanish to show the linear independence of these vectors. We obtain $\det(|v_1\rangle, |v_2\rangle, |v_3\rangle) = 2 \neq 0$.

1.3 $\|x\| = \sqrt{\langle x|x\rangle} = \sqrt{7}$. $\langle x|y\rangle = 7 - 2i$. $\langle y|x\rangle = 7 + 2i$, where the result of Exercise 1.4 may be used.

1.5 $c_1 = \langle e_1|v\rangle = 5/2$ and $c_2 = \langle e_2|v\rangle = 1/2$.

1.6 (1) $|e_1\rangle = |v_1\rangle/\|v_1\| = (-1, 2, 2)^t/3$. Then we obtain $\langle e_1|v_2\rangle = 0$ and $|e_2\rangle = |v_2\rangle/\|v_2\| = (2, -1, 2)^t/3$. Next, we find $\langle e_1|v_3\rangle = -3$ and $\langle e_2|v_3\rangle = 0$, from which we obtain $|f_3\rangle = (2, 2, -1)^t$ and $|e_3\rangle = (2, 2, -1)^t/3$.
(2) $c_1 = \langle e_1|u\rangle = 3$. Similarly we obtain $c_2 = 6$ and $c_3 = -3$.

1.7 $|e_1\rangle = |v_1\rangle/\|v_1\| = (1, i, 1)/\sqrt{3}$. $|f_2\rangle = |v_2\rangle - |e_1\rangle\langle e_1|v_2\rangle = (2, 1-i, i-1)^t$, from which we obtain $|e_2\rangle = |f_2\rangle/\|f_2\| = (2, 1-i, i-1)^t/2\sqrt{2}$.

1.9 Eigenvalues are $\lambda_1 = 1$ and $\lambda_2 = -1$. Corresponding normalized eigenvectors are $|\lambda_1\rangle = (e^{\pi i/4}, 1)^t/\sqrt{2}$ $|\lambda_2\rangle = (1, -e^{-\pi i/4})^t/\sqrt{2}$. Similar to Example 1.3, the unitary matrix which diagonalizes A is found to be

$$U = (|\lambda_1\rangle, |\lambda_2\rangle) = \frac{1}{\sqrt{2}} \begin{pmatrix} e^{i\pi/4} & 1 \\ 1 & -e^{-i\pi/4} \end{pmatrix}.$$

1.10 (2) Let $U|\lambda\rangle = \lambda|\lambda\rangle$. Hermitian conjugation yields $\langle\lambda|U = \lambda^*\langle\lambda|$. Multiplying these two equations, we obtain

$$\langle\lambda|U^\dagger U|\lambda\rangle = \langle\lambda|\lambda\rangle = |\lambda|^2\langle\lambda|\lambda\rangle,$$

from which we find $|\lambda| = 1$.

1.11 Use $U = i\sigma_x$ to obtain eigenvalues $\pm i$. Alternatively we may use $\operatorname{tr} U =$

$\lambda_1 + \lambda_2 = 0$ and $\det U = \lambda_1 \lambda_2 = 1$ to show $\lambda_i = \pm i$.

1.12 $U^\dagger = (I - iH)^{-1\dagger}(I + iH)^\dagger = (I + iH)^{-1}(I - iH)$, where use has been made of the identity $A^{-1\dagger} = A^{\dagger -1}$. We show

$$U^\dagger U = (I + iH)^{-1}(I - iH)(I + iH)(I - iH)^{-1}$$
$$= (I + iH)^{-1}(I + iH)(I - iH)(I - iH)^{-1} = I.$$

1.13

$$A = \sum_i \lambda_i |e_i\rangle\langle e_i| = \begin{pmatrix} 1 & -2i \\ 2i & 1 \end{pmatrix}.$$

1.14 (1) It follows from the observation $A = 2I + \sigma_x$ that the eigenvalues are 2 ± 1 and the eigenvectors are the same as those of σ_x. We denote $\lambda_1 = 3$ and $\lambda_2 = 1$ and $|\lambda_1\rangle = (1,1)^t/\sqrt{2}$ and $|\lambda_2\rangle = (1,-1)^t/\sqrt{2}$.
(2)

$$A = \frac{3}{2} \begin{pmatrix} 1 & 1 \\ 1 & 1 \end{pmatrix} + \frac{1}{2} \begin{pmatrix} 1 & -1 \\ -1 & 1 \end{pmatrix}.$$

(3) $\exp(i\alpha A) = e^{i3\alpha}|\lambda_1\rangle\langle\lambda_1| + e^{i\alpha}|\lambda_2\rangle\langle\lambda_2| = e^{2i\alpha}\begin{pmatrix} \cos\alpha & i\sin\alpha \\ i\sin\alpha & \cos\alpha \end{pmatrix}$. This
is expected from the decomposition $A = 2I + \sigma_x$ and the factorization $\exp(i\alpha A) = \exp(2i\alpha I)\exp(i\alpha\sigma_x)$, which follows from $[2I, \sigma_x] = 0$.

1.15 (1) $\lambda_1 = \lambda_2 = 1, \lambda_3 = 10$. The first two eigenvalues are degenerate. The corresponding eigenvectors are obtained by Gram-Schmidt orthonormalization. We may take

$$|\lambda_1\rangle = \frac{1}{\sqrt{2}} \begin{pmatrix} 1 \\ 0 \\ 1 \end{pmatrix}, |\lambda_2\rangle = \frac{\sqrt{2}}{6} \begin{pmatrix} 1 \\ 4 \\ -1 \end{pmatrix}, |\lambda_3\rangle = \frac{1}{3} \begin{pmatrix} -2 \\ 1 \\ 2 \end{pmatrix},$$

for example.
(2) It is easier to use Eq. (1.39). We obtain

$$A = 1 \times P_1 + 10 \times P_2 = 1 \times \frac{1}{9} \begin{pmatrix} 5 & 2 & 4 \\ 2 & 8 & -2 \\ 4 & -2 & 5 \end{pmatrix} + 10 \times \frac{1}{9} \begin{pmatrix} 4 & -2 & -4 \\ -2 & 1 & 2 \\ -4 & 2 & 4 \end{pmatrix}.$$

(3)

$$A^{-1} = P_1 + \frac{1}{10}P_2 = \frac{1}{10} \begin{pmatrix} 6 & 2 & 4 \\ 2 & 9 & -2 \\ 4 & -2 & 6 \end{pmatrix}.$$

1.16 It follows from Proposition 1.2 that

$$f(\alpha\hat{n}\cdot\boldsymbol{\sigma}) = f(\alpha(P_1 - P_2)) = f(\alpha)P_1 + f(-\alpha)P_2$$
$$= f(\alpha)\frac{I+\hat{n}\cdot\boldsymbol{\sigma}}{2} + f(-\alpha)\frac{I-\hat{n}\cdot\boldsymbol{\sigma}}{2}.$$

1.17

$$A = \begin{pmatrix} -1 & 0 \\ 0 & i \end{pmatrix} \begin{pmatrix} \sqrt{2} & 0 & 0 \\ 0 & \sqrt{2} & 0 \end{pmatrix} \frac{1}{\sqrt{2}} \begin{pmatrix} -1 & 0 & -i \\ 1 & 0 & -i \\ 0 & -\sqrt{2} & 0 \end{pmatrix}.$$

1.21 To simplify our argument and notations, let us suppose all the vectors are elements of $\mathbb{C}^2$ and write $|a\rangle = (a_1, a_2)^t$, for example. Then

$$|a\rangle\langle b| = \begin{pmatrix} a_1 b_1, a_1 b_2 \\ a_2 b_1, a_2 b_2 \end{pmatrix}, \quad |c\rangle\langle d| = \begin{pmatrix} c_1 d_1, c_1 d_2 \\ c_2 d_1, c_2 d_2 \end{pmatrix}.$$

while

$$|a\rangle \otimes |c\rangle = \begin{pmatrix} a_1 c_1 \\ a_1 c_2 \\ a_2 c_1 \\ a_2 c_2 \end{pmatrix}, \quad \langle b| \otimes \langle d| = (b_1 d_1, b_1 d_2, b_2 d_1, b_2 d_2).$$

By comparing the components of $(|a\rangle\langle b|) \otimes (|c\rangle\langle d|)$ and $(|a\rangle \otimes |c\rangle)(\langle b| \otimes \langle d|)$, we readily verify the equality of these matrices.

Chapter 2

2.2 The time development operator is

$$U(t) = e^{i\omega t \sigma_y/2} = \begin{pmatrix} \cos\omega t/2 & \sin\omega t/2 \\ -\sin\omega t/2 & \cos\omega t/2 \end{pmatrix}.$$

(1) $|\psi(t)\rangle = U(t)|\psi(0)\rangle = (\sin\omega t/2, \cos\omega t/2)^t$.
(2) $p(t) = |\langle\sigma_z = +1|\psi(t)\rangle|^2 = \sin^2(\omega t/2)$.
(3) $p(t) = |\langle\sigma_x = +1|\psi(t)\rangle|^2 = \frac{1}{2}(\cos\omega t/2 + \sin\omega t/2)^2 = \frac{1}{2}(1 + \sin\omega t)$.

2.4 Let $\rho = \sum_i \lambda_i |\lambda_i\rangle\langle\lambda_i|$ be the sepctral decomposition of ρ. Then

$$\rho^2 = \sum_i \lambda_i^2 |\lambda_i\rangle\langle\lambda_j|,$$

from which we obtain

$$0 = \operatorname{Tr} \rho - \operatorname{Tr} \rho^2 = \sum_i \lambda_i (1 - \lambda_i).$$

Since each summand is non-negative, we must have $\lambda_i = 0$ or $\lambda_i = 1$. From $\operatorname{Tr} \rho = \sum_i \lambda_i = 1$, we find only one of λ_i is 1 and the rest are 0. The converse is trivial.

2.5 We find

$$\rho_1^{\text{pt}} = \begin{pmatrix} \frac{p+1}{4} & 0 & 0 & 0 \\ 0 & \frac{1-p}{4} & \frac{p}{2} & 0 \\ 0 & \frac{p}{2} & \frac{1-p}{4} & 0 \\ 0 & 0 & 0 & \frac{p+1}{4} \end{pmatrix}.$$

The eivgenvalues are

$$\frac{p+1}{4}, \quad \frac{p+1}{4}, \quad \frac{p+1}{4}, \quad \frac{1}{4}(1 - 3p).$$

Therefore one of the eigenvalues is negative for $p > 1/3$ and the negativity does not vanish.

2.6 We find

$$\rho_2^{\text{pt}} = \begin{pmatrix} \frac{p}{2} & 0 & 0 & \frac{1-p}{2} \\ 0 & \frac{1-p}{2} & \frac{p}{2} & 0 \\ 0 & \frac{p}{2} & \frac{1-p}{2} & 0 \\ \frac{1-p}{2} & 0 & 0 & \frac{p}{2} \end{pmatrix}.$$

The eivgenvalues are

$$\frac{1}{2}, \quad \frac{1}{2}, \quad \frac{1}{2}(1 - 2p), \quad \frac{1}{2}(2p - 1).$$

We find one of the eigenvalues is always negative except at $p = 1/2$, where two of the eigenvalues vanish.

2.7 We obtain the corresponding density matrix

$$\begin{aligned}
\rho' &= |\psi'\rangle\langle\psi'| \\
&= \frac{1}{2}(|e_1\rangle\langle e_1| \otimes |e_2\rangle\langle e_2| + |e_2\rangle\langle e_2| \otimes |e_1\rangle\langle e_1| \\
&\quad - |e_1\rangle\langle e_2| \otimes |e_2\rangle\langle e_1| - |e_2\rangle\langle e_1| \otimes |e_1\rangle\langle e_2|).
\end{aligned}$$

Partial trace over the first Hilbert space yields

$$\rho = \operatorname{Tr}_1 \rho' = \frac{1}{2}(|e_1\rangle\langle e_1| + |e_2\rangle\langle e_2|) = \frac{1}{2}I.$$

It is a uniformly mixed state.

2.8

$$|\Psi\rangle = \frac{1}{2}(|\psi_1\rangle \otimes |\phi_1\rangle + \sqrt{3}|\psi_2\rangle \otimes |\phi_2\rangle).$$

2.11

$$\sqrt{\rho_1}\rho_2\sqrt{\rho_1} = \frac{1}{4}\begin{pmatrix} 1 & 0 & 0 & 1 \\ 0 & 0 & 0 & 0 \\ 0 & 0 & 0 & 0 \\ 1 & 0 & 0 & 1 \end{pmatrix}, \quad \sqrt{\sqrt{\rho_1}\rho_2\sqrt{\rho_1}} = \frac{1}{2\sqrt{2}}\begin{pmatrix} 1 & 0 & 0 & 1 \\ 0 & 0 & 0 & 0 \\ 0 & 0 & 0 & 0 \\ 1 & 0 & 0 & 1 \end{pmatrix}.$$

Therefore $F(\rho_1, \rho_2) = 1/\sqrt{2}$.

Chapter 3

3.2

$$\rho(\theta, \phi) = |\psi(\theta, \phi)\rangle\langle\psi(\theta, \phi)| = \begin{pmatrix} \frac{1+\cos\theta}{2} & \frac{1}{2}e^{-i\phi}\sin\theta \\ \frac{1}{2}e^{i\phi}\sin\theta & \frac{1-\cos\theta}{2} \end{pmatrix}$$

$$= \frac{1}{2}\left(I + \sin\theta\cos\phi\,\sigma_x + \sin\theta\sin\phi\,\sigma_y + \cos\theta\,\sigma_z\right).$$

3.3 By noting the identity $\operatorname{tr}\sigma_i\sigma_j = 2\delta_{ij}$, we show

$$\operatorname{tr}(\rho\sigma_k) = \frac{1}{2}\sum_i u_i \operatorname{tr}(\sigma_k\sigma_i)$$

$$= \frac{1}{2}\sum_i u_i 2\delta_{ik} = u_k.$$

3.4 We have to fix the order of the Bell basis vectors $\{|\text{Bell}_i\rangle\}$. We take $\{|\Phi^+\rangle, |\Psi^+\rangle, |\Psi^-\rangle, |\Phi^-\rangle\}$ as our order. Let $|\text{Bell}_i\rangle = \sum_j V_{ij}|j\rangle$, where $|j\rangle$ is the binary basis vectors in decimal expression. We have, for example, $|\text{Bell}_0\rangle \equiv |\Phi^+\rangle = \frac{1}{\sqrt{2}}|00\rangle + \frac{1}{\sqrt{2}}|11\rangle$. Repeating this for other basis vectors, we obtain

$$V = \frac{1}{\sqrt{2}}\begin{pmatrix} 1 & 0 & 0 & 1 \\ 0 & 1 & 1 & 0 \\ 0 & 1 & -1 & 0 \\ 1 & 0 & 0 & -1 \end{pmatrix}.$$

3.5 They all vanish.

3.6 (1) $|11\rangle|76\rangle$, (2) $|23\rangle|76\rangle$, (3) $|35\rangle|76\rangle$. Note that the second register is in the common state $|76\rangle$ for all the three cases.

Chapter 4

4.1 Suppose $U_{\text{CNOT}} = U_1 \otimes U_2$, where $U_k \in U(2)$. Then, for any tensor product state, we find $U_{\text{CNOT}}|\psi_1\rangle|\psi_2\rangle = (U_1|\psi_1\rangle)\otimes(U_2|\psi_2\rangle)$ is another tensor product state. If, however, U_{CNOT} acts on $|\psi\rangle = \frac{1}{\sqrt{2}}(|0\rangle+|1\rangle)\otimes|0\rangle$, we obtain the entangled state

$$U_{\text{CNOT}}|\psi\rangle = \frac{1}{\sqrt{2}}(|00\rangle + |11\rangle),$$

which is a contradiction.

4.2 $a|00\rangle + b|11\rangle$.

4.3 (1)

$$I \otimes |0\rangle\langle 0| + \sigma_x \otimes |1\rangle\langle 1| = \begin{pmatrix} 1 & 0 & 0 & 0 \\ 0 & 0 & 0 & 1 \\ 0 & 0 & 1 & 0 \\ 0 & 1 & 0 & 0 \end{pmatrix}$$

(2)

$$(I \otimes |0\rangle\langle 0| + \sigma_x \otimes |1\rangle\langle 1|)(|0\rangle\langle 0| \otimes I + |1\rangle\langle 1| \otimes \sigma_x)$$

$$= |00\rangle\langle 00| + |01\rangle\langle 10| + |10\rangle\langle 11| + |11\rangle\langle 01| = \begin{pmatrix} 1 & 0 & 0 & 0 \\ 0 & 0 & 1 & 0 \\ 0 & 0 & 0 & 1 \\ 0 & 1 & 0 & 0 \end{pmatrix}.$$

It leaves the basis vector $|00\rangle$ unchanged, but it produces the cyclic change of other basis vectors as $|01\rangle \to |11\rangle \to |10\rangle$.
(3) The result of (2) multiplied by CNOT from the left yields the matrix

$$|00\rangle\langle 00| + |01\rangle\langle 10| + |10\rangle\langle 01| + |11\rangle\langle 11| = \begin{pmatrix} 1 & 0 & 0 & 0 \\ 0 & 0 & 1 & 0 \\ 0 & 1 & 0 & 0 \\ 0 & 0 & 0 & 1 \end{pmatrix}.$$

It swaps the basis vector of the first qubit and that of the second qubit. Therefore it maps $|\psi_1\rangle|\psi_2\rangle \mapsto |\psi_2\rangle|\psi_1\rangle$.

4.5 The LHS is

$$(U_{\text{H}} \otimes U_{\text{H}})(I \otimes |0\rangle\langle 0| + X \otimes |1\rangle\langle 1|)(U_{\text{H}} \otimes U_{\text{H}})$$
$$= I \otimes \frac{1}{2}(|0\rangle+|1\rangle)(\langle 0|+\langle 1|) + U_{\text{H}}XU_{\text{H}} \otimes \frac{1}{2}(|0\rangle-|1\rangle)(\langle 0|-\langle 1|)$$
$$= \frac{1}{2}[I \otimes (I+X) + Z \otimes (I-X)] = \frac{1}{2}[(I+Z) \otimes I + (I-Z) \otimes X]$$
$$= U_{\text{CNOT}} = \text{RHS}.$$

4.6 $|00\rangle \mapsto |\Phi^+\rangle, |01\rangle \mapsto |\Psi^+\rangle, |10\rangle \mapsto |\Phi^-\rangle, |11\rangle \mapsto |\Psi^-\rangle$.

4.9 A NAND gate is obtained if NOT is applied after AND. Both NOT and

AND may be implemented using CCNOT gates. NAND is also obtained using the CCNOT gates.

4.10 (1) $\langle \Psi | \Phi \rangle = \langle \psi | \langle 0 | U^\dagger U | \phi \rangle | 0 \rangle = \langle \psi | \phi \rangle = (\langle \psi | \phi \rangle)^2$.
(2) If such U existed, we would have either $\langle \psi | \phi \rangle = 1$ or $\langle \psi | \phi \rangle = 0$ for arbitrary $|\psi\rangle$ and $|\phi\rangle$. This is a contradiction.

4.11 $U_H |\psi\rangle = \frac{1}{\sqrt{2}}(a|00\rangle + a|10\rangle + b|01\rangle - b|11\rangle)$. Suppose 0 is obtained upon the measurement of the first qubit. Then the wave function immediately after the measurement is $|0\rangle \otimes (a|0\rangle + b|1\rangle)$ and the second qubit is in the state $a|0\rangle + b|1\rangle$. Similarly, if 1 is obained upon the first qubit measurement, the second qubit state is $a|0\rangle - b|1\rangle$.

4.12 Let

$$U = \begin{pmatrix} a & e & j & n \\ b & f & k & p \\ c & g & l & q \\ d & h & m & r \end{pmatrix}$$

and take

$$U_1 = \begin{pmatrix} a^*/u & b^*/u & 0 & 0 \\ -b/u & a/u & 0 & 0 \\ 0 & 0 & 1 & 0 \\ 0 & 0 & 0 & 1 \end{pmatrix}.$$

Then it follows

$$U_1 U = \begin{pmatrix} a' & e' & j' & n' \\ 0 & f' & k' & p' \\ c' & g' & l' & q' \\ d' & h' & m' & r' \end{pmatrix}.$$

The third and the fourth components of the first column may be deleted similarly. (Use U_2 whose components are different from those of the unit matrix only in $(1,1), (1,3), (3,1)$ and $(3,3)$ matrix elements. See the next exercise.)

4.13 Use

$$U_1 = \begin{pmatrix} \frac{1}{\sqrt{2}} & \frac{1}{\sqrt{2}} & 0 & 0 \\ -\frac{1}{\sqrt{2}} & \frac{1}{\sqrt{2}} & 0 & 0 \\ 0 & 0 & 1 & 0 \\ 0 & 0 & 0 & 1 \end{pmatrix}, \quad U_2 = \begin{pmatrix} \sqrt{\frac{2}{3}} & 0 & \frac{1}{\sqrt{3}} & 0 \\ 0 & 1 & 0 & 0 \\ -\frac{1}{\sqrt{3}} & 0 & \sqrt{\frac{2}{3}} & 0 \\ 0 & 0 & 0 & 1 \end{pmatrix}, \quad U_3 = \begin{pmatrix} \frac{\sqrt{3}}{2} & 0 & 0 & \frac{1}{2} \\ 0 & 1 & 0 & 0 \\ 0 & 0 & 1 & 0 \\ -\frac{1}{2} & 0 & 0 & \frac{\sqrt{3}}{2} \end{pmatrix}$$

to reduce U to

$$U_3U_2U_1U = \begin{pmatrix} 1 & 0 & 0 & 0 \\ 0 & -\frac{1-i}{2\sqrt{2}} & -\frac{1}{\sqrt{2}} & -\frac{1+i}{2\sqrt{2}} \\ 0 & -\frac{3+i}{2\sqrt{6}} & \frac{1}{\sqrt{6}} & -\frac{3-i}{2\sqrt{6}} \\ 0 & -\frac{i}{\sqrt{3}} & -\frac{1}{\sqrt{3}} & \frac{i}{\sqrt{3}} \end{pmatrix}.$$

Then apply the prescription for a 3×3 matrix given in Lemma 4.1 for the rest of the nontrivial elements.

4.14 (1) $g_0 = 000, g_1 = 010, g_2 = 110$, for example.
(2)

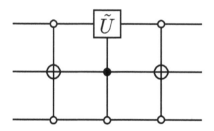

4.15

$$U_{CV}U_{CU} = (|0\rangle\langle 0| \otimes I + |1\rangle\langle 1| \otimes V)(|0\rangle\langle 0| \otimes I + |1\rangle\langle 1| \otimes U)$$
$$= |0\rangle\langle 0| \otimes I + |1\rangle\langle 1| \otimes (VU) = U_{C(VU)}.$$

4.16 The unitary matrix corresponding to the RHS of Fig. 4.12 is written as

$$(|0\rangle\langle 0| \otimes I \otimes I + |1\rangle\langle 1| \otimes I \otimes V)[(|0\rangle\langle 0| \otimes I + |1\rangle\langle 1| \otimes X) \otimes I]$$
$$\times [I \otimes (|0\rangle\langle 0| \otimes I + |1\rangle\langle 1| \otimes V^\dagger)][(|0\rangle\langle 0| \otimes I + |1\rangle\langle 1| \otimes X) \otimes I]$$
$$\times [I \otimes (|0\rangle\langle 0| \otimes I + |1\rangle\langle 1| \otimes V)]$$
$$= (|00\rangle\langle 00| + |01\rangle\langle 01| + |10\rangle\langle 10|) \otimes I + |11\rangle\langle 11| \otimes V^2$$
$$= (|00\rangle\langle 00| + |01\rangle\langle 01| + |10\rangle\langle 10|) \otimes I + |11\rangle\langle 11| \otimes U.$$

4.17 Instead of giving a general proof, we justify this statement by examining a few examples. (i) Let the input state of the first three qubits be 101. Then the first controlled-V gate is active and so is the controlled-$V^\dagger$ gate. The second controlled-V gate is not active since the second qubit is 0. The fourth qubit is then acted by $V^\dagger V = I$. (ii) Suppose the input state of the first three qubits is 111, next. Then both controlled-V gates are active while the controlled-$V^\dagger$ gate is not active. The fourth qubit is acted by $V^2 = U$.

Chapter 5

5.1 (1)
$$|\psi_3\rangle = |00\rangle \frac{1}{\sqrt{2}}(|0\rangle - |1\rangle).$$

The measurement outcome is 00 with a probability 1.

(2)
$$|\psi_3\rangle = |10\rangle \frac{1}{\sqrt{2}}(|0\rangle - |1\rangle).$$

The measurement outcome is 11 with a probability 1.

(3)
$$|\psi_3\rangle = \frac{1}{2}(-|00\rangle + |01\rangle + |10\rangle + |11\rangle)\frac{1}{\sqrt{2}}(|0\rangle - |1\rangle).$$

Chapter 6

6.2 (1) $\langle\psi|\psi\rangle = \mathcal{N}^2 \sum_{x=0}^{N-1} \cos^2(2\pi x/N)$. The summation is evaluated as

$$\sum_{x=0}^{N-1} \cos^2\left(\frac{2\pi x}{N}\right) = \sum_{x=0}^{N-1} \frac{\cos\left(\frac{4\pi x}{N}\right) + 1}{2} = \frac{N}{2} = 2^{n-1}.$$

Therefore $\mathcal{N} = 2^{-(n-1)/2}$.

(2) The x component is

$$\frac{1}{\sqrt{N}\sqrt{N/2}} \sum_{y=0}^{N} e^{-2\pi i x y/N} \cos\frac{2\pi y}{N}$$

$$= \frac{1}{\sqrt{2}N} \sum_y e^{-2\pi i x y/N}(e^{2\pi i y/N} + e^{-2\pi i y/N})$$

$$= \frac{1}{\sqrt{2}N} \sum_y (e^{-2\pi i(x+1)y/N} + e^{-2\pi i(x-1)y/N})$$

$$= \frac{1}{\sqrt{2}N} N(\delta_{x,1} + \delta_{x,N-1}) = \frac{1}{\sqrt{2}}(\delta_{x,1} + \delta_{x,N-1}),$$

from which we obtain

$$U_{\mathrm{QFT}n}|\psi\rangle = \frac{1}{\sqrt{2}}(|1\rangle + |N - 1\rangle).$$

6.3 Let $2^n/P = m \in \mathbb{N}$. Then we have a vector, after the application of QFT,

$$|\Psi'\rangle = \frac{1}{2^n} \sum_{x,y=0}^{2^n-1} e^{-2\pi i x y/2^n}|y\rangle|f(x)\rangle.$$

Let us separate the summation over x as

$$\sum_{x=0}^{2^n-1} h(x) \rightarrow \sum_{l=0}^{P-1} \sum_{k=0}^{m-1} h(kP+l),$$

where $h(x)$ is an arbitrary function. We obtain, after this replacement,

$$|\Psi'\rangle = \frac{1}{2^n} \sum_{l=0}^{P-1} \sum_{k=0}^{m-1} \sum_{y=0}^{2^n-1} e^{-2\pi i l y/2^n} e^{-2\pi i k y/m} |y\rangle |f(kP+l)\rangle$$

$$= \frac{1}{2^n} \sum_{y=0}^{2^n-1} \sum_{k=0}^{m-1} e^{-2\pi i k y/m} \sum_{l=0}^{P-1} e^{-2\pi i l y/2^n} |y\rangle |f(l)\rangle,$$

where use has been made of the periodicity $|f(kP+l)\rangle = |f(l)\rangle$.

Suppose $y = qm$ $(0 \leq q \leq P-1)$. Then

$$\sum_{k=0}^{m-1} e^{-2\pi i k y/m} = \sum_{k=0}^{m-1} e^{-2\pi i k q} = m.$$

If $y \neq qm$, on the other hand, we obtain

$$\sum_{k=0}^{m-1} e^{-2\pi i k y/m} = \frac{1 - e^{-2\pi i y}}{1 - e^{-2\pi i y/m}} = 0.$$

Accordingly,

$$|\Psi'\rangle = \frac{m}{2^n} \sum_{l=0}^{P-1} \sum_{q=0}^{P-1} e^{-2\pi i l q/P} |qm\rangle |f(l)\rangle.$$

The outcome $qm = q2^n/P$ $(0 \leq q \leq P-1)$ is obtained upon measurement of the first register.

6.4

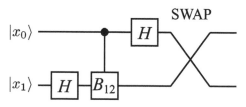

6.7 $K_n^{-1}(x,y) = e^{-i\theta_x} \delta_{xy}$.

6.8 Let $U = \left(e^{i\theta_x} \delta_{xy}\right)$ be a selective phase rotation transform matrix with $n = 3$. It is put in a block diagonal form

$$U = \begin{pmatrix} U_0 & 0 & 0 & 0 \\ 0 & U_1 & 0 & 0 \\ 0 & 0 & U_2 & 0 \\ 0 & 0 & 0 & U_3 \end{pmatrix},$$

where U_k is of the form $\mathrm{diag}(e^{i\theta_a}, e^{i\theta_b})$. Note that

$$U = |00\rangle\langle 00| \otimes U_0 + |01\rangle\langle 01| \otimes U_1 + |10\rangle\langle 10| \otimes U_2 + |11\rangle\langle 11| \otimes U_3$$
$$= A_0 A_1 A_2 A_3,$$

where $A_0 = |00\rangle\langle 00| \otimes U_0 + (|01\rangle\langle 01| + |10\rangle\langle 10| + |11\rangle\langle 11|) \otimes I$, for example. A quantum circuit implementing this gate is obtained from these observations as

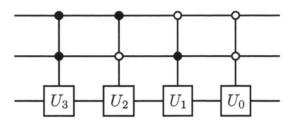

A filled circle in the figure is an ordinary control node, while a white circle is a negated control node, which may be implemented as in Fig. 6.5.

Chapter 8

8.2 It follows from $441 \leq 2^n < 882$ that $n = 9$. The period is 6 since $11^6 \equiv 1 \bmod 21$.

8.3 $61/45 = [1, 2, 1, 4, 3]$, $121/13 = [9, 3, 4]$.

8.4 The continued fraction expansion of $37042/Q$ is $[0, 28, 3, 4, 88, 1, 4, 3]$. We find for $k = 3$ that $p_3 = 13$, $q_3 = 368$ and $|13/368 - 37042/Q| \simeq 8.293 \times 10^{-8} \leq 1/(2Q)$. We have found $P = q_3 = 368$. The continued fraction expansion of $65536/Q$ is $[0, 16]$, and it fails to give the correct order.

8.5 The intermediate state of the bottom qubit in Fig. 8.6, for the input state $|c, a, b, c'\rangle$, is $|ab \oplus c'\rangle$. The last CCNOT gate adds $(a \oplus b)c \bmod 2$ to the bottom qubit to yield $|ab \oplus c' \oplus (a \oplus b)c\rangle = |ab \oplus ac \oplus bc \oplus c'\rangle$.

8.7 We find from $7 = 111$ that -7 is expressed as $1000 + 1 = 1001$.

Chapter 9

9.3 We evaluate

$$-i[H, \rho] = \frac{i\omega_0}{2}\begin{pmatrix} 0 & c_x - ic_y \\ -c_x - ic_y & 0 \end{pmatrix},$$

$$\sum_k L_k \rho L_k^\dagger = \begin{pmatrix} (\Gamma_+ + \Gamma_z)/2 + (\Gamma_z - \Gamma_+)c_z/2 & -\Gamma_z(c_x - ic_y)/2 \\ -\Gamma_z(c_x + ic_y)/2 & (\Gamma_- + \Gamma_z)/2 + (\Gamma_- - \Gamma_z)c_z/2 \end{pmatrix}.$$

$$-\frac{1}{2}\sum_k \left\{ \rho, L_k^\dagger L_k \right\}$$

$$= \begin{pmatrix} -(\Gamma_- + \Gamma_z)(1 + c_z)/2 & -(\Gamma_+ + \Gamma_- + 2\Gamma_z)(c_x - ic_y)/4 \\ -(\Gamma_+ + \Gamma_- + 2\Gamma_z)(c_x + ic_y)/4 & -(\Gamma_+ + \Gamma_z)(1 - c_z)/2 \end{pmatrix}.$$

Adding these terms yields the RHS of the Lindblad equation,

$$\begin{pmatrix} (\Gamma_+ - \Gamma_-)/2 - (\Gamma_+ + \Gamma_-)c_z/2 & (2i\omega_0 - \Gamma_+ - \Gamma_- - 4\Gamma_z)(c_x - ic_y)/4 \\ -(2i\omega_0 + \Gamma_+ + \Gamma_- + 4\Gamma_z)(c_x + ic_y)/4 & -(\Gamma_+ - \Gamma_-)/2 + (\Gamma_+ + \Gamma_-)c_z/2 \end{pmatrix},$$

from which the equations of motion for c_k are derived as

$$\frac{dc_x}{dt} = \omega_0 c_y - \left(\frac{\Gamma_+ + \Gamma_-}{2} + 2\Gamma_z \right) c_x,$$

$$\frac{dc_y}{dt} = -\omega_0 c_x - \left(\frac{\Gamma_+ + \Gamma_-}{2} + 2\Gamma_z \right) c_y,$$

$$\frac{dc_z}{dt} = (\Gamma_+ - \Gamma_-) - (\Gamma_+ + \Gamma_-)c_z.$$

By introducing the constants

$$c_z^{eq} = \frac{\Gamma_+ - \Gamma_-}{\Gamma_+ + \Gamma_-}, \quad \frac{1}{T_1} = \Gamma_+ + \Gamma_-, \quad \frac{1}{T_2} = \frac{\Gamma_+ + \Gamma_-}{2} + 2\Gamma_z,$$

the equations of motion are put in compact forms

$$\frac{dc_x}{dt} = \omega_0 c_y - \frac{c_x}{T_2}, \quad \frac{dc_y}{dt} = -\omega_0 c_x - \frac{c_y}{T_2}, \quad \frac{dc_z}{dt} = \frac{c_z^{eq} - c_z}{T_1}.$$

Chapter 10

10.1 The probability with which k bits are flipped in the received five bits is $\binom{5}{k} p^k (1 - p)^{5-k}$, $(0 \le k \le 5)$. The received five bits can be corrected if at most two bits are flipped. Therefore the success probability is

$$p_0 = (1 - p)^2 + 5p(1 - p)^4 + 10p^2(1 - p)^3 = (1 - p)^3(1 + 3p + 6p^2).$$

p_0 is as large as 0.99144 for $p = 0.1$.

10.3 Suppose U_β occurs in the first qubit. Encoding circuit outputs the state $a|+++\rangle + b|111\rangle$ for an input $|\psi\rangle = a|0\rangle + b|1\rangle$. The actions of U_β on the encoded state yields

$$a(\cos\beta|+++\rangle + i\sin\beta|-++\rangle) + b(\cos\beta|---\rangle + i\sin\beta|+--\rangle).$$

Action of the Hadmard gates in the error syndrome detection circuit maps these vectors to

$$a(\cos\beta|000\rangle + i\sin\beta|100\rangle) + b(\cos\beta|111\rangle + i\sin\beta|011\rangle).$$

The outputs of the error syndrome detection circuit for these vectors are

$$\cos\beta(a|000\rangle + b|111\rangle)|00\rangle + i\sin\beta(a|100\rangle + b|011\rangle)|11\rangle.$$

Bob will get $a|000\rangle + b|111\rangle$ when his measurement outcomes of the ancilla qubits are 00, which will happen with probability $\cos^2\beta$, while he will get $a|100\rangle + b|011\rangle$ when he observes ancilla qubits are 11, which will happen with probability $\sin^2\beta$. Bob applies X gate on the first qubit in the latter case.

10.5 (1) $A_1 = B_1 = 1$. (2) $A_1 = 1, B_1 = 0$. (3) $A_1 = 0, B_1 = 1$.

10.7 Let $a|+++\rangle + b|---\rangle$ be a codeword to be sent. The state after the action of the noise is

$$a|++\rangle\frac{1}{\sqrt{2}}(|100\rangle - |011\rangle) + b|--\rangle\frac{1}{\sqrt{2}}(|100\rangle + |011\rangle).$$

(1) It is easy to see $A_1 = B_1 = A_2 = B_2 = 0$. The bit-flip error syndrome detection circuit outputs the third group qubits in the state

$$\left[a|++\rangle\frac{1}{\sqrt{2}}(|100\rangle - |011\rangle) + b|--\rangle\frac{1}{\sqrt{2}}(|100\rangle + |011\rangle)\right]|11\rangle.$$

Therefore $A_3 = B_3 = 1$.
(2) Bob applies σ_x to the first qubit of the third group to obtain $a|++-\rangle + b|--+\rangle$. The action of $U_H^{\otimes 3}$ on $|\pm\rangle$ is

$$U_H^{\otimes 3}|+\rangle = \frac{1}{2}(|000\rangle + |011\rangle + |101\rangle + |110\rangle) \equiv |E\rangle$$

$$U_H^{\otimes 3}|+\rangle = \frac{1}{2}(|001\rangle + |010\rangle + |100\rangle + |111\rangle) \equiv |O\rangle.$$

Note that there are an even number of 1 in $|E\rangle$ and odd number in $|O\rangle$. Therefore the output of the phase-flip error syndrome detection circuit is

$$(a|EEO\rangle + b|OOE\rangle)|01\rangle.$$

We obtain $A_4 = 0, B_4 = 1$.
(3) is solved following Exercise 10.3.

10.8 It follows from Eq. (10.21) that

$$\begin{array}{cc}
\text{received code} & \text{syndrome} \\
(0001110) \to & (1,1,1)^t \\
(1101000) \to & (1,1,1)^t \\
(1100111) \to & (1,1,1)^t.
\end{array}$$

Chapter 12

12.1 It follows from Eq. (12.30) that

$$R\left(\frac{\pi}{2},\frac{\pi}{4}\right) = \begin{pmatrix} \dfrac{1}{\sqrt{2}} & -\dfrac{1}{2}(1+i) \\ \dfrac{1}{2}(1-i) & \dfrac{1}{\sqrt{2}} \end{pmatrix}.$$

$R(\pi/2,\pi/4)|0\rangle = (1/\sqrt{2},(1-i)/2)^t$, $R(\pi/2,\pi/4)|1\rangle = (-(1+i)/2,1/\sqrt{2})^t$.

12.3 An SU(2) matrix equivalent with U is

$$U' = \begin{pmatrix} e^{i\theta/2} & 0 \\ 0 & e^{-i\theta/2} \end{pmatrix}.$$

Comparing U' with Eq. (12.34), we identify $\beta = \theta$, $\alpha+\gamma = 0$ and $\alpha-\gamma = -\pi$, i.e., $\alpha = -\pi/2, \beta = \theta$ and $\gamma = \pi/2$.

12.4

$$e^{-i\delta\sqrt{1+\epsilon^2}\hat{n}\cdot I t}\begin{pmatrix} 1 \\ 0 \end{pmatrix} = \begin{pmatrix} \cos\Lambda - \dfrac{i\sin\Lambda}{\sqrt{1+\epsilon^2}} \\ -\dfrac{i\epsilon\sin\Lambda e^{-i\phi}}{\sqrt{1+\epsilon^2}} \end{pmatrix},$$

where $\Lambda = \delta\sqrt{1+\epsilon^2}t/2$. The state is $-|\uparrow\rangle$ for $\Lambda = \pi$, i.e., at $t = 2\pi/\delta\sqrt{1+\epsilon^2}$.

12.5 One is required simply to switch subscripts 1 and 2 in Eq. (12.43) to obtain the inverted CNOT gate;

$$U_{\text{CNOT}'} = Z_2\bar{Z}_1X_1U_J(\pi/J)Y_1.$$

Z_2 and $\bar{Z}_1$ are implemented with $\tilde{H}$ by making use of the relation (12.31).

12.7 $U_Z = \text{diag}(1,1,1,-1)$. We evaluate

$$U_B = Q^\dagger U_Z Q = \begin{pmatrix} 0 & 0 & 0 & i \\ 0 & 1 & 0 & 0 \\ 0 & 0 & 1 & 0 \\ -i & 0 & 0 & 0 \end{pmatrix}, \quad U_B^t U_B = \begin{pmatrix} -1 & 0 & 0 & 0 \\ 0 & 1 & 0 & 0 \\ 0 & 0 & 1 & 0 \\ 0 & 0 & 0 & -1 \end{pmatrix}.$$

The eigenvalues are $(-1,1,1,-1)$ and corresponding eigenvectors are

$$(1,0,0,0)^t, (0,1,0,0)^t, (0,0,1,0)^t, (0,0,0,1)^t,$$

from which we find $O_1 I$ and $h_D = \text{diag}(i, 1, 1, i) = e^{i\pi I_z \otimes I_z}$. O_2 is found as

$$O_2 = U_B (h_D O_1)^{-1} = \begin{pmatrix} 0 & 0 & 0 & 1 \\ 0 & 1 & 0 & 0 \\ 0 & 0 & 1 & 0 \\ -1 & 0 & 0 & 0 \end{pmatrix}.$$

Cartan decomposition is then

$$k_1 = Q O_1 Q^\dagger = I, h = Q h_D Q^\dagger = \begin{pmatrix} i & 0 & 0 & 0 \\ 0 & 1 & 0 & 0 \\ 0 & 0 & 1 & 0 \\ 0 & 0 & 0 & i \end{pmatrix} = e^{i\pi/4} e^{i\pi I_z \otimes I_z}$$

and

$$k_2 = Q O_2 Q^\dagger = \begin{pmatrix} -i & 0 & 0 & 0 \\ 0 & 1 & 0 & 0 \\ 0 & 0 & 1 & 0 \\ 0 & 0 & 0 & i \end{pmatrix} = \begin{pmatrix} e^{-i\pi/4} & 0 \\ 0 & e^{ipi/4} \end{pmatrix} \otimes \begin{pmatrix} e^{-i\pi/4} & 0 \\ 0 & e^{ipi/4} \end{pmatrix}.$$

The NMR pulse sequence is trivial for k_1 while we find $k_2 = (XY\bar{X}) \otimes (XY\bar{X})$ and $h = (X^2 \otimes I) e^{-i\pi I_z \otimes I_z} (\bar{X}^2 \otimes I)$. The obtained pulse sequence is

$$U_Z = e^{i\pi/4}[(XY\bar{X}) \otimes (XY\bar{X})](X^2 \otimes I) e^{\, i\pi I_z \otimes I_z} (\bar{X}^2 \otimes I)$$
$$= e^{i\pi/4}[(XYX) \otimes (XY\bar{X})] e^{-i\pi I_z \otimes I_z} (\bar{X}^2 \otimes I).$$

The sequence is further simplified by noticing that the overall phase can be dropped and XYX is replaced by a single pulse $R(\pi, \pi/4) = e^{-i\pi(I_x + I_y)/\sqrt{2}}$. In conclusion, the simplest pulse sequence is

$$U_Z = [R(\pi, \pi/4) \otimes (XY\bar{X})] e^{-i\pi I_z \otimes I_z} (\bar{X}^2 \otimes I).$$

12.8 We need to redefine U_{CNOT} so that it becomes an element of SU(4),

$$U_{\text{CNOT}} = e^{i\pi/4} \begin{pmatrix} 1 & 0 & 0 & 0 \\ 0 & 1 & 0 & 0 \\ 0 & 0 & 0 & 1 \\ 0 & 0 & 1 & 0 \end{pmatrix}.$$

It follows from

$$U_B = Q^\dagger U_{\text{CNOT}} Q = \frac{e^{i\pi/4}}{2} \begin{pmatrix} 1 & i & -1 & i \\ -i & 1 & -i & -1 \\ -1 & i & 1 & i \\ -i & -1 & -i & 1 \end{pmatrix}, \quad U_B^t U_B = \begin{pmatrix} 0 & 0 & -i & 0 \\ 0 & 0 & 0 & -i \\ -i & 0 & 0 & 0 \\ 0 & -i & 0 & 0 \end{pmatrix}$$

that $U_B^t U_B$ has eigenvalues $i, i, -i, -i$ and corresponding normalized eigenvectors $(0, -1, 0, 1)^t/\sqrt{2}, (-1, 0, 1, 0)^t/\sqrt{2}, (0, 1, 0, 1)^t/\sqrt{2}, (1, 0, 1, 0)^t/\sqrt{2}$. Now

$$O_1 = \frac{1}{\sqrt{2}} \begin{pmatrix} 0 & -1 & 0 & 1 \\ -1 & 0 & 1 & 0 \\ 0 & 1 & 0 & 1 \\ 1 & 0 & 1 & 0 \end{pmatrix}, \quad h_D^2 = \mathrm{diag}(i, i, -i, -i),$$

from which we take $h_D = \mathrm{diag}(e^{i\pi/4}, e^{i\pi/4}, e^{-i\pi/4}, e^{-i\pi/4})$. The other matrix O_2 is found as

$$O_2 = U_B (h_D O_1)^{-1} = \frac{1}{\sqrt{2}} \begin{pmatrix} 0 & -1 & -1 & 0 \\ -1 & 0 & 0 & 1 \\ 0 & 1 & -1 & 0 \\ 1 & 0 & 0 & 1 \end{pmatrix}.$$

The Cartan decomposition is found as

$$k_1 = Q O_1 Q^\dagger = \frac{1}{\sqrt{2}} \begin{pmatrix} 0 & i & 0 & i \\ -i & 0 & -i & 0 \\ 0 & i & 0 & -i \\ -i & 0 & i & 0 \end{pmatrix} = \frac{-i}{\sqrt{2}} \begin{pmatrix} 1 & 1 \\ 1 & -1 \end{pmatrix} \otimes \begin{pmatrix} 0 & -1 \\ 1 & 0 \end{pmatrix},$$

$$k_2 = Q O_2 Q^\dagger = \frac{1}{2} e^{i\pi/4} \begin{pmatrix} 1 & i & 1 & i \\ -i & -1 & -i & -1 \\ -i & 1 & i & -1 \\ -1 & i & 1 & -i \end{pmatrix}$$

$$= \frac{1}{2} \begin{pmatrix} 1-i & 1-i \\ -1-i & 1+i \end{pmatrix} \otimes \frac{1}{\sqrt{2}} \begin{pmatrix} i & -1 \\ 1 & -i \end{pmatrix}$$

and

$$h = Q h_D Q^\dagger = \frac{1}{\sqrt{2}} \begin{pmatrix} 1 & 0 & 0 & i \\ 0 & 1 & i & 0 \\ 0 & i & 1 & 0 \\ i & 0 & 0 & 1 \end{pmatrix} = e^{i\pi I_x \otimes I_x}.$$

To implement NMR pulse sequence, we notice $k_1 = U_H \otimes Y^2$, where $Y = e^{-i(\pi/2)I_y}$ and U_H is the Hadamard gate whose NMR implementation is given in Fig. 12.7. As for k_2, we find by inspecting Eq. (12.34), we immediately find

$$k_2 = \left[e^{i(\pi/2)I_y} e^{-i(\pi/2)I_x} \right] \otimes \left[e^{-i\pi I_y} e^{-i(\pi/2)I_x} \right].$$

Finally the Cartan subgroup element h is implemented with $JI_z \otimes I_z$ as

$$h = (\bar{Y} \otimes Y) e^{-i\pi I_z \otimes I_z} (Y \otimes \bar{Y}).$$

A naive pulse sequence obtained is

$$[(\bar{Y}X\bar{Y}) \otimes (Y^2 XY)]U_J(\pi/J)[(Y\bar{X}^2Y) \otimes (\bar{Y}Y^2)].$$

However, we notice the following simplifications:

$$\bar{Y}Y^2 = Y, \ Y^2 XY = \bar{X}\bar{Y}, \ Y\bar{X}^2Y = \bar{X}^2$$

up to an overall phase. Then the pulse sequence above is simplified as

$$[(\bar{Y}X\bar{Y}) \otimes (\bar{X}\bar{Y})] \cdot U_J(\pi/J) \cdot [\bar{X}^2 \otimes Y].$$

Compare this result with Eq. (12.43).

12.9 (1)

$$\tilde{\rho}_Y = \frac{1}{2}\begin{pmatrix} 1 - c_x & c_z - ic_y \\ c_z + ic_y & 1 + c_x \end{pmatrix}.$$

(2)

$$\rho_Y = \frac{1}{2}\begin{pmatrix} 1 - c_x & (c_z - ic_y)e^{-i\omega_0 t} \\ (c_z + ic_y)e^{i\omega_0 t} & 1 + c_x \end{pmatrix}.$$

Chapter 13

13.2 Use the following facts:

$$R\left(\frac{\pi}{2}, -\frac{\pi}{2}\right) = \begin{pmatrix} \frac{1}{\sqrt{2}} & \frac{1}{\sqrt{2}} \\ -\frac{1}{\sqrt{2}} & \frac{1}{\sqrt{2}} \end{pmatrix}, \ R\left(\frac{\pi}{2}, 0\right) = \begin{pmatrix} \frac{1}{\sqrt{2}} & -\frac{i}{\sqrt{2}} \\ -\frac{i}{\sqrt{2}} & \frac{1}{\sqrt{2}} \end{pmatrix},$$

$$R\left(\frac{\pi}{2}, \pi\right) = \begin{pmatrix} \frac{1}{\sqrt{2}} & \frac{i}{\sqrt{2}} \\ \frac{i}{\sqrt{2}} & \frac{1}{\sqrt{2}} \end{pmatrix}, \ R\left(\frac{\pi}{2}, \frac{\pi}{2}\right) = \begin{pmatrix} \frac{1}{\sqrt{2}} & -\frac{1}{\sqrt{2}} \\ \frac{1}{\sqrt{2}} & \frac{1}{\sqrt{2}} \end{pmatrix},$$

where $R_2(\theta, \phi) = R(\theta, \phi) \otimes I$ and $R(\theta, \phi)$ has been defined in Eq. (13.59).

Index